Computing with Bio-Molecules

Springer

Singapore
Berlin
Heidelberg
New York
Barcelona
Budapest
Hong Kong
London
Milan
Paris
Tokyo

Computing

with

Bio-Molecules

Theory
and
Experiments

Gheorghe Păun
(Editor)

 Springer

Dr. Gheorghe Păun
Institute of Mathematics
Romanian Academy of Sciences
PO Box 1-764
70700 Bucharest
Romania
E-mail: gpaun@imar.ro

Library of Congress Cataloging-in-Publication Data

Computing with bio-molecules : theory and experiments / editor,
    Gheorghe Păun.
        p.   cm. -- (Springer series in discrete mathematics and
    theoretical computer science)
        Includes bibliographical references (p.).
        ISBN 9814021059
        1. Molecular computers.   I. Păun, Gheorghe, 1950- .
II.   Series.
QA76.887.C66   1998
511.3—dc21   98-33723
                                                              CIP

ISBN 981-4021-05-9

© Springer-Verlag Singapore Pte. Ltd. 1998
Printed in Singapore

Typesetting: Camera-ready by Editor
SPIN 10676837   5 4 3 2 1 0

Preface

Molecular computing (especially DNA computing) means using bio-molecules as a support for computations and for devising computers. Contrast this with the opposite direction of research, the classic one, where computers are used in studying molecules (especially DNA).

The DNA molecule is one of the most compact support of information and has a very important feature from a computational point of view: it is a double stranded sequence (based on the Watson-Crick complementarity of the four nucleotides forming the two strands). This, on one hand, introduces a new data structure (the double tape, with a complementarity relation between the corresponding cells), on the other hand, makes available the power of the so-called twin-shuffle language, which can directly lead to characterizations of recursively enumerable languages, those recognized by the Turing machines. Moreover, DNA is a stable molecule, it can be handled in many ways already well controlled in the laboratory, and can be exponentially duplicated by the Polymerase Chain Reaction; this makes possible devising computations of a huge parallelism, a lot over the parallelism possible in electronic media. However, many processes involving DNA are still error-prone, they are nondeterministic and difficult to control and to scale-up, although in the last years a lot of information about DNA has been gained (mainly in the framework of the genome project).

Using DNA as a support for computation is not a new idea. Speculations about this possibility were made already in the fifties (R. P. Feynman), with details in the seventies (Ch. Bennett) and eighties (M. Conrad), but the feasibility of such computations was only recently proved. In 1994, L. Adleman reported a successful experiment of solving the Hamiltonian Path Problem in a graph by using only biochemical lab techniques. This experiment has raised both enthusiastic and not-so-enthusiastic (not to say pessimistic) comments. Anyway, the idea is revolutionary, because it makes possible the construction of computers of huge parallelism, thus letting us hope that problems which are intractable for the silicon computers can be solved in the new framework. Moreover, this attempt opens exciting theoretical research vistas. Basically, new data structures and new operations on them and on the usual data structures (strings, arrays, trees) are introduced, such as matching, splicing (crossing-over), insertion, deletion, duplication, etc. One of the most interesting theoretical developments is rooted in the splicing operation introduced already in 1987 by T. Head – the theory of the so-called H systems.

After Adleman's experiment several research groups were constituted in USA, Canada, Europe, Japan, workshops on DNA based computers were annualy organized in Princeton (1995, 1996) and Philadelphia (1997, 1998), many computer science conferences have invited plenary speakers or have organized sessions on this topic. The first European meeting devoted to DNA computing was the workshop organized in Mangalia, Romania, in August 1997. Similar meetings or sections of larger conferences were devoted to DNA computing. A PhD thesis has been recently presented in England, others are in preparation in Italy, Spain, The Netherlands, and maybe also in other countries. An important recent event was the Auckland conference *Unconventional Models of Computation* (January 1998), where DNA computing and quantum computing were the two main topics.

The field is definitely in a rapid expansion, due both to its theoretical appeal and to the promising applications. Actual computations were reported, but only referring to toy-problems; however, the speculations are spectacular: "how to break Data Encryption Standard in two months working in your basement", "how to construct a memory larger than the brain", "how to solve NP-complete problems in polynomial or even linear time". It is also strongly believed that "by-product" results will be important for the very field of biochemistry and for the understanding of the way life "computes" through evolution. Still, a lot of work remains to be done, both from a mathematical point of view and from a biochemical point of view. Up to now, biochemists have hardly considered the questions raised by using DNA molecules as computing "chips". Once the problem is posed, the progress is expected to be rapid.

DNA computing is a domain of a clear interdisciplinary nature. Meetings bringing together mathematicians, computer scientists, and biochemists are essential for the development of this domain. So are the books collecting the works of scientists working in these areas in search of the bio-computer (science).

*

The present book contains contributions by several active researchers in DNA computing, mainly mathematically oriented but also taking into consideration the lab reality. Some of these papers were presented during the Mangalia workshop mentioned above.

The volume starts with a general discussion about the interplay between linguistics, language theory, semiotics and biology, by S. Marcus, a "senior" of these fields, followed by two other "strategic papers" (by V. Manca and G. Ciobanu), making use of tools from logics, concurrent systems theory, λ- and π-calculus, etc. (Some of their central notions are those of multi-sets, membranes, solutions, adopted from biology as working metaphors in concurrent systems theory, and returning back to DNA computing as very powerful concepts.)

Then, we proceed along a series of "practical" issues: a suggestion of using double stranded molecules for solving the Hamiltonian Path Problem (T. Head), solutions to 3-SAT and 3-vertex colorability problems (N. Jonoska, St. A. Karl, M. Saito), Boolean circuit evaluation (M. Ogihara, A. Ray), and Boolean transitive closure (P. E. Dunne, M. Amos, A. Gibbons). All these are approached by DNA techniques, ...in spite of the thermodynamic constraints (R. Deaton, M. Garzon) or the combinatorial constraints (S. Biswas) on DNA-based computing. A possibly better implementation of DNA-like computation in silicon media is discussed by Gh. Ştefan.

A series of subsequent papers is devoted to some of the main models in (theoretical) DNA computing: H systems, sticker systems, Watson-Crick automata. One inquiries about possibilities of characterizing recursively enumerable languages (R. Freund), one presents universality results for finite automata and Watson-Crick finite automata (C. Martín-Vide, Gh. Păun, G. Rozenberg, A. Salomaa), as well as explicit constructions of universal H systems based on multisets (P. Frisco, G. Mauri, Cl. Ferretti); the possibility of restricting the length of patterns in splicing rules without losing the universality is examined by A. Păun and M. Păun.

The papers which follow become more and more theoretically oriented: one deals with the influence of the representation of splicing rules on the results concerning generative power (H. J. Hoogeboom, N. van Vugt), one side splicing systems (T. Head), self-splicing (J. Dassow, V. Mitrana). The splicing operation is then extended to array languages (K. Krithivasan, S. R. Kaushik) and to a variant of the shuffle operation on routes (A. Mateescu). The last two papers of the volume can be considered as pieces of pure mathematics: algebraic approaches to various DNA operations (Z. Li), fixed-point approaches to cut and paste operations (R. Ceterchi). It is worth mentioning that such algebraic tools are very effective and useful in formal language theory.

The contents of this volume confirms the appreciation that theory is well ahead of practice in this area. DNA computing *in info* is well ahead of DNA computing *in vitro*. This situation is often encountered in computer science. Remember only that computers themselves have first existed as mathematical models, as – universal – Turing machines, and only later as electronic physical devices. Let us hope that history will repeat itself in the case of DNA based computers...

*

Many thanks are due to all contributors to this book, as well as to Ian Shelley from Springer, Singapore, for the very efficient and pleasant cooperation.

Gheorghe PĂUN
May 1998

Contents

Language at the Crossroad of Computation and Biology

Solomon MARCUS

Romanian Academy, Mathematical Sciences
Calea Victoriei 125, 71102 Bucharest, Romania
smarcus@stoilow.imar.ro

The research in the recently initiated field of molecular computation cannot be understood in its deep significance without considering its broad scientific and historical context. The natural cognitive framework of this concern should be the triangle MIND-LANGUAGE-BIOLOGY. Strongly associated with MIND are BRAIN and COMPUTATION, while LANGUAGE has two basic aspects: natural and artificial, the latter having as a particular important case the formal languages. In respect to biology, the most important aspects related to the above concern are: neuroscience, molecular biology and cellular biology.

Attempts to improve computation by using the structure and behavior of DNA, i.e., of heredity, were preceded for about fifty years, by the opposite itinerary, trying to improve the understanding of heredity by using computational models. Despite their opposition in respect to direction (from DNA to computation in the first case, from computation to DNA in the second case), these two approaches are both based on what heredity and computation have in common. The research done so far shows that this common denominator includes as a basic pattern the human language, in its both hypostases: natural language and formal language. We reach in this way the cognitive triangle COMPUTATION-LANGUAGE-HEREDITY, which is a specific form of the above considered triangle. On the other hand, we can consider the brain as a biological concern, while computation is, in some theories, a legitimate substitute of mind; so, we reach the cognitive triangles COMPUTATION-LANGUAGE-BRAIN and COMPUTATION-LANGUAGE-NERVOUS ACTIVITY.

The presence of language (be it natural or formal) in all these triangles is motivated by the fact that the research done so far pointed out the linguistic paradigm as a basic link between mind and biology and more specifically

1

between computation and heredity, between computation and brain and between computation and nervous activity. This link takes various forms, from analogy and contiguity to cognitive modeling and cognitive metaphor.

So, we organize our presentation as follows: a) biological metaphors in historical linguistics; b) computation and formal languages as models of natural languages; c) natural languages as a model of computation; d) natural languages as metaphors and models of heredity; e) formal languages and formal grammars as models of heredity; f) DNA recombination as a source of computation models, via formal languages; g) computational biology and the corresponding information metaphors; h) bridging experimental and theoretical biological computation; i) computational brain theories of language. Since language is a particular case of a sign system, we consider also: j) sign systems and living systems.

a) Biological metaphors in historical linguistics. The first important metaphor orienting historical linguistics was the Darwinian metaphor, proposed by August Schleicher [118]. It gave birth to the "tree-model", as an explanation of the mechanism of language evolution. The other model was the "wave-model" due to J. Schmidt (1872). (A review of them belongs to Goebl, 1883.) Schleicher, author of the family tree-theory, thought of language as an organism which could grow and decay, and whose changes could be analyzed using the methods of the natural sciences. The family tree metaphor determined a sequence of other correlated metaphors. In respect to Romance family, Latin is the parent language, Italian, French, Spanish, Romanian, Portuguese, Catalan are daughter languages, Portuguese is a sister of Spanish and Romanian is a sister of Italian, etc. There are some limits of validity of these metaphors, for example, in contrast with what happens in a human family, a parent language does not survive after a daughter language is born, while a language is never born in a definite moment, like a human being, and never dies at a fixed moment.

The wave model of linguistic change assumes that a change starts at point X and moves simultaneously through a geographical area and through the strata of society. The further it travels, the less effect the change has. If many changes were taking place at once, the speakers furthest away from X would gradually lose their linguistic identity with those at X. They would become different dialects, and in due course perhaps even different languages (Bailey [4]). However, in the last part of XIXth century it was believed that a sound change affects the whole of a language simultaneously. One sound system would smoothly develop into the next, and all words which contained a particular sound would be affected in the same way

(Crystal [25], p. 332). The metaphor of a wave has proved attractive since the late XIXth century: a change spreads through a language in the same way as a stone sends ripples acrosss a pool (Crystal [25], p. 332). This model seemed to be better than the family-tree model, with its clean splitting off of branches (Crystal [25], p. 332). A major progress in this direction is due to Labov [75] with his sociolinguistic studies of linguistic variation. His assumption was that the variation in language use which is found in any community is evidence of change in progress in a language.

An interesting viewpoint in this respect is brought by Forster [39] and Forster et al. [40], [41]. They unite the tree model and the wave model of language evolution into one geometric network model. This approach has previously been used for reconstructing DNA evolution and now it is applied to vocabulary lists of closely related languages. Forster ([39], p. 184) starts with the remark that it would be desirable to visualise both tree aspects and wave aspects of language evolution in a single diagram and observes that this type of problem is perfectly tailored to network methods originally developed for reconstructing phylogenetic relationships from DNA sequences (Bandelt [5]; Bandelt et al. [6]). During evolution, a given DNA sequence acquires mutations at random positions, causing the progeny sequences to become more and more dissimilar from one another and from their ancestral sequence as time passes, yielding the treelike aspect of DNA evolution. Some times, mutations at the identical position in two differènt DNA sequences will cause these two sequences to become more similar, resulting in convergence (Forster [39], p. 184).

In order to establish which form of linguistic information is most amenable to phylogenetic analysis, Forster refers to the one hundred word list proposed by Swadesh [127] as the basic vocabulary of a given language. Then, Forster combines the tree model and the wave model: word stem replacements as well as uptake of ancient idiosyncratic substrate lead to divergence between originally related vocabularies (the tree aspect of language evolution), while the wave aspect of language evolution is given by the fact that convergence of vocabularies is determined by loan events of word stems. Forster observes that although the mechanisms of DNA and vocabulary evolution are not analogous, the effects seem to be. The treelike and the wavelike events of vocabulary evolution can be united and visualised in a phylogenetic network. Finally let us observe that, in the second and third decades of XXth century, Roman Jakobson, one of the most important linguists of the XXth century, trusted the naturalistic metaphor in linguistics, while later, as we will see, he pointed out the opposite itinerary, leading to linguistic metaphors in biology (more exactly, in the study of heredity).

b) Computation and formal languages as models of natural languages. Computation became a cognitive model for natural languages with the pioneering work of Chomsky [16], [17]. Chomsky's hierarchy of formal grammars and languages was motivated by needs of the study of syntax of natural languages. Since each type of formal grammars in Chomsky's hierarchy is equivalent to a specific type of automata, which, in their turn, are specific types of a Turing machine, it follows that Chomsky's hierarchy leads to a hierarchy of computational models for natural languages.

Strongly associated with the computational-formal linguistic approach to natural languages is the computational approach to nervous activity (McCulloch & Pitts [93], Kleene [73]), where logical calculus and finite automata simulate the behavior of nerve nets.

The linguistic explanatory capacity of formal grammars and of various types of automata was discussed in a very large number of papers; it is no need to insist in this respect.

c) Natural languages as a model of computation (via programming languages). Short time after the introduction of ALGOL 60 by means of Backus-Naur normal forms (BNNF) it is proved by Ginsburg & Rice [47] that BNNF are equivalent with Chomsky's context-free grammars. This revealed that a tool born in the study of syntactic structures of natural languages remains adequate and relevant for the study of the syntax of programming languages. In this way, it is proved that natural languages are a term of reference for programming languages, in other words, linguistics is able to give cognitive models for computation (because a programming language is a tool to implement an algorithm, i.e., a specific type of computation). However, ALGOL cannot remain at BNNF; its syntax is supplemented with some nonformalized conditions concerning the requirement to declare any identifier, any simple variable, any array, etc. Do these semantic conditions have some influence on the generative type of ALGOL ? Floyd [38] has shown that the answer is affirmative, so, the semantics of ALGOL 60 is not context-free, it is context sensitive. The same is true for the semantics of FORTRAN IV and of COBOL. Păun [101] has shown that ALGOL 60, FORTRAN IV and COBOL are neither simple matrix nor finite index matrix languages; the conditions leading to non-context-free rules are always of a semantic type. Similarity with natural languages is given by the fact that the latter are also somewhere between context-free and context sensitive; however, non-context-free rules are required in natural languages already at their syntactic level (this very controversial statement seems to be now generally accepted), while in programming languages they appear only at the semantic level. This discrepancy may have the following explanation: in programming languages, the distinction between semantics and syntax is purely conventional and it is very sharp, while in natural languages there is no such sharp distinction; some semantic factors are involved in some syntactic aspects.

d) Natural languages as metaphors and models of heredity. The idea to look at heredity through the glasses of natural languages and of linguistics came first as a metaphor, then as a model. Watson [128] tells us that heredity is a form of communication. For Asimov [3], the nucleotide bases are letters (or words) and they form an alphabet. Jacob [61] considers that the sense of the genetic message is given by the combination of its signs in words and by the arrangement of words in phrases. A more systematic linguistic view of the genetic code is proposed by Jakobson [62], [63], whose starting point was just the remark that discoveries in the field of genetics are presented in a linguistic terminology. He observes that the dictionary of the genetic code encompasses 64 distinct words defined as triplets. To Jakobson belongs the idea to take the nucleotide bases as phonemes of the genetic code. He argues that "since our letters are mere substitutes for the phonemic pattern of language, and the Morse alphabet is but a secondary substitute for letters, the subunits of the genetic code are to be compared directly with phonemes". Jakobson ([62], p. 438) calls attention on the essential binary character of both the phoneme and the nucleotide base: "All the interrelations of phonemes are decomposable into several binary oppositions of the further indissociable distinctive features. In an analogous way, two binary oppositions underlie the four 《letters》 of the nucleic code: thymine T, cytosine C, guanine G, and adenine A. A size relation [...] opposes the two *pyrimidines* T and C to the larger *purines*, G and A. On the other hand, the two pyrimidines (T vs C) and, equally, the two purines (G vs A) [...] present two contrary orders of the donor and acceptor: T/C = G/A and T/G = C/A."

Verbal and genetic code show the same hierarchical design: from lexical to syntactic units of different grades, in natural languages; from codons to cistrons and operons in the genetic language. See also, in this respect, Ratner [112]. Like words in natural languages, genes are transmitted not in isolation but in sequences whose structure (chromosomes) could be compared to the structure of utterances or sentences. Like the meaning of words in language, the meaning of some codons depends on the context, on their position in the genetic message; see Clark & Marcher [22]. Like in natural languages, we have in heredity phenomena of synonymy: some different codons may have the same genetic meaning. Jacob [61] believes that this phenomenon gives some suppleness to the writing of heredity. However, sometimes things work differently. As Monod [94] pointed out, the translation from RNA language into the protein language is essentially unidirectional, in contrast with speech, where translation at any level is bidirectional.

We adopt now the convention proposed by Marcus [87]. We distinguish between the genetic code (GC) and the genetic language (GL), the former being a part of the later. GL has two strata. The inferior stratum of GL, interpreted as its syntactic level, is the stratum of nucleic acids. The superior stratum of GL, interpreted as its semantic level, is that of proteins and metabolism. Syntax has two substrata: DNA and RNA. Semantics has two substrata: proteins and metabolism.

The principle of duality of patterning, asserted by Martinet [89] for natural languages, claims that natural languages are based on two articulations: a first one, in minimal syntagmatically meaningful linguistic units, called morphemes, and a second one, in syntagmatically meaningless linguistic units, called phonemes. Morphemes convey semantic meaning (be it lexical or grammatical), while phonemes convey only phonological meaning; see, for more, Beadle & Beadle [10], Masters [90], and Marcus [87]. Similarly, nucleotides convey only chemical meaning and are syntagmatically minimal, while codons are syntagmatically minimal units conveying genetic meaning (due to their correspondence with amino acids).

Alternative proposals for the interpretation of duality of patterning in GL belong to Grassi [50] and Tulio de Mauro [92]. Grassi proposes to consider nucleotides as units of first articulation and sugars, phosphate groups and bases as units of second articulation. Tulio de Mauro uses the term *moneme* as a unit of first articulation which is not obligatory provided with a meaning. Nucleotides would be the monemes of GL and they are articulated in three types of units of second articulation. We reject these interpretations. The principle of duality of patterning asserts in essence the possibility to obtain a very large number of syntagmatically minimal meaningful units by syntagmatic combination of a very small number of meaningless units. So, any interpretation of this principle in GL should contrast genetically meaningful units with genetically meaningless units; this condition is not satisfied in Grassi's and in Tulio de Mauro's proposals.

The analogy between nucleotides and phonemes is motivated by three reasons: 1) nucleotides cannot be syntagmatically decomposed in smaller units; 2) they have no genetic meaning, but only a chemical meaning (for instance, thymine and cytosine consist of hydrogen, oxygen, carbon, and nitrogen) and 3) they can be paradigmatically analyzed in binary distinctive features (which here are chemical features). So, *chemical* corresponds to *phonological*, while *genetic* corresponds to *semantic*. There are however some quantitative and structural differences: 1) we have only four types of nucleotides, while the number of phonemes is larger than 20; 2) a phoneme is an *unordered* set of distinctive features (for instance: P = {non-vocalic, consonantal, grave, non-compact, non-strident, non-nasal, non-continuant, non-voiced}), while a nucleotide is an *ordered* set of distinctive features, because chemical elements are arranged in different graphs (see, for instance, Hartman & Suskind [54]); 4) in contrast with phonemes, which are *abstract* entities, nucleotides are *material* entities. However, the type-token distinction is valid for both phonemes and nucleotides; the same type of phoneme may have different occurrences in a statement and the same type of nucleotide may have different occurrences in a DNA or RNA; 5) syntax of phonemes in language has only one level (substratum), while syntax of nucleotides has two substrata, DNA and RNA; 6) syntax of nucleotides in DNA is double stranded, while syntax of phonemes in language is single stranded (like in RNA); 7) The length of a RNA molecule is generally much smaller than the length of a DNA molecule, but much larger than the length of a statement in language (the length being measured by the number of nucleotides-occurrences in the first case and by the number of phonemes-occurrences in the second case); 8) there is a process of *transcription* from DNA to RNA (A $\to$ U, T $\to$ A, G $\to$ C, C $\to$ G), while no similar operation in language exists (because the syntax of phonemes does not have two substrata).

We described in details the relation nucleotides-phonemes, in order to understand, in this specific case, what does it mean that B is a cognitive model of A (in our case, the phoneme is adopted as a cognitive model of the nucleotide); B shares with A some basic features, giving to the results obtained in the investigation of B some relevance in respect to A; on the other hand, B fails to share with A some important features of A, so B is only an approximation of A, B is in some respect heterogeneous in respect to A and this fact makes possible the existence of some methods that can be applied to B, but not to A. Cognitive models, like cognitive metaphors, have a conflictual nature, they always include a tension of two opposite forces: similarity and dissimilarity. This fact has to be kept in mind in what follows.

The next level of RNA substratum is the level of *codons* which share with the morphemes of natural languages the following basic properties: 1) each codon is a sequence of nucleotides (i.e., of genetic phonemes); 2) they have a genetic meaning (to a given codon corresponds one or several amino acids); 3) there is no syntagmatic decomposition of a codon in shorter genetically meaningful sequences of nucleotides. In contrast with morphemes (in natural languages), whose length is variable, the length of any codon is the same: equal to three. In contrast with the number of morphemes in a natural language, which is variable and very large (between thousand and ten thousands), the number of codons in GL is fixed and relatively small (equal to 64). Concatenation of codons leads to finite, but very large strings called *polynucleotide chains*. Concatenation of morphemes in a natural language leads to *utterances* and *texts*, usually much shorter than polynucleotide chains.

The genetic code (GC) is only a part of GL, understood as the correspondence existing between the codons of RNA and the different types of amino acids; we can interpret these codons as lexical morphemes. Monod [94] defines the GC as the rule associating to a given polynucleotide sequence a polypeptidic sequence. It is equivalent to the preceding one. The considered sequence is like a string of phonemes where we have already introduced a morphemic segmentation. We replace each codon by its corresponding amino acid. No ambiguity may appear, because the possible homonymy is solved by means of the context.

It is necessary to distinguish between the theoretical genetic language (TGL) and the experimental genetic language (EGL), in respect to which three types of properties can be considered: 1) properties belonging to both TGL and EGL; 2) properties that belong only to TGL; 3) properties that belong only to EGL. All the properties discussed so far are of first type.

These distinctions change with the time. For instance, DNA-sequencing is known for longtime, but only in 1975 it is invented experimentally, by Frederick Sanger (who in 1950 invented experimentally protein sequencing).

Modern linguistics has developed the study of some basic dichotomies in respect to which natural languages are analyzed. An interesting problem is to check their validity for GL. Grassi proposes to interpret TGL as equivalent to language and EGL as equivalent to speech. If we interpret the same dichotomy as general-individual or as systematic-accidental, we can take as language the GC and as speech the possible combinatorial structures of DNA (and their various mutations) in the way heredity is realized by various living beings. The evolution of the GL seems to be too slow to permit observation of its history, so only synchronic aspects are investigated. Diachrony remains beyond our means of research. Paradigmatic aspects of GL are less important than in natural languages. Syntagmatic aspects are predominant. The genetic competence is obtained by means of some generative devices, as they will be described in the following. The genetic performance is the actualization of this competence. Similarly, we distinguish between the genetic deep structure and the genetic surface structure. Grassi proposes to take as deep structure the syntax of amino acids and as surface structure the DNA and the RNA. This choice is not consistent with the Chomskian view of this dichotonomy, where the deep structure is always given by a generative grammar. The distinction grammatical-lexical can be realized at the level of codons, by considering as grammatical codons those which are associated with no amino acid, but used merely to signal the end of a genetic message; there are only three grammatical codons: UAA, UAG, UGA. All the other 61 codons are lexical.

There is a possibility to define, for each codon, its degree of synonymy and, by means of it, the concept of a degenerate genetic code. Beisson [12] claims that EGL has no ambiguity, while other authors, such as Clark & Marcher [22] claim the existence of some ambiguities: some codons change their meaning in respect to their position in the RNA strings. On the other hand, there are contexts where the grammatical codons are transformed in lexical codons. Anyway, GL is predominantly devoid of ambiguity. Roman Jakobson points out the opposition between the context sensitive nature of natural languages and of GL, on the one hand, and the context-free nature of formalized languages, on the other hand. He is right in respect to the language of propositional calculus or other simple formal languages, but he is wrong in respect to many formal, or semi-formal non-context-free languages occurring in computation and in the modeling of programming languages.

Another class of problems is related to the degree of stability of GL: to what extent to small modifications of the codons correspond small modifications of the associated amino acids ? ("Small" is understood here in respect to a notion of distance between two different strings.) An answer is given by Marcus [87], who shows, in the same order of ideas, that GL fulfils to a large extent the principle ruling natural languages: the most important information is carried by the first part of a word. In respect to the property of continuity of GL (to what extent a small difference in

meaning is associated with a small difference in expression ?), it is proved
that GL is very discontinuous.

Going further, to larger units, it is shown that cistrons in GL belong
to utterances in natural languages (an utterance is, following Harris [53],
p. 14, any stretch of talk by one person, before and after which there is
silence on the part of the person). Segmentation in cistrons is simpler than
segmentation in utterances, because the former are separated by grammatical codons, while the latter are usually separated by no material border.
Operons in GL correspond to discourses in natural languages. There is in
GL a one-to-one correspondence between citrons and polypeptide chains;
it replaces the old correspondence between genes and proteins.

e) Formal languages and formal grammars as models of heredity.
GL shares with natural languages a very important feature:
the impossibility to give an explicit necessary and sufficient condition for
a string (of nucleotides, of phonemes, resp.) to be well-formed. What
structure should have a string of nucleotides to define the heredity of a living
being ? In absence of an answer to this question, one can look for answers
to less difficult questions, trying to reach only an approximate idea of the
process of protein formation. It seems that the first attempt in this respect
belongs to Pawlak [100], who tries to exploit the constant length of codons,
taking them (and not the nucleotides) as basic bricks of his approach. Since
there are 64 types of codons (possible strings) on the alphabet of four
elements, denoted by 0, 1, 2, and 3, and since only twenty of these strings
of length 3 over the alphabet $\{0, 1, 2, 3\}$ are associated with some amino
acids (whose number is just 20), Pawlak looks for a procedure to select
from the possible 64 strings of length 3 only 20 of them. Representing each
amino acid by a regular triangle, Pawlak proposes a recursive procedure to
define well-formed strings of such triangles (i.e., of amino acids). For each
triangle we choose a base and, correspondingly, a left side and a right side.
Let i, j and k be the numbers associated to the left side, to the base and
to the right side, respectively, in such a way that i is strictly smaller than
j, while k is not larger than j. There are exactly 20 types of triangles one
can get in this way, by means of the numbers 0, 1, 2, 3. They correspond,
for Pawlak, to the 20 types of amino acids. Here is the weak point of his
approach, because it seems that there is no chemical or genetic motivation of
his procedure. However, the simple fact that this combinatorial restriction
leads to exactly 20 possibilities may be provisionally enough to accept it, in
absence of another better choice. Now a recursive definition of well-formed
strings of Pawlak triangles is adopted, in the following way: 1) every Pawlak

triangle is a well-formed string; 2) given a well-formed string x, the following string y is also well-formed: y is obtained from x by adding only one new triangle β, whose base is either the left or the right side of a triangle α belonging to x, in such a way that the number associated to this base in β is also associated to the corresponding side in α; no side of β is either a base or a side of a triangle in x; 3) only those strings are well-formed, which are obtained by means of rules 1 and 2. A well-formed string to which no new triangle can be added, in order to get again a well-formed string, is said to be a *terminal string*. Terminal strings are associated with proteins. Pawlak writes a grammar (which is not a Chomskian, but a dependency one) in which every rule corresponds to one of his triangles: $j \rightarrow ik$. From this representation a simpler one, by means of graphs, can be obtained. If we order the triangles using the lexicographic order (0 precedes 1, which precedes 2, which precedes 3), then we denote them by the letters from a to t, in their alphabetic order. Bernard Vauquois (one of the authors of ALGOL 60), proposed the following context-free grammar corresponding to Pawlak's dependency grammar (personal communication, 1971), where S, A, B, and C are nonterminal symbols:

$$S \rightarrow a, S \rightarrow bA, A \rightarrow a, A \rightarrow bA, S \rightarrow c, S \rightarrow dA, B \rightarrow c, B \rightarrow dA,$$
$$S \rightarrow eB, S \rightarrow fA, B \rightarrow eB, B \rightarrow fA, S \rightarrow gAA, S \rightarrow hAB, B \rightarrow gAA,$$
$$B \rightarrow hAB, S \rightarrow hBA, B \rightarrow hBA, S \rightarrow i, C \rightarrow i, S \rightarrow jA, C \rightarrow jA,$$
$$S \rightarrow kB, C \rightarrow kB, S \rightarrow lC, C \rightarrow lC, S \rightarrow mA, C \rightarrow mA, S \rightarrow nAA,$$
$$C \rightarrow nAA, S \rightarrow oAB, C \rightarrow oAB, S \rightarrow oBA, C \rightarrow oBA, S \rightarrow pAC,$$
$$C \rightarrow pAC, S \rightarrow pCA, C \rightarrow pCA, S \rightarrow qB, C \rightarrow qB, S \rightarrow rBA,$$
$$SC \rightarrow rBA, \rightarrow rAB, C \rightarrow rAB, S \rightarrow sBB, C \rightarrow sBB, S \rightarrow tBC,$$
$$C \rightarrow tBC, S \rightarrow tCB, C \rightarrow tCB.$$

The protein formation process is simulated by the language of derivations in the above grammar and this language is no longer context-free.

The Pawlak-Vauquois approach to the genetic syntax presents some shortcomings, four of which will be mentioned here: a) with every well-formed string of nucleotide bases, the string obtained by replacing A by T, T by A, C by G, and G by C is also well-formed (in view of law of complementarity of the bases); this requirement was not taken into consideration in the procedure defining the well-formed strings; b) there is no biological motivation of the order and hierarchy of the nucleotides constituting a codon; c) the formation of proteins seems to be a process in which configurations of amino acids are expanding simultaneously in various directions, in contradiction with the above generative devices, in which, at every step, only one term of the string under consideration is affected; d) the double helicoidal structure of DNA, discovered in 1953 by Watson and Crick, is completely ignored in the procedure above.

Maybe the above shortcomings could be avoided by considering some alternative generative devices, such as picture grammars (see, for instance, Rosenfeld [114]) and Lindenmayer systems (Lindenmayer [81]); the former allow polydimensional structures, like the double helix, while the latter, as parallel rewriting systems, could allow simultaneous expansion in various directions. Starting from Marcus [87], Creangă, Filipoiu & Petrescu [23] made the first attempt to model the double helix by means of picture grammars (Kaneff [67], Rosenfeld [114]). A first attempt, very introductory, to use Lindenmayer's approach in the study of the protein formation is due to Marcus [87], who uses in this respect what he calls a semi-Lindenmayer system. Both these directions deserve to be deepen.

Many other problems deserve attention. For instance, how does the GL behave as an error detecting and error correcting code (if it does) ? A similar investigation for natural languages is due to Beckmann [11]. The problem of genetic mutations could be a part of a connotative, or metaphorical, genetics.

In respect to the genetic-linguistic similarity, let us quote some remarks by Manca [84], as a reaction to Marcus [87]: "[...] the paper seems to anticipate a deep understanding of a strong similarity between biomolecular systems and symbolic systems, that is the starting point of new perspectives in formal language theory: DNA computing, molecular computing, artificial life and new computational models based on biochemical metaphors. (By the way, I think that Turing, in his studies about morphogenesis, was implicitly aware of this connection)."

"However, apart the analogy analysis, the principal question of the paper seems to me the search for well-formedness conditions. You write: ⟪No necessary and sufficient conditions are known for the well-formedness of a string of nucleotide bases... This is just what happens in natural languages, where many necessary conditions are known for the well-formedness of a string, but none of these conditions is both necessary and sufficient for the well-formedness of any string.⟫ It seems to me that this is the crucial point in any notion of 'grammaticality' for very complex symbolic systems. As they are complex and open to signification processes that cannot be established in some prefixed and definitive manner, at the same time, well-formedness is intrinsically forced to be only partially defined. Of course, this fact does not imply that in these cases any search for well-formedness is vacuous, but only that we need to cope with phenomena where theoretical tools are requested that could be adequate for describing soundness in some partial, gradual, vague, or may be, analogical sense."

"[...] It is very interesting to compare all the ways Pawlak's language can be described: by triangles, by trees, by Chomskian non-context-free rules, by Vauquois's context-free rules, by L rules. But, as you pointed out, some expressive limitations are present in all these approaches. It would be interesting to investigate whether some new kinds of symbolic systems can present more expressive adequacy. For example, some sound form can be obtained by combining two unsound forms (this is a reasonable situation from a biological point of view). How can you manage such a phenomenon, within a traditional generative device ?"

"At present time, I am studying some extensions of symbolic systems aimed to express formally typical phenomena of biological reactions. In these systems two important aspects are intended to be developed: 1) distribution of information expressed by rewriting at any step finite languages, rather than single strings; 2) regulation expressed by syntactical conditions in terms of logical formulae similar to 'rudimentary predicates' considered in Salomaa's book [116] (technically, a restricted class of predicative formulae I call Sigma-1-star, in order to guarantee some natural effectivity)."

"For example, you can represent in a reasonable manner the first well-formed string of fig. 3 in Marcus [87] by the set of strings $\{t011, t'010, Ltt'\}$ meaning intuitively that you have two triangles, say t, t'; the triangle t has the left side labeled by 0, the base by 1, and the right side by 1, analogously the triangle t'; moreover, they are connected in L manner, that is, by making the left of the first one (t) coincide with the basis of the second one (t'). In general, the notion of well-formed Pawlak configuration can be formalized by a (finite) language over the alphabet $\{0, 1, 2, L, R, t, t', \ldots\}$, where a suitable logical condition holds. In this manner, you can combine two non-Pawlak configurations by generating a Pawlak configuration. I think that in a similar way it is possible to add information expressing geometrical or pictorial aspects (e.g., helicoidal structure). [...] It would be interesting to compare my ideas with your approach. I wrote a short paper about 《metabolic systems》 (I sent a copy to George Păun), where only basic definitions and some examples are considered in order to explain the idea of language rewriting systems. In another recent paper, I studied logical representability of syntactical notions. It is my intention to join these approaches in order to give a more powerful notion of metabolism in terms of symbolic systems." (Manca is referring here to [85], [86].)

f) DNA recombination as a source of computation models, via formal languages. The point of departure here is the recombinant behavior of double stranded DNA molecules, that can be regarded as strings over the alphabet consisting of the four compound two-level symbols

$$
\begin{array}{cccc}
A & C & G & T \\
T & G & C & A
\end{array}
$$

Taking an example from Head, Păun & Pixton ([57], p. 295), let us consider two molecules, molecule 1:

$$
\begin{array}{l}
A\,A\,A\,A\,A,\,G\,A,\,T\,C,\,A\,A\,A\,A\,A \\
T\,T\,T\,T\,T,\,C\,T,\,A\,G,\,T\,T\,T\,T\,T
\end{array}
$$

and molecule 2:

$$
\begin{array}{l}
C\,C\,C\,C\,C\,C,\,T\,G\,G,\,C\,C\,A,\,C\,C\,C\,C\,C\,C \\
G\,G\,G\,G\,G\,G,\,A\,C\,C,\,G\,G\,T,\,G\,G\,G\,G\,G\,G
\end{array}
$$

(Commas have no material presence; they belong to the explanation that will follow.) Suppose these molecules are situated in an apppropriate aqueous solution in which the restriction enzymes *Dpn*I and *Bal*I are present and a ligase enzyme is present. *Dpn*I will cut the first molecule between $\frac{A}{T}$ and $\frac{T}{A}$ and *Bal*I will cut the second molecule between $\frac{G}{C}$ and $\frac{C}{G}$. By ligation, the molecule

$$
\begin{array}{l}
A\,A\,A\,A\,A,\,G\,A,\,C\,C\,A,\,C\,C\,C\,C\,C\,C \\
T\,T\,T\,T\,T,\,C\,T,\,G\,G\,T,\,G\,G\,G\,G\,G\,G
\end{array}
$$

may arise. More generally, if two appropriate restriction enzymes and a ligase are present in an appropriate aqueous solution, from any two double stranded DNA molecules of the forms $x_1 u_1 u_2 x_2$ and $y_1 u_3 u_4 y_2$, the molecule $x_1 u_1 u_4 y_2$ may arise. This is the kernel of the so-called splicing operation.

Important recent developments in formal language theory provide new generative devices that allow the modeling of molecular recombination processes by corresponding generative processes acting on strings. The language of all possible strings that may be generated in the above way represents the set of all possible molecules that may be generated by the biochemical recombination processes.

This type of approach was initiated by Head [56]. For a general survey, see Head, Păun & Pixton [57]. The splicing operation, which is at the origin of this approach, reaches, in some of its variants, the power of

Turing machines, so, in certain conditions, one can compute by splicing exactly what one can compute by Turing machines. So, computation by splicing is situated at the midway between computational biology and biological computation. It belongs to computational biology by the first part of its itinerary, resulting from the attempt to build a computational model of recombinant behavior of double stranded DNA molecules; it belongs to biological computation by the second part of its itinerary, leading to a type of computing equivalent to Turing machines. In this respect, computing by splicing meets Adleman's [1] pioneering work starting the so-called molecular computation.

g) Computational biology and the corresponding information metaphors. Lander, Langridge & Saccocio, [77] call attention on the need to understand the specific aims of some neighbor, strongly correlated fields, such as genetics, biochemistry and molecular biology.

Genetics. The increasing concern with the molecule basis of biological function is genuinely related to some fundamental questions such as: How do cells "know" when to divide ? How does memory work ? How do we see ? Starting from observation and experiment, genetics developed the idea that discrete entities – later called genes – are the basic units of inheritance of biological function. The existence of genes was first deduced by means of experimental evidence; only almost hundred years later genes were physically isolated.

Biochemistry is another important approach to biological function, interested in the study of actual physical molecules, as opposed to the abstract concepts of genetics. In opposition with genetics, interested in the study of the whole organism, biochemistry takes individual components, each of which being studied separately. So, biochemistry and genetics are complementary disciplines, contrasting as concrete vs abstract and individual vs global. Biochemistry discovered: that some specific molecules (called proteins) are responsible for almost all the functions in a living organism; that proteins are linear chains of building blocks (called amino acids); that there are only 20 types of amino acids; that typically 50 to 500 of these amino acids connect together to make one protein; that proteins fold into unique shapes based on their amino acid sequence.

Molecular biology links genetics and biochemistry, pointing out: the relationship between genes and proteins; the composition of genes, which is made of DNA, proved to be the heredity material of all species; the double helix structure of DNA, fundamental for its function as the agent of storage and transfer of genetic information. This fact is symptomatic for the manner in which, in biology, shape determines properties, that is, structure usually determines function. The double helix structure, with its complementarity principle, is a redundancy phenomenon, helping to face loss or damage of information during the life of a cell. When such a loss or damage occurs, there is a basis for biological mutations. Developing further the information metaphor in biology, we can look at the DNA double helix as a "clever and robust information storage and transmission system[...] Biology ⟪reads⟫ DNA (actually, the RNA copy of it) like a Turing machine tape" (Lander, Langradge & Saccocio, [77], p. 35). Biologists are not yet able to predict accurately the shape in which a protein will fold. Now it is possible to read the specific sequence of individual genes and to predict the sequence of the proteins that they encode. The problem is to use this information to predict the actual biological function of these proteins. Up through 1955, mapping the shape of a molecule was a complicate matter of hand calculations. This explains why crystalographers were among the first users of electronic computers for scientific purposes. Yet, today, even with the most modern technology, it is extremely difficult to establish the structure of a molecule. Until 1991, only 50 or so protein structures were determined per year. The itinerary from DNA and RNA to the final molecular structure or shape is a big challenge of computational molecular biology. We know how to convert into amino acids an arbitrary stretch of DNA, but we do not as yet know *which* stretch should be converted (i.e., most of the DNA does not, in fact, code for particular proteins). We don't know either what the amino acids, once converted, do as proteins. The approach to these problems requires both experimental and theoretical investigations, both mathematical and computational. As Lipton observes, computer scientists are likely to make the greatest contributions to molecular biology not as programmers, but rather as highly trained problem solvers.

An important source of problems in computational biology is the human genome project, which began about ten years ago (in 1988) and "is aimed at determining the location of the estimated 100,000 human genes that comprise the entire human genome. The purpose is also to analyze the structure of DNA, the famous double helical strand of base pairs discovered by James D. Watson and Francis Crick in 1953" (Frenkel, [43], p. 41). When this project will be accomplished, we will be able to decode, in a routine way, the genes that make life possible. "Moving from symbolic to numerical processing, the computational requirements include developing the best and fastest algorithms for comparing sequence data and for the prediction of protein folding, which involves combinatorics, parallel processing and supercomputing" (Frenkel, [43], p. 42). As Frenkel points out, this project provides a setting for AI and neural networks applications, for instance, in defining splice junctions between pieces of DNA. In respect to the language paradigm, central in our interest here, we mention that there are attempts to look at the genetic code as sentences of paragraphs to be deciphered, drawing on pattern recognition and linguistic analysis techniques.

Let us conclude with the remark that computational biology is the concern of a special "Journal of Computational Biology" started in 1994 by Mary Ann Liebert, Inc. The ACM Press has published RECOMB 97, Proceedings of the First Annual International Conference on Computational Molecular Biology (January 20 – 23, 1997, Santa Fe, New Mexico). A special issue of the journal *Discrete Applied Mathematics* (published by North-Holland, Amsterdam), volume 71, numbers 1-3, December 1996, was dedicated to computational molecular biology.

As we have seen, computational biology is concerned with computational models of biological phenomena, the aim of using the former being the knowledge of the latter. The next section is concerned with biological computation, trying the opposite itinerary: how can we improve computation, starting from biological phenomena ?

h) Bridging experimental and theoretical biological computation.

The starting remark for Adleman, [1], was that DNA stores information using a four-letter alphabet and this information is manipulated by living organisms in a similar way computers work through strings over a two-letter alphabet. Adleman's question is simple: Could DNA be made to function like a computer ? The machines around us are becoming more lifelike. Adleman takes the opposite direction: What if life itself could be used to solve problems ? Can DNA shift from reproducing life to *thinking* about it ? (Bass, [7]). It is difficult to overestimate the semiotic weight of these questions. Towards the end of 1993, Adleman had already a design for the world's first molecular computer and its first problem to solve: the so-called Hamiltonian path problem. Taking an arbitrary set of cities a salesman has to travel between, exactly once to each of them, what is the shortest route linking those cities ? When the number of cities grows, this problem becomes hard, it cannot be solved by algebraic equations. Adleman took the case of seven cities connected by fourteen possible routes. The problem was to find an itinerary connecting the cities by a path beginning in Atlanta, ending in Detroit and passing

through each intervening city only once. In order to simplify things, we
will take only four cities from the seven considered by Adleman: Atlanta,
Baltimore, Chicago, and Detroit. Nonstop flights can be taken between
Atlanta-Chicago, Chicago-Detroit, Chicago-Baltimore, Baltimore-Detroit.
It is easy to see that starting in Atlanta and ending in Detroit one can find
three sequential flights that pass through all four cities: from Atlanta to
Chicago, from Chicago to Baltimore, and from Baltimore to Detroit. Adleman's concern was whether DNA can solve such problems. Let us assign
to each city a DNA name comprising six letters, selected at random from
the four-letter DNA alphabet (as a matter of fact, Adleman used 20-letter
names). Atlanta: ATGCGA, Baltimore: CGATCC, Chicago: GCTTAG,
Detroit: GTCCGG. In this way, flights between cities were assigned a DNA
flight number. This number was obtained by combining the last three letters from the DNA name of the originating city with the first three letters
from the DNA name of the destination city. For instance, the flight from
Atlanta to Chicago is from ATGCGA to GCTTAG, so its DNA flight number is CGAGCT. Now every strand of DNA has a complementary strand
obtained by substituting T for A, G for C, A for T, and C for G. When these
complementary strands get near each other, they stick together. This fact
gives DNA its double-stranded spiral structure. In this way, each city has
a complementary DNA name: Atlanta: TACGCT, Baltimore: GCTAGG,
Chicago: CGAATC, Detroit: CAGGCC.

Genetic engineering makes possible to design molecules with a desired
sequence. Employing these techniques, Adleman synthesized 30 trillion
molecules for the complement of each of his seven cities and another 30
trillion molecules for each of the 14 routes. Bass When thrown into a test
tube holding a fiftieth of a teaspoon of aqueous solution, these 60 trillion
DNA sequences stuck together; they formed long chains of DNA flight numbers "splinted" together by DNA complements. After every genetic strand
is splinted with its complement, the answer is contained in the string of
DNA letters (a molecule) that meets the following requirements: it begins
with the DNA flight number for a connecting flight out of Atlanta (CGA),
ends with the DNA flight number for a connecting flight into Detroit (GTC)
and contains the DNA names of all the intervening cities, in a sequence that
solves the problem. However, there is still a difficulty, expressed metaphorically by Adleman (as Bass [7] tells us) in this form: "Hey, waiter, there's a
Hamiltonian path in my soup !" So, the problem is how to fish this winning
molecule out of his test tube. This difficulty was transgressed and Adleman identified strings of DNA molecules in whose alphabet was encoded

the solution to his problem.

In terms of the number of operations performed, Adleman's DNA computer, due to the massively parallel reactions involved in biochemistry (large number of molecules working together) was a hundred times faster than the best of today's serial supercomputers.

Molecular computation is a marvelous example of a challenge where efforts come concomitantly from both experimental and theoretical directions. Adleman's successful experiment showed how standard methods of molecular biology can be used to solve (a small instance of) a computationally intractable problem. Are we able to construct a universal machine out of biological macromolecular components ? Not yet and may be the moment when this goal will be achieved is far away. In the meantime, from a theoretical direction we have good news. As we pointed out, computing by splicing is equivalent to computing by a Turing machine. Can we bridge this gap between Adleman's experimental approach and the theoretical approach ? Splicing is related to ideas and methods of formal language theory, while Adleman's approach is different. There are, however, strong reasons to expect that this gap can be reduced, because universality in respect to computation should match the well-known universality of the genetic alphabet and of the genetic code. As we have seen, Adleman's experiment was based on the double-stranded and complementarity of the DNA and of the massively parallel reactions involved in biochemistry. Similarly, Rozenberg & Salomaa, [115], explain the universality of DNA computing by the fact that Watson-Crick complementarity is in some form present in all models of DNA computing and, at the same time, complementarity has the same features as the twin-shuffle language, known to be a basis of universality (Engelfriet & Rozenberg, [36]). A further step in this investigation is proposed by Salomaa, [117], taking also into account the works by Kari, Păun, Rozenberg, Salomaa & Yu, [69] and by Freund, Păun, Rozenberg & Salomaa, [44]. The paradigm of complementarity yields computational power. Suppose we may encode some things as words x and some other things as words y and assume that some agent leads to a relation $R(x, y)$ (in the case of DNA the agent is just the nature and $R(x, y)$ is just the double strand $A \to T, G \to C, T \to A, C \to G$). Then we can obtain conclusions about x if we know that $R(x, y)$ holds for some y. Salomaa observes that such a conclusion is made by Adleman [1]: the sequence of vertices encoded by x is actually a path in the graph considered (related to the salesman problem).

In order to obtain the twin-shuffle language, TS, consider first the alphabet $\{0, 1, \bar{0}, \bar{1}\}$ (interpretation: $0 = $ A, $1 = $ G, $\bar{0} = $ T, $\bar{1} = $ C), so $\bar{\bar{0}} = 0$

and $\bar{\bar{1}} = 1$. Given a word z over $\{0, 1, \bar{0}, \bar{1}\}$, we can decide in two steps whether z belongs or not to TS: erase first all the barred letters from z, leaving a word z'; then erase all the non-barred letters from z, as well as the bars from the barred letters, leaving a word z''; the word z is in TS if and only if $z' = z''$. Some kind of interplay takes place between complementarity and TS.

The universality of a model of DNA computing means that it can produce the family of recursively enumerable languages (which are just the languages generated by type-0 grammars in Chomsky hierarchy); this family is nothing else than the family of languages that are Turing computable. As examples of universal models of DNA computing we mention the family of acceptors or automata and the family of grammar-like generator devices. Informally speaking, a Watson-Crick finite automaton has a DNA-like alphabet, a finite set of states among which one is initial and some of them are final states, and a finite set of transitions. The inputs are double strands. Two reading heads read the upper and the lower strands, respectively. A double strand is accepted by the automaton if it can be computed by means of available transitions. The language accepted by the Watson-Crick automaton consists of all upper strands of the double strand accepted by the automaton. It is proved that every recursively enumerable language is a weak coding of a language accepted by a Watson-Crick automaton. (For a rigorous version of these concepts see Freund, Păun, Rozenberg & Salomaa, [44], where these automata were introduced.)

In a third step, sticker systems were introduced, as a computational model which is an abstraction of the way Watson-Crick complementarity is used in DNA computing (Păun & Rozenberg, [103]). Watson-Crick complementarity proves to be a powerful tool to bridge these important areas of computability: automata theory and molecular computing (Freund, Păun, Rozenberg & Salomaa, [44]; Kari, Păun, Thierrin & Yu, [70]). Challenged by a new successful experiment (Ouyang et al., [96]), Păun introduced what he calls DNA computing by carving ([102]).

We can infer from the considerations above that automata and formal grammars, on the one hand, and molecular computation, on the other hand, share two important features: massive parallelism (in computation, in the first case, in biochemical reactions, in the second) and complementarity (under the form of the twin-shuffle language, in the first case, and under the form of Watson-Crick complementarity in the second case). This common denominator is the source of hope to reduce more and more the gap between the experimental and the theoretical approaches to molecular computation.

i) Computational brain theories of language. Our guides in this respect will be Chomsky [20] and Schnelle [122], [123]. It is known for longtime that language has a biological foundation; see Lenneberg [78] as a basic reference in this respect. The development of AI and of cognitive science is looking now for its prehistory. For Chomsky, the first cognitive revolution is associated with the Cartesian theory of mind. Schnelle [123] believes that the next step consists of two other theories of mind, both of them due to Leibniz. Cartesian intuitionism is replaced by a symbolically computational theory of the mind which explains human rationality as determined and guided by calculatory combinations of symbols; this is the second cognitive revolution. Chomsky remains within this theoretical framework, valid for most modern approaches to computational and cognitive science, based on logic, mathematics, and computer science. Schnelle refers also to Leibniz' dynamical theory of mind, the theoretical core of Leibniz' monadology; the calculatory theory of mind is only a part of its dynamical theory, the latter generating a third cognitive revolution, whose implications remain to be developed. This is the way Leibniz tries to solve the so-called unification problem of the analysis of language as a mental phenomenon: "How can empirically correct results of linguistics be unified with the other sciences concerned with the pertinent objects of nature, i.e., with the physics, chemistry, and neurology of the brain ?" Two lines of thought were developed in this respect: some researchers, like Chomsky, believe that the unification problem is exclusively the task of neurosciences, while other researchers, like Schnelle, believe that linguistics too has a basic role in this respect. But both Chomsky and Schnelle agree that the limits of the XVIIth century cognitive revolution have not yet been transcended. See, for more, Schnelle, [119].

There is a strong link between the computational mind theories of language and the computational brain theories of language. We will follow the development of the latter, in order to better understand the development of the former. The main shift of focus in this respect was brought by Chomsky's generative viewpoint (for Schnelle, 'generative' is equivalent to 'explicit'; however, the former is more than the latter, because analytic mathematical models of language are also explicit, but they are opposite, complementary to the generative viewpoint). This shift was "from behavior or the product of behavior to states of the mind/brain that enter into behavior" (Chomsky [18], p. 3). 'Behavior' is understood here as externally observable behavior (of human beings or animals), contrasting with the internal behavior of mind/brain.

Let us first observe that already in the XIXth century neurologists developed models of language representation in the brain, motivated by the discovery of language deficits produced by lesions called aphasia. At that time it was implicitly assumed that "the brain parts involved in language production and comprehension are the sites where language is represented in the brain" (Pulvermüller [111], p. 283).

The brain is conceived as a cooperative system of systems. Following Arbib et al. [2], chapters 4 and 5), *cooperative* computation is "the style of the brain". It is important to characterize the architecture of the interacting areas functionally and neuro-anatomically. Linguistics may serve as a pattern in this respect, because, as Schnelle ([122], p. 51) observes, "the idea of an architecture in which complexity is distributed over various components is similar to the linguistic idea of levels and sub-levels of linguistic description (phonology, morphology, syntax, semantics)."

A framework for computational brain theory is due to Churchland & Sejnowski [21]; for them, such a theory is about the central nervous system, its subsystems, neural networks, neurons, synapses, cell membranes, molecules, and ions etc. So, there are many levels of organization. The dynamic properties, laws, or algorithms remain to be specified for each level. These authors support the "cognitivist pecking order", holding that "until cognitive faculties, such as languages, high-level vision etc have been described completely in their symbolic and intentional aspects, modeling these tasks in terms of interactive brain systems is premature". This strategy is opposed to the "realism pecking order", holding that "until the whole neuron is thoroughly and completely modeled, modeling even a small circuit... is premature". These two opposite orders are usually adopted at the same time and they both agree that brain systems modeling is premature and hence computational neuroscience presently impossible. For Churchland & Sejnowski, nervous systems are vast networks of networks, with various regions specializing for various tasks. This ideal recalls the idea of a universal grammar conceived as a hypothetical brain, proposed by Calude, Marcus & Păun [14]. Each particular human competence has its particular generative grammar. How so many grammars find room in our brain, where are they located, how do we succeed to select and identify at any time the grammar we need and, after this, to put it back in its previous place and to use it again when necessary ? We assume the existence of a universal, global competence, a competence of the second order, of a general nature, whose activity is just to manage, to generate particular, individual competences; it is like a meta-grammar depending on one or several parameters; as soon as we choose particular values for these parameters we get a particular grammar, coping with a particular human competence.

Kosslyn & Koenig [74] are primarily concerned with finding out where the components which implement language and other cognitive faculties are located in the brain and how they depend on perceptual preprocessing and motor organizational postprocessing. Controversies in this respect (for instance, how syntactic operations are performed in the brain ?) occur, leading to alternative views by Deane [29], Kean [72], Kosslyn & Koenig [74] and Schnelle [120]. Pinker [106], [107] compares associationism (which describes the brain as a homogeneous network of interconnected units modified by the learning mechanism that records correlations among frequently co-occurring input patterns) and rule-and-representation theories (which describe the brain as a computational device in which rules and principles operate on symbolic data structures). Schnelle [120] proposes a distribution of tasks over brain systems, taking into account that "we need a stratification of interactive areas which cooperate distinctively with parts of the limbic system". See also Schnelle [121] and Fadiga & Gallese [37].

Neurobiologists see cognitive processes in terms of cell assemblies. Karpf [71] proposes a reconsideration of the cognitive hypothesis of modularity: accepted for low-level cognition, modularity should be substituted, for high level cognition, by an interface of systems and subsystems implementing the modules postulated by linguistic grammar theory. If for Chomsky the essential properties of language belong to the innate human nature, for Karpf language is a product of nurture (the distinction "nature – nurture" describes for Karpf the distinction between "innate" and "aquired"). Damasio [26] introduces the concept of *convergence zones* (which relate the classical levels of brain processing) and the notion of *regionalization* (meaning that different domains are formed, which are dedicated to different cognitive tasks). The main idea of Pulvermüller [109] is that neither single neurons nor states of the entire brain, but groups of neurons, Hebbian cell assemblies, correspond to linguistic units or psychological entities.

In a more recent contribution, Pulvermüller [111] gives clear (provisional) answers to the following linguistically relevant questions: 1) which parts of the brain are involved in representing and processing language ? 2) why are these brain regions involved ? 3) which properties of language elements determine their cortical representations ? 4) which neuronal processes underlie the processing of language elements ? Unanswered remains the question 5) how are sequencing rules laid down in the brain and which biological principles govern these rules ? The author's view point is based on large-scale neuronal theories of language having the origin in the Hebbian associationist learning principles (Hebb [58]); his last word is: "A more detailed neuronal grammar is clearly one of the most important targets of future research addressing neurobiology of language."

An important contribution belongs to Edelman [30], who proposes a neo-Darwinian program, trying to show how development (embriology) is related to evolution, how genes affect form through development; he develops an epigenetic theory of language and speech, following which language can only be understood as a system added to and integrated into a sequence of genetically prior biological systems of animal cognition. Edelman is referring to *qualia*, which for him constitute the personal or subjective experiences, feelings, and sensations that accompany experience. "The qualia problem is a semantic problem: the relation of words resp. phrases to experiences is not a relation between word-forms and *entities* (external or internal for the individual) *which are the same for all individuals*, but a relation between word-forms resp. phrases (which are quasi identical for all people who communicate) and experiences which are *essentially different and only globally similar among different individuals*." Schnelle puts in contrast Edelman's view with "a formal theory, in which the meanings of the terms and sentences are strictly defined and therefore identical for all users" (Schnelle [122], p. 92) and observes that in a behaviorist approach to ordinary language like that of Wittgenstein the considered problem does not appear, because it cannot be explained in ordinary language. Things are different in Edelman's approach: the existence of qualia can be proved as well as the impossibility of identical language reference to qualia. Higher order consciousness based on language is essentially fuzzy in respect to our own experience.

The authors quoted so far aimed to show which areas in the brain cooperate to reach certain type of internal and external behavior, how features of intention and understanding of words and sentences are represented in activated cell assemblies distributed over functionally and anatomically different areas and how the network of connectivities emerges genetically and epigenetically (Schnelle [122], p. 94). By means of these achievements, the task to build a computational brain theory of language is formulated by Schnelle as follows (ibidem): a) define formally the "space" S of possible *representational states* (one for each feature representation of partially and completely understood words and sentences in the brain from the set recursively enumerated by a formal grammar); b) determine the *"kinetics" of language use* in the space S, in formally defining the sequence of partial states of representations (e.g., partially understood sentences) leading to complete representational states (e.g., complete understood sentences) or the sequence of complete representational states which cohere (e.g., sentences that make up a *coherent* text; c) define the "dynamics" of language use, i.e., the representational state space in terms of embodied laws, formal constraints, or rules which describe how the kinematically defined sequences (feature structures of partially understood words and sentences) can be computed [...] and, finally, d) define the *dynamics of language acquisition*, i.e., of the emergence of the law or rule embodiments.

Schnelle ([122], pp 95 – 96) formulates a basic requirement for computational cognitive science and linguistics: it has to cope simultaneously with the different levels of formal description, the numerical ones of dynamic systems theory and the symbolic ones familiar to formal linguist. These two types of mathematics, a continuous type in the first case, a discrete one in the second case, are rarely familiar to the same researcher. Following Schnelle (idem, pp. 101 – 102), there are two possible strategies: the first strategy is to develop computational fragments which illustrate the potentialities of the new methods; the second strategy is systematically genetic. The first strategy has in its turn two variants: to start from associationistic psychological frameworks and methods or to start from methods of symbolic representation applied in formal logic and formal linguistics. In the second strategy, primitive, and even proto-primitive, grammatical systems, structurally closely related to the sensorimotor stages of a child's development should first be formalized and specified in computational form.

We are back to the Leibnizian problem of relating mind, expressed in terms of calculi, to the body, specified in terms of physical dynamical systems, in the terms of which linguistic structure should be formulated.

j) Sign systems and living systems. All problems discussed so far can be considered in a semiotic setting. Every time we were confronted with mappings between different sign systems. Reference to language was sometimes by contigency, sometime by analogy and, in this latter case, sometimes by cognitive modeling, sometimes by cognitive metaphor. Important aspects were eluded. To give only one example, we paid no attention to L systems, introduced by Lindenmayer [81] in order to model the development of simple filamentous organisms. Beyond this biological function, L systems are a chapter of formal languages relevant for computation. At the same time, growth functions associated to sequences generated by L systems offer to Kari, Rozenberg & Salomaa ([68], pp. 288 – 301) the occasion of a theatrical scenario with three *dramatis personae*, a formal language theorist, a pure mathematician and a lonesome traveller; they talk about communication and commutativity, merging and stages of death, stagnation and malignancy; biological and medical terminology has here a delicate double status, it accepts both a literal reading and a metaphorical reading ("cancer is sometimes identified with exponential growth").

In respect to the theatrical scenario and the bio-medical metaphors adopted by Kari, Rozenberg & Salomaa, a natural question cannot be avoided. Do these procedures have a purely external, ornamental function, as in classical rhetorics, or are they cognitive means allowing a better, deeper understanding of L systems ? The importance of this question is related to the fact that science – and particularly biology, which is our concern here – are full of metaphorical transfers and we have to elucidate their cognitive function. In a recent paper ([88]), we have argued the importance of cognitive metaphors in today information sciences and their isomorphism with cognitive models; we have argued that such congnitive metaphors impose a kind of dictatorship, they dictate what problems will be investigated, what concepts will be introduced, what methods will be used, what aspects will be in attention and what other aspects will be left aside. However, as soon as such metaphors are no longer able to fulfil their task in respect to the new data and aspects available, they are replaced by other, more powerful metaphors. If, for instance, the steam-engine metaphor was powerful in the XIXth century, it had to be replaced in the second part of the XXth century by the computer metaphor in the field of biology.

A systematic study of the semiotic metaphor in biology belongs to Emmeche & Hoffmeyer [35]. They pay attention to the manner in which vitalistic or finalistic explanations never totally disappear, they reappear in new guises (i.e., metaphors). They show how the neo-Darwinian dillema: Do organisms and parts of them develop their specific forms just because such forms were the most functional, the most successful in reproduction ? was associated with some appropriate metaphors, having a constitutive role in the development of biology. Emmeche & Hoffmeyer [35] are concerned with basic aspects such as information sciences as a new source of metaphors in biology, linguistic and semiotic metaphors in biology, the metaphors of nature as language, the culture-as-nature analogy, life as learning and thought, life as memory systems, organisms as cognitive systems, life/organisms/genetic systems as computers, life as linguistic/semiotic system. A critical approach to the life-as-language metaphor of G. Forti leads, among other results, to the idea that neo-Darwinism cannot account for the role of code-duality between analog and digital codes in evolution. An interesting Peircean perspective on metaphors and genes is proposed. Reference is made to Bateson [8], Berlinsky [13], Campbell [15], Forti [42], Gould [49], Hansen [51], Hawkins [55], Kalmus [66], Löfgren [83], Pattee [98], [99], Picardi [105], Platnick & Cameron [108], Shanon [125], Stuart [126], Weisbuch [129], Zwick [130], to quote only works directly related to the linguistics-biology interference. In direct continuation of this research we have to quote Hoffmeyer & Emmeche [60]. The use and misuse of metaphor in biology, mainly related to the semantics of molecular evolution are the concern of Levinthal [80]. In the same area we mention the important works by Sebeok [124], on initial conditions, and Emmeche [33], on life as a semiotic phenomenon. We call attention on the linguistic-semiotic potential of population genetics (Lancaster [76]) and on the human genom project in its relation to computer science (Frenkel [43]) and on mapping and interpreting biological information (Lander, Langridge & Saccocio [77]). A general non-technical presentation of Adleman's approach is proposed by Bass [7]. Artificial life is considered by Emmeche [31], [32]; the same author is concerned with the computational notion of life [34]. Gerard, Kluckhohn & Rapoport [46] point out analogies between biological and cultural evolution. A review of the semiotic aspects of biology belongs to Hoffmeyer [59]. Jenkins [64] is concerned with language vs genetics. Levelt [79] considers speaking as brain activity. Lipton [82] completes Adleman's approach, Maturana [91] refers to biology of language, Paton [97] concerns computing with biological metaphors.

References

[1] L. Adleman, Molecular computation of solutions to combinatorial problems, *Science*, 226 (1994), 1021 – 1024.

[2] M. A. Arbib, E. J. Conklin, J. C. Hill, *From Schema Theory to Language*, Oxford Univ. Press., Oxford, New York, 1987.

[3] I. Asimov, *Il codice genetico* (transl. from English), Einaudi, Torino, 1968.

[4] C. J. Bailey, *Variation and Linguistic Theory*, Center for Applied Linguistics, Arlington, VA, 1973.

[5] H. S. Bandelt, Phylogenetic networks, *Verh. naturwiss. Ver Hamburg*, NF 34 (1994), 51 – 71.

[6] H. J. Bandelt, P. Forster, B. C. Sykes, M. B. Richards, Mitochondrial portraits of human populations using median networks, *Genetics*, 141 (1995), 743 – 753.

[7] Th. A. Bass, Gene, *Genie Wired*, August 1995, 114 – 117, 164 – 168.

[8] G. Bateson, *Steps to an Ecology of Mind*, Chandler, New York, 1972.

[9] E. Baum, R. Lipton, eds., *DNA Based Computers*, American Mathematical Society, DIMACS Series 27, 1996.

[10] G. Beadle, M. Beadle, *The Language of Life. An Introduction to the Science of Genetics*, New York, 1966.

[11] P. Beckmann, *The Structure of Language. A New Approach*, The Golem Press, Boulder, Colorado, 1972.

[12] J. Beisson, *La génétique*, "Qui sais-je ?" no. 113, Presses Universitaires de France, Paris, 1971.

[13] D. Berlinsky, The language of life, in *Complexity, Language, and Life: Mathematical Approaches* (= *Biomathematics*, 16) (J. L. Casti, A. Karlqvist, eds.), Springer-Verlag, Berlin, 1986, 231 – 267.

[14] C. Calude, S. Marcus, Gh. Păun, The universal grammar as a hypothetical brain, *Revue Roumaine de Linguistique*, 24, 5 (1979), 479 – 489.

[15] J. Campbell, *Grammatical Man. Information, Entropy, Language and Life*, Penguin, New York, 1982.

[16] N. Chomsky, Three models for the description of language, *IRE Trans. in Information Theory*, IT-2, 3 (1956), 113 – 124.

[17] N. Chomsky, *Syntactic Structures*, Mouton, The Hague, 1957.

[18] N. Chomsky, *Knowledge of Language*, Praeger, New York, 1986.

[19] N. Chomsky, *Language and Problems of Knowledge*, MIT Press, Cambridge, Mass., 1988.

[20] N. Chomsky, *Language and Thought*, Moyer Ball, Wakefield, RI, and London, 1993.

[21] P. S. Churchland, T. J. Sejnowski, *The Computational Brain*, MIT Press, Cambridge, Mass., 1992.

[22] B. F. C. Clark, K. A. Marcher, How proteins start, *Scientific American*, 218 (Jan. 1968).

[23] L. D. Creangă, Fl. Filipoiu, Fl. Petrescu, Picture grammars in molecular genetics, *Revue Roumaine de Linguistique – Cahiers de Linguistique Thèorique et Appliquée*, 17, 1 (1980), 49 – 57.

[24] F. H. C. Crick, The genetic code, *Scientific American*, 215 (Oct. 1966).

[25] D. Crystal, *The Cambridge Encyclopedia of Language*, Cambridge Univ. Press, Cambridge, 1987.

[26] A. R. Damasio, Time-locked multiregional retroactivation: A system-level proposal for the neural substrates of recall and recognition, *Cognition*, 33 (1989), 25 – 62.

[27] A. R. Damasio, H. Damasio, Brain and language, *Scientific American*, 267 (1992), 62 – 71.

[28] R. Dawkins, *The Selfish Gene*, Oxford Univ. Press, New York, Oxford, 1976.

[29] D. Deane, *Grammar in Mind and Brain*, Mouton de Gruyter, Berlin, New York, 1992.

[30] G. M. Edelman, *Bright Air, Brilliant Fire: On the Matter of Mind*, Basic Books, New York, 1992.

[31] C. Emmeche, Life as an abstract phenomenon: is Artificial Life possible ?, in *Toward a Practice of Autonomous Systems. Proc. of the First European Conf. on Artificial Life* (F. J. Vasela, P. Bourgine, eds.), MIT Press, Cambridge, Mass., 1992, 466 – 474.

[32] C. Emmeche, *The Garden in the Machine. The Emerging Science of Artificial Life*, Princeton Univ. Press, Princeton, 1994.

[33] C. Emmeche, The computational notion of life, *Theoria*, Segunda Epoca, 9, 21 (1994), 1 – 30.

[34] C. Emmeche, Defining life as a semiotic phenomenon, *VIth International Congress "Semiotics bridging nature and culture"*, International Association for Semiotic Studies, Guadalajara, Mexico, July 13 – 18, 1997.

[35] C. Emmeche, J. Hoffmeyer, From language to nature. The semiotic metaphor in biology, *Semiotica*, 84, 1/2 (1991), 1 – 42.

[36] J. Engelfriet, G. Rozenberg, Fixed point languages, equality languages and representations of recursively enumerable languages, *J. of the ACM*, 27 (1980), 499 – 518.

[37] L. Fadiga, V. Gallese, Action representation and language in the brain, *Theoretical Linguistics*, 23, 3 (1997), 267 – 280.

[38] R. W. Floyd, On the non-existence of a phrase structure grammar for Algol 60, *Comm. of the ACM*, 5 (1962), 483 – 484.

[39] P. Forster, Network analysis of word lists, *Third Intern. Conference on Quantitative Linguistics*, August 26 – 29, 1997, Research Institute for the Languages of Finland, 1997, 184 – 186.

[40] P. Forster, R. Harding, A. Torroni, H. J. Bandelt, Origin and evolution of Native American mtDNA variation: a reappraisal, *American J. of Human Genetics*, 59 (1996), 935 – 945.

[41] P. Forster, A. Toth, H.-J. Bandelt, Phylogenetic network analysis of word lists, *International Conf. on Quantitative Linguistics*, Helsinki, August 1997.

[42] G. Forti, Structure and evolution in language and in living beings, *Scientia*, 112 (1977), 69 – 79.

[43] K. A. Frenkel, The human genome project and informatics, *Comm. of the ACM*, 34, 11 (1991), 41 – 51.

[44] R. Freund, Gh. Păun, G. Rozenberg, A. Salomaa, Watson-Crick finite automata, *Proc. of the Third DIMACS Workshop on DNA Based Computers*, June 23 – 25, 1997, Philadelphia, 305 – 317.

[45] N. S. Galles, Biological foundations of linguistic diversity, *Theoretical Linguistics*, 23, 3 (1997), 251 – 266.

[46] R. Gerard, C. Kluckhohn, A. Rapoport, Biological and cultural evolution: some analogies and explorations, *Behavioral Science*, 1 (1956), 6 – 31.

[47] S. Ginsburg, H. G. Rice, Two families of languages related to ALGOL, *Technical Memo* 578/000/1, SDC, Santa Monica, Calif., 1961.

[48] H. Goebl, "Stammbaum" und "Welte", *Zeitschrift für Sprachwiss.*, 2 (1983), 3 – 44.

[49] S. J. Gould, Darwinism and the expansion of evolutionary theory, *Science*, 216 (1982), 380 – 387.

[50] L. Grassi, Il codice linguistico e altri codici: il codice genetico, *Versus. Quaderni di studi semiologici*, 3 (1972), 93 – 111.

[51] O. Hansen, Are the genes of universal grammar more than structural ? *Hereditas*, 95 (1981), 213 – 218.

[52] D. J. Haraway, *Crystals, Fabrics, and Fields; Metaphors of Organicism in Twentieth-Century Developmental Biology*, Yale Univ. Press, New Haven, 1976.

[53] Z. S. Harris, *Structural Linguistics*, Phoenix Books, The Univ. of Chicago Press, Chicago, 1961.

[54] P. E. Hartman, S. R. Suskind, *Gene Action*, Foundations of Modern Genetics Series, Prentice Hall, Englewood Cliffs, New Jersey, 1965.

[55] D. Hawkins, *The Language of Life. An Essay in the Philosophy of Science*, W. H. Freeman and Co., San Francisco, 1964.

[56] T. Head, Formal language theory and DNA: an analysis of the genetic capacity of specific recombinant behaviors, *Bull. Math. Biology*, 49 (1987), 737 – 759.

[57] T. Head, Gh. Păun, D. Pixton, Language theory and molecular genetics: generative mechanisms suggested by DNA recombination, in *Handbook of Formal Languages* (G. Rozenberg, A. Salomaa, eds.), vol. 2, Springer-Verlag, Berlin et al., 1997, 295 – 360.

[58] D. O. Hebb, *The Organization of Behavior. A Neuropsychological Theory*, Wiley, New York, 1949.

[59] J. Hoffmeyer, Semiotic aspects of biology: Biosemiotics, in *Semiotics. A Handbook on the Sign-Theoretic Foundations of Nature and Culture* (R. Possner, K. Robering, Th. A. Sebeok, eds.), Walter de Gruyter, Berlin, New York, 1997.

[60] J. Hoffmeyer, C. Emmeche, Code-duality and the semiotics of nature, in *On Semiotic Modeling* (M. Anderson, F. Merrell, eds.), Mouton de Guyter, Berlin, New York, 1991, 117 – 166.

[61] F. Jacob, *La logique du vivant. Une histoire de l'hérédité*, Gallimard, Paris, 1970.

[62] R. Jakobson, Linguistics, Chapter VI in *Main Trends of Research in the Social and Human Sciences*, I, Mouton Unesco, Paris, The Hague, 1970, 419 – 463.

[63] R. Jakobson, *Essais de linguistique générale*, vol. II, *Rapports internes et externes du langage*, Argument 57, Les Editions de Minuit, Paris, 1973.

[64] L. Jenkins, Language and genetics, *Theoretical Linguistics*, 5, 1 (1978), 77 – 82.

[65] N. K. Jerne, The generative grammars of the immune system, *Science*, 229 (1985), 1057 – 1059.

[66] H. Kalmus, Analogies of language to life, *Language and Speech*, 5, 1 (1962), 15 – 25.

[67] S. Kaneff, ed., *Picture Language Machines*, Academic Press, New York, 1970.

[68] L. Kari, G. Rozenberg, A. Salomaa, L systems, in *Handbook of Formal Languages* (G. Rozenberg, A. Salomaa, eds.), vol. 1, Springer-Verlag, Berlin et al., 1997, 253 – 328.

[69] L. Kari, Gh. Păun, G. Rozenberg, A. Salomaa, S. Yu, DNA computing, matching systems and universality, *TUCS Technical Report*, 49, 1996.

[70] L. Kari, Gh. Păun, G. Thierrin, S. Yu, Characterizing recursively enumerable languages using insertion-deletion systems, *Proc. of the Third DIMACS Workshop on DNA Based Computers*, June 23 – 25, 1997, Philadelphia, 318 – 333.

[71] A. Karpf, *Selbstorganisationsprozesse in der sprachlichen Ontogenese: Erst- und Freundsprache*, G. Narr, Tübingen, 1993.

[72] M. L. Kean, On the biologically real models or human linguistic capacity, in *Formal Grammars: Theory and Implementation* (R. Levin, ed.), Oxford Univ. Press, New York, Oxford, 1992.

[73] S. C. Kleene, Representation of events in nerve nets and finite automata, in *Automata Studies* (C. E. Shannon, McCarthy, eds.), Princeton Univ. Press, 1956.

[74] S. T. M. Kosslyn, O. Koenig, *Wet Mind – The New Cognitive Neuroscience*, Free Press, New York, 1992.

[75] W. Labow, Objectivity and commitment in linguistic science; the case of the Black English trial in Ann Arbor, *Language and Society*, 11 (1982), 165 – 201.

[76] P. Lancaster, Population genetics, in *Mathematical Models of the Real World*, Prentice Hall, Englewood Cliffs, New Jersey, 1976, 104 – 127.

[77] E. S. Lander, R. Langridge, D. M. Saccocio, Mapping and interpreting biological information, *Comm. of the ACM*, 34, 11 (1991), 33 – 39.

[78] E. H. Lenneberg, *Biological Foundations of Language*, Wiley, New York, 1967.

[79] W. Levelt, *Speaking*, MIT Press, Cambridge Mass., 1989.

[80] M. Levinthal, The use and misuse of metaphor in biology: Studies of molecular evolution, in *On Semiotic Modeling* (M. Anderson, F. Merrell, eds.), Mouton de Gruyter, Berlin, New York, 1991, 167 – 184.

[81] A. Lindenmayer, Mathematical models for cellular interaction in development; I and II, *J. of Theoretical Biology*, 18 (1968), 280 – 315.

[82] R. Lipton, DNA solution of hard computational problems, *Science*, 268 (1995), 542 – 545.

[83] L. Löfgren, Life as an autolinguistic phenomenon, in *Autopoiesis: A Theory of Living Organization* (M. Zeleny, ed.), North Holland, Amsterdam, 1981, 236 – 249.

[84] V. Manca, E-mail message to S. Marcus, 14 April 1997.

[85] V. Manca, String rewriting and metabolism: A logical perspective, in the present volume.

[86] V. Manca, D. Martino, From string rewriting to logical metabolic systems, in *Grammatical Models of Multi-Agent Systems* (Gh. Păun, A. Salomaa, eds.), Gordon and Breach, London, 1998.

[87] S. Marcus, Linguistic structures and generative devices in molecular genetics, *Cahiers de Linguistique Thèorique et Appliquée*, 11, 1 (1974), 77 – 104.

[88] S. Marcus, Metaphors as dictatorship, in *Welt der Zeichen-Welt der Dinge, World of Signs-World of Things* (J. Bernard, J. Wallmannsberger, G. Withalm, eds.), Akten der 8 Symposium der Österreichischen Geselschaft für Semiotik, Insbruck, 1993, ÖGS, Wien, 1997, 87 – 108.

[89] A. Martinet, *Élèments de linguistique générale*, Armand Colin, Paris, 1967.

[90] R. D. Masters, Genes, languages and evolution, *Semiotica*, 2, 4 (1970), 295 – 320.

[91] H. R. Maturana, Biology of language: the epistemology of reality, in *Psychology and Biology of Language and Thought: Essays in Honor of Eric Lenneberg* (G. A. Miller, E. Lenneberg, eds.), Academic Press, New York, 1978, 27 – 63.

[92] T. de Mauro, Modelli semiologici: l'arbitrarietà semantica, *Lingua e Stile*, 1, 1 (1966), 37 – 63.

[93] W. S. McCulloch, E. Pitts, A logical calculus of the ideas imminent in nervous activity, *Bulletin of Mathematical Biophysics*, 5 (1943), 115 – 133.

[94] J. Monod, *Le hasard et la nècesité. Essau sur la philosophie naturelle de la biologie moderne*, Seuil, Paris, 1970.

[95] W. Nöth, *Handbook of Semiotics*, Indiana Univ. Press, Bloomington, 1990.

[96] Q. Ouyang, P. D. Kaplan, S. Liu. A. Libchaber, DNA solution of the maximal clique problem, *Science*, 278 (October 1997), 446 – 449.

[97] R. C. Paton, ed., *Computing with Biological Metaphors*, Chapman and Hall, New York, 1994.

[98] H. H. Pattee, How does a molecule become a message ?, *Developmental Biology Supplement*, 3 (1969), 1 – 16.

[99] H. H. Pattee, Dynamic and linguistic modes of complex systems, *Intern. J. of General Systems*, 3 (1977), 259 – 266.

[100] Z. Pawlak, *Gramatyka i Matematika*, Panstwowe Zakady Wydawnietw Szkolnych, Warzsawa, 1965.

[101] Gh. Păun, On the place of programming languages in Chomsky's hierarchy, *Stud. Cerc. Matem.*, 33, 4 (1981), 455 – 466 (in Romanian).

[102] Gh. Păun, (DNA) Computing by carving, *Report CST – 97-17 of the Center for Theoretical Studies*, Charles Univ. and the Academy of Sciences of the Czech Republic, December 1997.

[103] Gh. Păun, G. Rozenberg, Sticker systems, *Theoretical Computer Sci.*, 205 (1998).

[104] Gh. Păun, G. Rozenberg, A. Salomaa, *DNA Computing. New Computing Paradigms*, Springer-Verlag, Berlin, New York, 1998.

[105] E. Picardi, Some problems of classification in linguistics and biology, 1800 – 1830, *Historiographia Linguistica*, 4, 1 (1977), 31 – 57.

[106] S. Pinker, Rules of language, *Science*, 25 (1991), 530 – 534.

[107] S. Pinker, Regular and irregular morphology and the psychological status of rules of grammar, *Proc. of the 17th Annual Meeting of the Berkeley Linguistic Society*, Berkeley Linguistic Society, Berkeley, CA, 1992.

[108] N. I. Platnick, H. Don Cameron, Cladistic methods in textual, linguistic, and phylogenetic analysis, *Systematic Zoology*, 26 (1977), 380 – 385.

[109] F. Pulvermüller, Constituents of a neurological theory of language, *Concepts of Neuroscience*, 3 (1992), 157 – 200.

[110] F. Pulvermüller, On connecting syntax and the brain, in *Brain Theory* (A. Aertsen, ed.), Elsevier Science Publ., Amsterdam, 1993.

[111] F. Pulvermüller, Brain-theoretical perspectives on language, *Theoretical Linguistics*, 23, 3 (1997), 281 – 302.

[112] V. Ratner, Lineinaja uporjadočennost geneticeskih soobscenii, *Problemy Kibernetiki*, 16 (1966).

[113] M. Richards, H. Corte-Real, P. Forster, V. Macaulay, H. Wilkinson-Herbots, A. Demaine, S. Papihar, R. Hedges, H. J. Bandelt, B. Sykes, Paleolitic and neolitic lineages in the European mitochondrial gene pool, *American J. of Human Genetics*, 59 (1996), 185 – 203.

[114] A. Rosenfeld, *Picture Processing by Computer*, Academic Press, New York, London, 1969.

[115] G. Rozenberg, A. Salomaa, Watson-Crick complementarity, universal computations, and genetic engineering, *Computer Sci. Technical Report* 28, Leiden University, 1996.

[116] A. Salomaa, *Formal Languages*, Academic Press, New York, 1973.

[117] A. Salomaa, Turing, Watson-Crick and Lindenmayer. Aspects of DNA complementarity, in *Unconventional Models of Computation* (C. Calude, J. Casti, M. J. Dinneen, eds.), Springer-Verlag, Singapore, 1998, 94 – 107.

[118] A. Schleicher, *Die Darwinsche Theorie und die Sprachwissenschaft*, Weimar, 1863.

[119] H. Schnelle, From Leibniz to Artificial Intelligence, in *Topics in the Philosophy of Artificial Intelligence* (L. Albertazzi, R. Poli, eds.), Mitteleuropäisches Kulturinstitute, Bozen, 1991.

[120] H. Schnelle, Sprache und Gehirn, *Kognitionswissenschaft*, 4 (1994), 1 – 16.

[121] H. Schnelle, Language and brain, in *Origins of Semiotics* (W. Nöth, ed.), Mouton-Walter de Gruyter, Berlin, 1993, 339 – 363.

[122] H. Schnelle, Approaches to computational brain theories of language. A review of recent proposals, *Theoretical Linguistics*, 22, 1/2 (1996), 49 – 104.

[123] H. Schnelle, Reflections on Chomsky's "Language and Thought", *Theoretical Linguistics*, 22, 1/2 (1996), 105 – 124.

[124] Th. A. Sebeok, Semiotics and the Biological Sciences: Initial Conditions, *Discussion Paper*, 17, 1D30, Collegium Budapest Institute for Advanced Study, Budapest, 1996.

[125] B. Shanon, The genetic code and human language, *Synthèse*, 39 (1978), 401 – 415.

[126] C. I. J. M. Stuart, Bio-informational equivalence, *J. of Theoretical Biology*, 113 (1985), 511 – 636.

[127] M. Swadesh, Towards a greater accuracy in lexicostatistics dating, *International J. of American Linguistics*, 21 (1955), 121 – 137.

[128] S. D. Watson, *Biologie molèculaire du gène*, Ediscience, Paris, 1968.

[129] G. Weisbuch, Networks of automata and biological organization, *J. of Theoretical Biology*, 121 (1986), 255 – 267.

[130] M. Zwick, Some analogies of hierarchical order in biology and linguistics, in *Applied General Systems Research* (= NATO Conference Series) (G. J. Klir, ed.), Plenum Press, New York, 1978, 521 – 529.

String Rewriting and Metabolism:
A Logical Perspective

Vincenzo MANCA

Dipartimento di Informatica, Università di Pisa
Corso Italia 40, 56125 Pisa, Italy
mancav@di.unipi.it

1   Introduction

String rewriting is the basis of formal language theory [15], [16], where symbolic systems introduced by Chomsky [3], Marcus [11], Lindenmayer [9], and Head [7] with their main variants and related formalisms [4], [5], [6], [14], seem to amount, according to reasonable taxonomic and enumerative criteria, to several hundreds of different types.

However, all these systems can be easily reduced to the following common structure:

$$S = (L_0 \, , \, \Rightarrow^* , \, L_f),$$

where L_0 is the *initial* (usually finite) language, $\Rightarrow^*$ is the reflexive and transitive closure of the yield relation $\Rightarrow$ (on the strings of some alphabet), and L_f is the *final* language (usually all the strings of a finite alphabet of terminal symbols).

The distinguishing feature of this system is the rewriting relation: its combinatorial schema (replacement, insertion, splicing), its sequential or parallel nature, its rules and regulation strategy, and its generative mechanism, usually expressed by the generated language $L(S)$:

$$L(S) = \{\alpha \, \in L_f \mid \exists \, \beta \in L_0 \, (\beta \Rightarrow^* \alpha)\}.$$

We indicate in this paper how a general and rigorous definition of a string generative system, based on the above triple, can be developed in logical terms. In such a perspective, we indicate how all the important aspects of the usual systems can be formally and uniformly described. A rewriting relation $\Rightarrow$ can be logically represented if it is possible to determine a suitable first order model $\mathcal{M}$ and a first order formula φ such that:

$$\alpha \Rightarrow \beta \;\; \text{iff} \;\; \varphi(\alpha, \beta) \;\; holds \;\; in \;\; \mathcal{M}.$$

This paper outlines some initial steps which derive from this definition. Section 2 introduces the notion of representability within first order models and gives some introductory examples. Section 3 presents the logical representation of typical syntactical relations and provides the logical characterization of classical classes of formal languages. Section 4 shows the computational universality of three different notions of logical representability. These notions are based on: i) the *standard syntactic model* SYN on the domain ω^* of finite strings of natural numbers, ii) the theory RRR inspired by Raphael Robinson's arithmetical theory RR, and iii) the $\Sigma(SYN)_1$-formulae resembling the Σ_1-formulae of arithmetical hierarchy. Due to the aforementioned universality results, it is reasonable to define a set or a relation as (*logically*) ω^*-*representable* iff it is representable according to one of these three methods. Therefore, ω^*-*representability* is the key concept for a general, logical definition of (string) generative system: it is a system $(L_0, \Rightarrow, L_f)$ where all three components are ω^*-representable. We will develop in a future work a detailed analysis of the most important string rewriting systems (not only generative), aimed at showing that each of them is completely and naturally described in logical terms. However, the examples and the results given in the present paper constitute a well founded basis for a rigorous, logical analysis of rewriting.

Our approach implies the possibility of unification and comparison of well-known symbolic systems, and provides a natural framework where different aspects of classical systems could be naturally combined and integrated into new computational formalisms and Artificial Life models [13]. Moreover, logical rewriting can be extended to deal with the new perspectives opened by DNA and molecular computing. To this end, we show that an interesting notion of *metabolism*, suggested by chemical and biochemical processes, can be formalized in terms of logical representability within a first order model which extends SYN. In this model, called $META$, several important phenomena, such as distribution, cooperation, and complex levels of syntactical aggregation, can easily be expressed. The final section introduces the basic definitions and presents some examples which illustrate this potentiality of logical rewriting.

2 Logical Representability

Assume the 'marvelous' 7 logical symbols:

$$\rightarrow, \neg, \wedge, \vee, \leftrightarrow, \forall, \exists$$

with the standard syntactical and semantical first order logical notions (equality predicate = is assumed in its usual usage) that can be found in introductory treatises or basic chapters of textbooks in mathematical logic (cf., for example, [1], [2], [8], or [17]).

We recall that $\models$ is the predicative formalization of proof consequence $\Rightarrow$, according to the specific sense developed by mathematical logic. More precisely, this symbol has two different, though related, meanings. We write:

$$\mathcal{M} \models \varphi$$

to mean *validity w.r.t. models*, inasmuch as it expresses that the formula φ holds in the model $\mathcal{M}$, assuming that all the symbols of φ can be interpreted in $\mathcal{M}$ (φ is a $\Sigma(\mathcal{M})$-formula, where $\Sigma(\mathcal{M})$ is the signature of the model $\mathcal{M}$). A natural extension of this meaning is:

$$\mathcal{M} \models \Phi,$$

where Φ is a set of formulae, which expressees that $\mathcal{M} \models \varphi$ holds for any formula φ of Φ.

The second meaning of $\models$ is the first order *logical validity*. In this sense:

$$\Phi \models \varphi$$

means that for all first order models (more exactly Σ-models, if Σ is the signature of the language of Φ):

$$\mathcal{M} \models \Phi \Rightarrow \mathcal{M} \models \varphi.$$

The two meanings are related, in fact if we put:

$$Th(\mathcal{M}) = \{\psi \mid \mathcal{M} \models \psi\},$$

then

$$(\mathcal{M} \models \varphi) \Leftrightarrow (Th(\mathcal{M}) \models \varphi).$$

Given a formula $\varphi(x)$ with a free variable x and an individual term t, we indicate by $\varphi(t)$ the formula obtained from $\varphi(x)$ by replacing in it all the occurrences of x with the individual term t.

Definition 2.1 *Let* Σ *be a signature and* $\mathcal{M}$ *be a* Σ*-model. A subset* S *of the domain of* $\mathcal{M}$ *is represented in* $\mathcal{M}$ *by the formula* $\varphi(x)$ *with a free variable* x *when:*

$$a \in S \Leftrightarrow \mathcal{M} \models \varphi(a).$$

Note that $\varphi(a)$ is generally a formula in the signature which extends Σ with the elements of the set S considered as new individual constants. In the same manner we could define the representability of properties and relations within a given model.

Notation 2.1 *When a set, or a relation, is representable within a model* $\mathcal{M}$ *by a formula* φ *that belongs to a class* Γ *of formulae, we say that it is* $\mathcal{M}$*-representable by* Γ.

Let AR be the *standard arithmetical model*:

$$AR = (\omega, +, \cdot, 0, 1),$$

where ω is the set of natural numbers, $+$ and $\cdot$ denote the plus and times operations on natural numbers respectively, and 0 and 1 denote numbers zero and one, respectively. As usual, we may use, ambiguously, the same symbols to indicate either the symbols of a signature Σ, or the corresponding functions, relations and individuals of a Σ-model.

Example 2.1 *The set of prime numbers is represented in* AR *by the following formula of the first order language of signature* $\{+, \cdot, 0, 1\}$:

$$\forall y\, z\, (\neg x = 1 \wedge (x = y \cdot z \rightarrow x = y \vee x = z)).$$

Notation 2.2 $S \equiv_{\mathcal{M}} \varphi$ *indicates that* φ *represents* S *within* $\mathcal{M}$. *When the model* $\mathcal{M}$ *is clearly understood, the subscript* $\mathcal{M}$ *may be omitted.*

The following example is a list of very simple formalizations within the standard arithmetical model AR, where $P(n)$ means that n is a prime number, $|P| = \aleph_0$ means that the set of prime numbers is denumerable, and $<, >$ is Cantor's pairing function (cf. [17]):

Example 2.2

$$x \leq y \equiv \exists z (x + z = y),$$
$$x = 0 \equiv \forall y (x + y = y),$$
$$|P| = \aleph_0 \equiv \forall x \exists y (P(y) \wedge y \geq x),$$
$$< x, y >= z \equiv (z + z = (x + y + 1) \cdot (x + y) + x + x).$$

In order to develop a logical representation of syntactical notions we introduce the following model SYN which we call the *standard syntactic model*:

$$SYN = (\omega^*, \; --, \; |\,|, 0, \lambda).$$

where ω^* is the set of finite strings of natural numbers, $\_\_(\alpha, \beta)$ abbreviated
by $\alpha\beta$ is the concatenation of strings α and β, and $|\alpha|$ is the length of string
α, and 0, λ are the constants for zero and the empty string. In this model
numbers can be conceived as symbols of an infinite alphabet ω, therefore
any language can be embedded in its domain (we may identify $0, 1, 2, \ldots$
with letters $a, b, c, \ldots$).

We write: $\alpha(i) = n$ to say that symbol (number) n occurs in the string
α at position i, $\alpha \preceq \beta$ to say that α is a substring of β, i.e., the string
constituted by the symbols of α which occur between two (not necessarily
distinct) positions; and $a \in \alpha$ iff $a \preceq \alpha$ and $|a| = 1$.

Example 2.3

$$y \in \omega \;\equiv\; (|y| = \||y\||),$$
$$(x(i) = y) \;\equiv\; \exists yz(x = zyw \;\wedge\; y \in \omega \;\wedge\; |zy| = i),$$
$$(x \preceq y) \equiv \exists zw(y = zxw),$$
$$(x \in y) \equiv (x \preceq y \wedge |x| = 1),$$
$$(x + y = z) \;\equiv\; \exists uv(x = |u| \;\wedge\; y = |v| \;\wedge\; z = |uv|).$$

Example 2.4 *The languages* $\{a^n b^n \mid n \in \omega\}, \{a^n b^n c^n \mid n \in \omega\}$ *are SYN-*
representable:

$$\alpha \in \{a^n b^n \mid n \in \omega\} \;\equiv\; \exists xy(\alpha = xy \wedge |x| = |y| \wedge,$$
$$\forall uv((u \preceq x \wedge |u| = 1 \to u = a) \wedge (v \preceq y \wedge |v| = 1 \to v = b)),$$
$$\alpha \in \{a^n b^n c^n \mid n \in \omega\} \;\equiv\; \exists xyz(\alpha = xyz \wedge |x| = |y| \wedge |x| = |z| \wedge,$$
$$\forall uvw((u \preceq x \wedge |u| = 1 \to u = a) \wedge$$
$$(v \preceq y \wedge |v| = 1 \to v = b) \wedge$$
$$(w \preceq z \wedge |w| = 1 \to w = c))).$$

Note that with the representation of $+$ in SYN we inherit in it all the
arithmetical relations whose *definiens* uses such a function symbol. The
following example is more complicated. How can we represent the product
between natural numbers in terms of relationships among strings? We
can get a solution to this question by representing the natural process of

calculating a product by means of iterated sums. In fact if $m \cdot k = n$, there exists a sequence of natural numbers $m, m \cdot m, m \cdot m \cdot m, \ldots$, of length k that ends with n.

Example 2.5

$$m \cdot k = n \equiv$$
$$\exists w(|w| = k \wedge w(1) = m \wedge$$
$$\forall x(x < k \rightarrow w(x+1) = w(x) + m) \wedge w(k) = n).$$

The idea of the previous example has a wide application. Moreover, notice that the simple structure of the formula representing multiplication is mainly due to the possibility of *SYN*-representing computations directly (thus avoiding all kinds of syntax encoding, e.g., Gódel's β-function, cf. [17]). In fact, we have:

Example 2.6 *The following relations are representable in the model* SYN, *where π_k is the k-th prime number, $G(n, m, k)$ indicates that m is a prime number, and m^k is the maximum power of m that is divisor of n:*

$$n = m^k,$$
$$n = m!,$$
$$n = \pi_k,$$
$$G(n, m, k).$$

3   Logical Syntax of Formal Languages

We assume as known the classical notions of formal language theory; for further details see [15], [16]. We only recall briefly some basic definitions in order to fix notations.

Given a finite alphabet A of symbols, $A = \{a_1, \ldots, a_n\}$, we can embed it into a finite subset N_A of natural numbers by means of a one-to-one mapping f between A and N_A. In this manner a corresponding one-to-one mapping f' between A^* and N_A^* is defined in the natural way:

$$f'(\lambda) = f(\lambda) = \lambda,$$
$$f'(a_i \alpha) = f(a_i) f'(\alpha) \text{ for each } a_i \in A, \ \alpha \in A^*.$$

Therefore, we can consider languages over alphabet A as particular subsets of the free monoid ω^*. This embedding allows us to represent syntactic properties in the model *SYN*.

Let us consider a Chomsky grammar $G = (T, N, P, S)$, where T is the finite alphabet of terminals, N is the finite set of nonterminals, P is the finite set of productions $\alpha \to \beta$, with $\alpha \in V^+$, $\beta \in V^*$ (V^* being the free monoid generated by V, and $V^+ = V^* - \{\lambda\}$) and, finally, $S \in N$ is the start symbol. From S and replacing iteratively in a string the left side of a production with the right one, a grammar G generates a type-0 language $L(G)$ constituted by all $\alpha \in T^*$ generated from S with a finite number of replacements. Type-0 languages coincide with the recursively enumerable subsets of T^*, i.e., the subsets generated by some process that is effective w.r.t. some universal computation formalism.

A *substitution* f is a mapping from a finite alphabet A to the powerset of A: $f : A \to 2^A$. Thus, f associates some language with each symbol of A. Such a mapping can naturally be extended to strings in A^* in the following manner (here juxtaposition indicates concatenation between languages):

$$f(\lambda) = \lambda,$$
$$f(ax) = f(a)f(x), \ a \in A, \ x \in A^*.$$

Further, we can also extend f to languages $L \subseteq A^*$ by defining

$$f(L) = \bigcup_{x \in L} f(x).$$

If $f(a)$ is a singleton for all $a \in A$, then we say that f is a *homomorphism* in A, and we identify $f(a)$ with its unique element.

Consider the following syntactical relation:

- $sub(\alpha, i, j, \beta)$ means that the string β is the substring of α starting at position $i + 1$ and ending at position j: $\alpha[i, j] = \beta$, $i < j$. For example:

$$\alpha = b \ a \ \overbrace{b \ c \ a \ d}^{\beta} \ b, \text{ with } i = 2 \text{ and } j = 6.$$

- $occur(\alpha, a, n)$ means that in the string α the symbol a occurs exactly n times: $|\alpha|_a = n$,

- $hom_f(\alpha, \beta)$ means: $f(\alpha) = \beta$, where f is a string homomorphism,

- $perm(\alpha, \beta)$ means that α is a permutation of β,

The following lemma will be useful in our discussion.

Lemma 3.1 *Syntactical relations: sub, occur, perm, and hom$_f$ are SYN-representable.*

Proof. We proceed as follows:
  1. *sub*

$$(\alpha[i, j] = \beta) \equiv (i = j \to \beta = \lambda) \land$$
$$(i < j \to (\exists \, \gamma, \delta \, (\alpha = \gamma\beta\delta \land \mid\gamma\mid = i \land \mid\gamma\beta\mid = j))).$$

  2. *occur*

$$(\mid\alpha\mid_a = n) \equiv \exists\nu(\mid\nu\mid = n \land \forall i(1 \leq i \leq n \leftrightarrow \alpha(\nu(i)) = a))$$
$$\land\forall i \leq \mid\alpha\mid(\alpha(i) = a \to \exists j \leq n(\nu(j) = i)).$$

  3. *perm*

$$perm(\alpha, \beta) \equiv \forall x(x \in \alpha \leftrightarrow x \in \beta) \land \mid\alpha\mid = \mid\beta\mid$$
$$\land\forall x(x \in \alpha \to \mid\alpha\mid_x = \mid\beta\mid_x).$$

  4. *hom$_f$*

$$f(\alpha) = \beta \equiv (\alpha = \lambda \to \beta = \lambda) \land (\alpha \neq \lambda \to$$
$$(\exists \nu \, \forall i \, (i \leq \mid\nu\mid \to \beta[\nu(i), \nu(i + 1)] = f(\alpha(i))))).$$

□

Example 3.1

homomorphism: $f(c) = ab,\ f(d) = ccc,\ f(e) = \lambda$

1	2	3	4	5	6	7	8		(*i*)
c	*d*	*e*	*c*	*d*	*e*	*e*	*c*		(*α*)

↓	↓	↓	↓	↓	↓ ↓	↓		↓

a b	*c c c*	*λ*	*a b*	*c c c*	*λ*	*λ*	*a b*	(*β*)
1 2	3 4 5		6 7	8 9 10			11 12	([$\nu(i)$, $\nu(i + 1)$])

0	2	5	5	7	10	10	10	12	(*ν*)
1	2	3	4	5	6	7	8	9	(*i*)

A *Dyck* language is a context-free language generated by a grammar

$$G_k = \{\{a_1, a_2, \ldots, a_k, b_1, b_2, \ldots, b_k\}, \{S\}, P, S\},$$

where P consists of productions $\{S \rightarrow SS, S \rightarrow \lambda\} \cup \{S \rightarrow a_i S b_i \mid 1 \leq i \leq k\}$. $L(G_k)$ can be considered as the language of correct strings over parentheses of k types (where a_i and b_i are each pair of left and right parentheses of type i).

Example 3.2 *Given the Dyck expression* $\alpha = (_3 \ (_1 \ )_1 \ )_3 \ (_2 \ )_2$ *we can associate with it a sequence of strings where the first one is* α, *and each of the others is obtained from the previous one by deleting a pair of parentheses:*

$$\alpha_0 = (_3 \ (_1 \ )_1 \ )_3 \ (_2 \ )_2$$
$$\alpha_1 = (_3 \ )_3 \ (_2 \ )_2$$
$$\alpha_2 = (_3 \ )_3$$

Put:

$$Corresp(p, q) \equiv (\ p =(_1 \ \wedge \ q =)_1 \ ) \ \vee$$
$$(\ p =(_2 \ \wedge \ q =)_2 \ ) \ \vee \ \ldots \ \vee \ (\ p =(_k \ \wedge \ q =)_k \ ),$$

where the formula $Corresp(p, q)$ *depends on the parentheses constituting the language* D *(k denotes the number of kinds of parentheses in the language D_k). The following instances of* δ *and* ν *relate to the Dyck sequence in the example:*

$\delta =$	1	2	3	4	5	6	7	8	9	10	11	12
	$(_3$	$(_1$	$)_1$	$)_3$	$(_2$	$)_2$	$(_3$	$)_3$	$(_2$	$)_2$	$(_3$	$)_3$

$$\nu = \begin{matrix} 0 & 6 & 10 & 12 \\ 1 & 2 & 3 & 4 \end{matrix} \quad (i)$$

$$\delta[\nu(1), \nu(2)] = \delta[0, 6] = \alpha_0$$
$$\delta[\nu(2), \nu(3)] = \delta[6, 10] = \alpha_1$$
$$\delta[\nu(3), \nu(4)] = \delta[10, 12] = \alpha_3$$

In general, the first order formalization of D_k *is the following:*

$$\alpha \in D_k \equiv (\alpha = \lambda) \vee (\exists p, q \ (\alpha = pq \wedge Corresp(p, q)) \vee$$
$$(|\alpha| > 2 \wedge \exists \delta, \nu \ 9\delta[\nu(1), \nu(2)] = \alpha \ \wedge$$
$$\exists p, q \ (\delta[\nu((|\nu| - 1)), \nu(|\nu|)] = pq \wedge Corresp(p, q)) \wedge$$
$$\forall i \ (1 \leq i \leq (|\nu| - 2) \rightarrow \exists \beta, \gamma, r, s \ (\delta[\nu(i), \nu(i+1)] = \beta r s \gamma \ \wedge$$
$$\delta[\nu(i+1), \nu(i+2)] = \beta \gamma \wedge Corresp(r, s)))))$$

The following theorem gives us a first result of logical representability for an important class of formal languages.

Theorem 3.1 *Any regular language is* SYN-*representable.*

Proof. Let us consider a regular language $L(e)$ described by means of a regular expression e. We give the formalization of relation $\alpha \in L(e)$ by induction over the structure of expression e:

1. if $e = \{\lambda\}$, then $\alpha \in L(e) \equiv \alpha = \lambda$;

2. if $e = \{a\}$, $a \in A$, then $\alpha \in L(e) \equiv \alpha = a$;

3. if $e = e_1 + e_2$, with relations $\alpha \in L(e_1)$ and $\alpha \in L(e_2)$ already formalized, then

$$\alpha \in L(e) \equiv \alpha \in L(e_1) \lor \alpha \in L(e_2);$$

4. if $e = e_1 e_2$, with relations $\beta \in L(e_1)$ and $\gamma \in L(e_2)$ already formalized, then

$$\alpha \in L(e) \equiv \exists \beta, \gamma \, (\alpha = \beta \gamma \land \beta \in L(e_1) \land \gamma \in L(e_2));$$

5. if $e = (e_1)^*$, then we build an auxiliary array of integers to express $\alpha = \alpha_1 \alpha_2 \ldots \alpha_n$; this allows us to formalize $\alpha \in L^*$, provided that the formalization of $\alpha_i \in L$ is already available:

$$(\alpha \in L(e)) \equiv \exists \nu \, (\forall i \, (1 \leq i \leq (|\nu| - 1) \to \alpha[\nu(i), \nu(i+1)] \in L(e_1))).$$

$\square$

From the examples and theorems given so far we can easily obtain the following theorem.

Theorem 3.2 *Any context-free language is* SYN-*representable.*

Proof. A language L is context-free if and only if there exists a homomorphism h, a *Dyck* language D and a regular language R such that $L = h(D \cap R)$. Therefore, it is sufficient to combine the representability into SYN of Dyck languages, of regular languages and of homomorphisms. $\square$

From this and previous results, we can deduce the representability of any type-0 language L in the model SYN. In fact, we know [16] that $L = h(L_1 \cap L_2)$ for a suitable homomorphism h and two context-free languages L_1, L_2. However, by using the techniques already introduced we also have a direct proof of this general theorem.

Theorem 3.3 *Any type-0 language is* SYN-*representable.*

Proof. (Sketch) Let L be a language generated by a type-0 grammar G. A string α belongs to L if there exist two strings β and ν, the first one constituted by the concatenation of all strings of a derivation of α, the second one (whose length is the length of the derivation $+ 1$) being a vector that allows us to extract from β, for every pair of values of consecutive indices, the derivation steps of α. In order to characterize all strings of L and only them, it is sufficient to express that, for any derivation of them, any step is obtained from the previous one by applying some production of G. This is easily representable by a simple logical condition expressed by concatenation and existential quantification. $\square$

Is any *SYN*-representable language a type-0 language? If not, which class of formulae represents type-0 languages? In order to answer these questions, we will extend the notion of first order representability.

4   Three Logical Characterizations of Computational Universality

In this section we extend the notion of logical representability by introducing: i) representability within a theory and ii) axiomatic representability. We then show that there is a class of formulae, indicated by $\Sigma(SYN)_1$ such that SYN-representability by this class is computationally universal, i.e., identifies the class of type-0 languages. Moreover, this representability is equivalent to the representability within a particular theory called RRR by $\Sigma(SYN)_1$, and is also equivalent to the axiomatic representability within RRR.

Let us recall some basic notions about first order theories. A set Φ of Σ-formulae is *recursively enumerable* (r.e.) or *semidecidable* if we have an algorithm that will effectively generate all the formulae of Φ. It is *decidable* or *recursive* iff we have an algorithm for deciding when a given Σ-formula belongs to Φ or not, or equivalently, iff we can effectively generate all the formulae of Φ and all the Σ-formulae that do not belong to Φ. A Σ-*theory* Φ is a set of formulae closed w.r.t. the first order logical consequence, that is, for every Σ-formula: $\Phi \models \varphi \Rightarrow \varphi \in \Phi$. If AX is a decidable (recursive) set, then Φ is an *axiomatic* theory (if AX is finite, Φ is finitely axiomatizable). Given the effective nature of first order logical consequence (for the deductive completeness of proof systems), the set of theorems of an axiomatizable theory is recursively enumerable. A theory Φ is *axiomatizable* iff $\Phi = \{\varphi \mid AX \models \varphi\}$ for some r.e. set of formulae AX called *axioms* of Φ. A Σ-theory Φ is *complete* iff, for any Σ-formula φ, $\Phi \models \varphi$ or $\Phi \models \neg\varphi$ holds (Φ is *sound* iff it is not the case that both conditions hold). A very simple, but fundamental result establishes that any axiomatizable and complete theory is also decidable.

We can extend the notion of logical representability of sets (and relations) by the following definition. Let T_Σ be the set of terms without variables on the signature Σ.

Definition 4.1 *Let Φ be a theory over the signature Σ. A subset S of T_Σ is represented in Φ, by the formula $\varphi(x)$ with a free variable x, when:*

$$a \in S \Leftrightarrow \Phi \models \varphi(a).$$

Notation 4.1 *When a set is representable within a theory Φ by a formula φ that belongs to a class Γ of formulae, we say that it is Φ-representable by Γ.*

An interesting case of representability within a theory is *axiomatic* representability:

Definition 4.2 *A subset S of T_Σ is Φ-representable by a finite set of axioms AX if for some formula φ in the signature of $\Phi \cup AX$:*

$$a \in S \Leftrightarrow \Phi \cup AX \models \varphi(a).$$

A set S is axiomatically representable when it is AX-representable by some finite set of axioms AX. Of course, representability within a theory and axiomatic representability can naturally be extended to any relation.

Remark 4.1 *We use symbols $\leq$ and $\preceq$ (meaning the usual order on natural numbers and the substring inclusion respectively) as abbreviations, according to the formal representations, in terms of $+$, $-$, given in Examples 2.2 and 2.3.*

Consider the following theory RRR on the signature $\{\omega, --, \|, \lambda\}$, where ω indicates a denumerable set of constants for natural numbers (denoted by the usual symbols). This theory is strongly related to Raphael Robinson's arithmetical theory RR (R^- in [17]), and consists of the following axiom schemata, where given k constants $n_1, \ldots, n_k \in \omega$ with $k \geq 1$, then $(n_1, \ldots, n_k)$ stands for the left-associated term $((\ldots(n_1 n_2)\ldots)n_k)$, and if $k = 1$, then $(n_1, \ldots, n_k)$ has to be considered equivalent to n_1. That is, we identify left-associated terms (without variables) of RRR with the elements of ω^*.

- $(n_1 \ldots n_k)\,(m_1 \ldots m_h) = (n_1 \ldots n_k m_1 \ldots m_h)$,

- $(n_1 \ldots n_k)\,\lambda = \lambda\,(n_1 \ldots n_k) = (n_1 \ldots n_k)$,

- $|\lambda| = 0$,

- $|(n_1 \ldots n_k)| = k$,

- $\neg(n_1, \ldots, n_k) = (m_1 \ldots m_h)$, for any $(n_1, \ldots, n_k) \neq (m_1 \ldots m_h) \in \omega^*$,

- $\forall x\,(x \leq n \leftrightarrow (x = 0 \vee x = 1 \vee \ldots \vee x = n))$,

- $\forall x\,(x \preceq (n_1 \ldots n_k) \leftrightarrow \bigvee_{i \leq j \leq k} x = (n_i \ldots n_j) \vee x = \lambda)$.

The following are interesting examples of first order axiomatic representabilities, where a, b, and c stand for any three natural numbers.

Example 4.1 $\{a^n b^n | n \in \omega\}$ *is RRR-representable by the following axioms:*

- $L(\lambda)$,

- $\forall x\,(L(x) \rightarrow L(axb))$

In fact: $\alpha \in \{a^n b^n \mid n \in \omega\} \leftrightarrow RRR \cup AX \models L(\alpha)$.

Example 4.2 $\{a^n b^n c^n \mid n \in \omega\}$ *is RRR-representable by the following axioms:*

- $L(\lambda)$,

- $\forall x\,y\,(L(xby) \rightarrow L(axbbyc))$.

In fact: $\alpha \in \{a^n b^n c^n \mid n \in \omega\} \leftrightarrow RRR \cup AX \models L(\alpha)$.

In general we can prove that:

Theorem 4.1 *Any type-0 language is axiomatically RRR-representable.*

Proof. Let G be the grammar of a type-0 language L with terminal symbols $a_1, \ldots, a_n$ and start symbol S (write $\alpha \rightarrow \beta \in G$ to say that $\alpha \rightarrow \beta$ is a production of G). Consider the following axioms AX in a signature which extends that of RRR with a unary predicate D:

$$AX = \{D(S)\} \cup \{D(x\alpha y) \rightarrow D(x\beta y) \mid \alpha \rightarrow \beta \in G\}$$

In this manner we have:

$$\alpha \in L \leftrightarrow RRR \cup AX \models D(\alpha) \wedge \forall x(x \in \alpha \rightarrow x = a_1 \vee \ldots \vee a_n).$$

$\square$

The next example shows the expressive power of RRR-axiomatic representability.

Example 4.3

$$\alpha \in \{a^p \mid p \text{ is a prime number}\} \Leftrightarrow RRR \cup AX \models L(\alpha).$$

where AX are the following axioms:

- $\forall x \ R(x, x)$,

- $\forall x \ y \ (R(x, y) \rightarrow R(x, xy))$,

- $\forall x \ (L(x) \leftrightarrow \forall z(z \in x \leftrightarrow z = a) \wedge \forall x \ y(y \preceq x \rightarrow (\neg R(y, x))))$.

Consider the following abbreviations:

$$\forall x \leq t \,.\varphi \ \equiv \ \forall x(x \leq t \rightarrow \varphi),$$

$$\forall x \preceq t \,.\varphi \ \equiv \ \forall x(x \preceq t \rightarrow \varphi),$$

$$\exists x \leq t \,.\varphi \ \equiv \ \exists x(x \leq t \wedge \varphi),$$

$$\exists x \preceq t \,.\varphi \ \equiv \ \exists x(x \preceq t \wedge \varphi),$$

where the term t is said to bound the quantifiers (a normal quantification considered as an *unbounded* quantification). A formula φ of the signature $\Sigma(AR)$ of AR is a $\Sigma(AR)_1$-formula, or in brief, according the standard usage, a Σ_1-formula iff it is constructed starting from atomic formulae by means of connectives, bounded universal quantifications and either unbounded or bounded existential quantification. Analogously, a $\Sigma(SYN)$-formula formula φ over the signature $\Sigma(SYN)$ of SYN is a $\Sigma(SYN)_1$-formula iff any universal quantification that occurs in it is a bounded quantification. For the sake of brevity, we refer to Σ_1 and to $\Sigma(SYN)_1$ as the set of Σ_1-formulae and the set of $\Sigma(SYN)_1$-formulae respectively.

The most important fact about Σ_1-formulae is a theorem strictly connected to Kleene's Normal Form Theorem in Computability theory (cf. Th. I.11.7 and related topics in [17]): *A set L of natural numbers is recursively enumerable iff it is AR-representable by some Σ_1-formula.*

Σ_1-formulae and $\Sigma(SYN)_1$-formulae, are intrinsically related to the RR theory, which we report below, and to the already considered RRR theory. These theories allow us to analyze fundamental arithmetical and syntactical notions in terms of logical representability.

Robinson's RR first order arithmetical theory of signature $\{\omega, +, \cdot\}$, where as for RRR ω is a denumerable set of constants, consists of the following axiom schemata:

- $n + k = m$ iff $AR \models n + k = m$,

- $n \cdot k = m$ iff $AR \models n \cdot k = m$,

- $\neg n = m$, for all $n \neq m \in \omega$,

- $\forall x (x \leq n \leftrightarrow (x = 0 \vee x = 1 \vee \ldots \vee x = n))$, for all $n \in \omega$.

Definition 4.3 *A theory Φ is Σ_1-sound iff for every Σ_1-formula φ:*

$$\Phi \models \varphi \Rightarrow AR \models \varphi.$$

Definition 4.4 *A theory Φ is $\Sigma(\text{SYN})_1$-sound iff for every $\Sigma(\text{SYN})_1$-formula φ:*

$$\Phi \models \varphi \Rightarrow \text{SYN} \models \varphi.$$

Theorem 4.2 (Σ_1-*completeness of RR*) *Let φ a Σ_1-formula with k free variables and let $n_1, \ldots, n_k \in \omega$; then:*

$$AR \models \varphi(n_1, \ldots, n_k) \Leftrightarrow RR \models \varphi(n_1, \ldots, n_k).$$

Proof. By induction on the complexity of φ (see Th. III.6.13 in [17]).　□

Corollary 4.1 *If Φ is a Σ_1-sound extension of theory RR, then it is also a Σ_1-complete theory.*

Theorem 4.3 ($\Sigma(\text{SYN})_1$-*completeness of RRR*) *Let φ a $\Sigma(\text{SYN})_1$-formula with k free variables and let $\alpha_1, \ldots, \alpha_k \in \omega^*$; then:*

$$\text{SYN} \models \varphi(\alpha_1, \ldots, \alpha_k) \Leftrightarrow RRR \models \varphi(\alpha_1, \ldots, \alpha_k).$$

Proof. By induction on the complexity of φ, analogously to the proof of the Σ_1-completeness of RR.　□

Corollary 4.2 *If Φ is a $\Sigma(\text{SYN})_1$-sound extension of theory RRR, then it is also a $\Sigma(\text{SYN})_1$-complete theory.*

Now we can prove that the three notions of: SYN-representability by $\Sigma(SYN)_1$-formulae, RRR-representability by $\Sigma(SYN)_1$-formulae, and axiomatic RRR-representability are equivalent each to other inasmuch as all of them are universal, i.e., a language is type-0 iff it is logically representable according to one of these methods.

Theorem 4.4 *A language L is type-0 iff it is* SYN-*representable by a* $\Sigma(SYN)_1$-*formula.*

Proof. A recursively enumerable set L of strings is generated by some effective device (a Turing Machine or a type-0 grammar). We avoid going into details, because, in any case, we can express any step of the computation process which generates a string $\alpha \in L$ by some string, and thus we can express the generation process by the sequence of these strings. This means that we have a string γ and a string ν such that the length of ν is equal to the steps of the computation, γ is the concatenation of all the sequences representing the steps, and for each i such that $1 \leq i \leq |\nu|$, the substring representing the i-th step is located between positions $\nu(i), \nu(i+1)$ of γ. Moreover, the string representing $(i+1)$-th step has to be obtained from the one representing step i-th by some symbolic transformations specified by the particular generation method. All these properties of γ and ν can easily be expressed formally in the model SYN by some $\Sigma(SYN)_1$-formula. Conversely, iff L is SYN-represented by some $\Sigma(SYN)_1$-formula, then for $\Sigma(SYN)_1$-completeness of RRR it is also RRR-representable by this formula. Therefore, for the recursive enumerability of the theorems of an axiomatizable theory, L is recursively enumerable. $\qquad\square$

Theorem 4.5 *A language L is type-0 iff it is RRR-representable by some* $\Sigma(SYN)_1$-*formula.*

Proof. It follows from the previous theorem and from the $\Sigma(SYN)_1$-completeness of RRR. $\qquad\square$

Theorem 4.6 *A language is axiomatically RRR-representable iff it is type-0.*

Proof. We have already shown that any type-0 language is axiomatically RRR-representable. Conversely, if the language is RRR-representable by some finite set of axioms AX, then theory $RRR \cup AX$ has a recursively enumerable set of axioms, hence the language is recursively enumerable, therefore it is a type-0 language.   □

The previous three theorems suggest the following definition.

Definition 4.5 *A language L is (logically) ω^*-representable if one of the following conditions holds:*

- *L is SYN-representable by $\Sigma(\text{SYN})_1$,*

- *L is RRR-representable by $\Sigma(\text{SYN})_1$,*

- *L is axiomatically RRR-representable.*

Theorem 4.7 *All syntactical relations considered in the previous section are logically ω^*-representable.*

Proof. It is sufficient to check that all the formulae considered are or can be equivalently transformed into $\Sigma(SYN)_1$-formulae, and therefore, for the previous theorems, they are ω^*-representable.   □

Our logical characterizations of type-0 languages are very close to an analogous representation theorem formulated in terms of *rudimentary predicates* (cf. [16] Ch. III, Th. 12.5, and [18]). A rudimentary predicate is essentially determined by a $\Sigma(SYN)_1$-formula where no symbol $| \; |$ occurs. The representation of a language via rudimentary predicates is equivalent to its representability by $\Sigma(SYN)_1$, with no occurrence of $| \; |$ within the following model:

$$(\omega^*, --, \varpi).$$

where ϖ denotes a predicate that holds on atomic strings (i.e. single numbers). This equivalence is a straightforward consequence of the next theorem.

Theorem 4.8 $|\alpha| = |\beta|$ *is representable in the model $(\omega^*, --, \varpi)$ by a $\Sigma(\text{SYN})_1$-formula with no occurrence of $| \; |$.*

Proof. (Informal) We can express that two strings have the same length by using an auxiliary symbol, say $\sharp$. Assume that α and β are not empty and not atomic. Let us say that a string is *monic* iff it is constituted only by occurrences of the symbol $\sharp$, and that a monic string γ is a *full substring* of a string δ when it occurs in δ as a substring that is not a proper substring of another monic string. Under these assumptions $|\alpha| = |\beta|$ iff there are two strings α', β' that satisfy the following requirements:

- both α' and β' begin with one symbol followed by $\natural$, and end with a monic substring;

- a monic string γ can occur only once in α' as its full substring;

- a monic string γ is a substring of α' iff it is a substring of β';

- iff $\gamma\natural$ is a full substring of α', then also γ is a full substring of α' which precedes $\gamma\natural$ (from the left);

- given a monic string γ, let $\gamma\natural\natural$, $\gamma\natural$, and γ be full substrings of α' and β', and let α_1, α_2 and β_1, β_2 be the pairs of strings such $\gamma\alpha_1\gamma\natural\alpha_2 \preceq \alpha'$ and $\gamma\beta_1\gamma\natural\beta_2 \preceq \beta'$. In this case: α_2 is obtained from α_1 by adding to it only one (rightmost) symbol; analogously, β_2 is obtained from β_1 by adding to it only one (rightmost) symbol;

- α and β are the two strings that in α' and β' are between the two rightmost (monic) full substrings;

- the two ending monic strings of α' and β' are equal.

For example, if $\alpha = abc$ and $\beta = cbb$, the strings α' and β' considered above are:
$$\alpha' = a\natural ab\natural\natural abc\natural\natural\natural,$$
$$\beta' = c\natural cb\natural\natural cbb\natural\natural\natural.$$

It is easy to verify that we can put together all these conditions in a $\Sigma(SYN)_1$-formula with no occurrences of $||$. $\square$

5   Logical Metabolism

In this section we show that logical rewriting can be extended to *metabolic rewriting*, if a model $META$, more powerful than SYN, is considered, where not only the syntactical structure of strings, but also localization relations (distribution and encapsulation) between strings can be represented. The most important mathematical aspect of metabolic rewriting is that at any step a finite language, rather than a single string, is rewritten. This means that classical generative systems are particular metabolic systems, and in turn, it is possible to associate a *canonic* metabolic system with any generative system. It is sufficient that at any step of a generative process, we keep all generated strings in a language L, and consider a rewriting step from a string $\alpha \in L$ to a string β as the passage from L to $L \cup \{\beta\}$. Model $META$ allows us to express the abstract form of transformations that holds in chemical and biochemical processes. The reader can find in [10] the original motivations to logical metabolism along with examples showing its chemical and biochemical relevance. Here we only want to stress its natural relationship with logical rewriting, therefore, we consider the basic definitions and revisit in metabolic terms some classical examples. Consider the following structure:

$$META = (\mathcal{F}, \sqsubset, 0, \lambda, \emptyset, \cdot, |\,|, \subseteq, \cup, \|\,\|),$$

where:

1. $\mathcal{F}$ is the set of finite subsets of strings constituted by natural numbers and the special symbol (control character) $\sqsubset$ which we call also *separator*. Numbers and $\sqsubset$ are identified with the strings constituted by only one element, and strings are identified with sets constituted by only one string.

2. $0, \lambda, \emptyset$ are respectively the number zero, the empty string and the empty set.

3. $\cdot, |\,|$ are respectively the catenation between languages (as usual, indicated only by juxtaposition) and the length of the longest string of a given language, where $|\lambda| = 0$, and $|\sqsubset| = 0$ (λ is the neutral element w.r.t. catenation).

4. $\subseteq, \cup$ and $\|\,\|$ are respectively the usual set inclusion relation, the union between languages, and the operation yielding the cardinality of a language.

Definition 5.1 (Metabolite) *A metabolite or metabolic state is an element of the domain $\mathcal{F}$ of the model $META$.*

The model $META$ is an extension of the model SYN, i.e. all sets, relations, and operations representable in SYN are also representable in $META$, therefore all the results of representability that hold in this model continue to hold in $META$. In particular, in the model $META$ we can express all the usual arithmetical operations and relations, and all the classical rewriting relations of formal language theory (of Chomsky grammars, Lyndenmayer Systems, contextual grammars, splicing systems). Let us recall some important results which follow directly from the representability properties of SYN and from the fact that $META$ is an extension of SYN.

Theorem 5.1 *Standard arithmetical relations are representable in META.*

Proof. It follows from the representability of standard arithmetic in *SYN*.

□

As an example, we can consider the following simple way to represent in *META* the sum of natural numbers:

$$m + k = n \iff META \models \exists LM(|L| = m \land |M| = k \land |LM| = n).$$

Extend the signature $\Sigma(META)$ so that it includes the symbols $\leq$, $\preceq$ defined in the usual way. A $\Sigma(META)_1$-formula is a $\Sigma(META)$-formula existentially bounded by a binary predicate belonging to $\{\leq, \preceq, \subseteq\}$.

Theorem 5.2 (Universality of *META*) *The class of type-0 languages coincides with the class of languages that are META-representable by $\Sigma(META)_1$-formulae.*

Proof. It follows from the main representability result proved for *SYN*. □

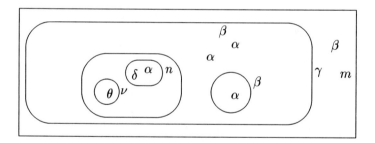

A Metabolite

If we use the separator $\sqsubset$, we can distinguish the notion of a membrane from the abstract notion of a set. In fact, a membrane α of the metabolite M is viewed as a particular string that *envelops* other strings, i.e., for some finite language L, the language $L \sqsubset \alpha = \{\beta \sqsubset \alpha \mid \beta \in L\}$ is a subset of M. This means that we encode relations of localization between membranes and their internal objects, by means of string suffixes. This simple trick allows us to express, as finite languages, complex structures, with many aggregation levels. The localization path of any single component is traced in a sort of hierarchical address: if α, β, γ are strings that do not contain occurrences of the separator $\sqsubset$, and $\alpha \sqsubset \beta \sqsubset \gamma$ belongs to the metabolite M, then a molecule α is internal to the membrane β which, in turn, is internal to the membrane γ. More generally, any finite structure can be expressed by a language where other separators for expressing particular relations are allowed.

In order to describe the evolution of a metabolic state, let us consider a finite set Φ of first order *metabolic* formulae, with two free variables:

$$\Phi = \{\, \varphi_i(X,Y) \,\}_{i \in 1,\ldots,n}$$

We say that a metabolite M *evolves* in one step to the other metabolite M' according to φ_i, and we write $M \xrightarrow{\varphi_i} M'$, if and only if $\varphi_i(M, M')$ holds in the model $META$ (in $\varphi_i(M, M')$ the metabolites M, M' replace the free variables X, Y of φ_i):

$$M \xrightarrow{\varphi_i} M' \ \ \textit{if and only if} \ \ M \models \varphi_i(M, M').$$

Let us consider the binary relation $\xrightarrow{\Phi}$ obtained from the union of all $\xrightarrow{\varphi_i}$. By means of this relation, language evolution can be conceived as a metabolite succession $M_0 \to M_1 \to \ldots \to M_n \to \ldots$, where each metabolite is obtained from the previous one, by making the transformations described in the theory Φ.

Definition 5.2 (Logical Metabolic System) *A metabolic system $\mathcal{M}$ is a triple:*

$$(M_0, \Phi, M_f),$$

where M_0 is the initial *metabolite, Φ is a finite class of metabolic formu-lae each of them being a metabolic agent, and M_f is the metabolic filter. The initial metabolite M_0 and the filter M_f are languages representable in $META$. At any step i of a metabolic evolution, the set $M_i \cap M_f$ is the anabolism at the step i, and the set difference M_i / M_f is the catabolism at the step i.*

The language $L(\mathcal{M})$ generated by $\mathcal{M}$ is the union that collects the an-abolism of any step, where $\xrightarrow{\Phi}_*$ denotes the reflexive and transitive closure of $\xrightarrow{\Phi}$:

$$L(\mathcal{M}) = \bigcup_{M_0 \xrightarrow{\Phi}_* L_i} L_i \cap M_f.$$

Theorem 5.3 *Any Chomsky grammar G determines a logical metabolic system $\mathcal{M}(G)$ generating the same language.*

Proof. For any production $\alpha \to \beta$ of a grammar G consider the following formula ($\gamma \in X$ stands for $\gamma \subseteq X \wedge \|\gamma\| = 1$):

$$\exists xy(x\alpha y \in X \ \wedge \ x\alpha y \in Y \ \wedge \ x\beta y \in Y).$$

Let Φ be the set of all formulae associated with the productions of G. This is the set of metabolic formulae of the system $\mathcal{M}(G)$. The starting symbol S of G is the initial metabolite of $\mathcal{M}(G)$, and if T is the set of terminal symbols of G, then T^* is the filter of $\mathcal{M}(G)$. For example, the metabolic system associated with the simple grammar $\{S \to ab/aSb\}$ provides such a kind of metabolic evolution (subscripts and superscripts in the arrows is avoided):

$$L_0 = \{S\} \to L_1 = \{S, aSb\} \to L_2 = \{S, ab, aSb\} \to L_3$$
$$= \{S, ab, aabb, aSb\} \ldots$$

where $L_i \to L_{i+1}$ iff for some $\alpha S\beta \in L_i$, either $L_{i+1} = L_i \cup \{\alpha ab\beta\}$, or $L_{i+1} = L_i \cup \{\alpha aSb\beta\}$. $\square$

Similar theorems can be proved for any classic string rewriting system, after representing its rewriting into the model SYN, and then providing its translation within $META$.

In our formalization a metabolite, inasmuch as it is a set, represents a natural way to distribute information in different components, and according to our formalization of metabolic evolution driven by first order formulae, represents a system where evolution is intrinsically multi-agent and cooperative. In fact any formula can be considered as a (metabolic) agent, which may be triggered by some specific enzymes and completely identified in its behaviour by the condition expressed by the formula. The possibility of using the expressive power of first order logic allows us to define complex regulation mechanisms and cooperation patterns among agents. Of course, we cannot hope to express by such a logical machinery all the complexity of real metabolic processes. The main inadequacy is in the connections that real metabolic processes have with many different levels, e.g., physical, chemical, genetic, epigenetic, or embryological. A comprehension of how these levels could communicate is a challenge when attempting to obtain a deep understanding of living organisms. Our model is too idealized because metabolites are considered in their pure aspect of aggregation, classification and localization. However, despite its intrinsic limitations, we think this could be a useful step in the analysis of many natural processes. A very general structure of a metabolic formula $\varphi(X, Y)$ is the following:

$$\exists Z, M_1, \ldots, M_k(X = Z \cup A(M_1, \ldots, M_k) \;\land\; \psi(X, M_1, \ldots, M_k) \;\land$$
$$Y = Z \cup B(M_1, \ldots, M_k)),$$

where $A(M_1, \ldots, M_k), B(M_1, \ldots, M_k)$ are terms of variables $M_1, \ldots, M_k$. We abbreviate such a formula in the premises/conclusions format, where the free variables X, Y are implicit ($M_1, \ldots, M_k$ are the *existential* variables of the rule):

$$\frac{A(M_1, \ldots, M_k), \psi([X], M_1, \ldots, M_k)}{B(M_1, \ldots, M_k)}$$

In the formula $\psi([X], M_1, \ldots, M_k)$ the brackets in $[X]$ mean that X is a *tacit* variable, i.e., whenever a formula $L \subseteq X$ occurs in $\psi(X, M_1, \ldots, M_k)$, then it is replaced by L in $\psi([X], M_1, \ldots, M_k)$ (reading: L is a subset of the metabolite to which the rule is applied). The intuitive meaning of such a rule is the following procedure: *Consider a metabolite, say M. If, for some $M_1, \ldots, M_k$, it includes the subset $A(M_1, \ldots, M_k)$ (determined by the subsets $M_1, \ldots, M_k$) and if M verifies the condition ψ then transform M into a metabolite, say M', where the subset $B(M_1, \ldots, M_k)$ replaces the subset $A(M_1, \ldots, M_k)$.*

Example 5.1 *The language*

$$L = \{a^n b^n c^n \mid n \in \omega\}$$

is generated by a logical metabolic system where the initial metabolite is abc, the filter is $\{a, b, c\}^*$, and there are the metabolic formulae (γ, β are existential variables):

$$\frac{\gamma b \beta}{\gamma b \beta, a\gamma bb\beta c}.$$

The next example presents a system with only three formulae which is able to generate a very complex language. Its expressive power is due to the use of non trivial logical conditions.

Example 5.2 *The language*

$$L = \{a^p \mid p \text{ is a prime number}\}$$

is generated by the logical metabolic system with the initial metabolite

$$\{aa, aaa \sqsubset m, aaaa \sqsubset aa\}$$

with the filter $\{a\}^*$, and with the following metabolic formulae (μ, π, M, L are existential variables):

$$\frac{\pi \sqsubset m, \forall \beta, \gamma (\gamma \sqsubset \beta \wedge \beta \neq m \rightarrow \pi \prec \gamma)}{\pi, a\pi \sqsubset m, \pi\pi \sqsubset \pi} \tag{1}$$

$$\frac{\begin{array}{c} M, \\ \mu \sqsubset m, \\ \forall \beta (\mu \sqsubset \beta \wedge \beta \neq m \leftrightarrow \mu \sqsubset \beta \subseteq M), \\ \forall \beta (\mu \sqsubset \beta \subseteq M \leftrightarrow \beta\mu \sqsubset \beta \subseteq L) \end{array}}{L, a\mu \sqsubset m} \tag{2}$$

The generation of the first four prime numbers (*a number p stands for the string a^p*):

1   $\{2, 3 \sqsubset m, 4 \sqsubset 2\} \xrightarrow{(1)}$

2   $\{2, 3, 4 \sqsubset m, 4 \sqsubset 2, 6 \sqsubset 3\} \xrightarrow{(2)}$

3   $\{2, 3, 5 \sqsubset m, 6 \sqsubset 2, 6 \sqsubset 3\} \xrightarrow{(1)}$

4   $\{2, 3, 5, 6 \sqsubset m, 6 \sqsubset 2, 6 \sqsubset 3, 10 \sqsubset 5\} \xrightarrow{(2)}$

5   $\{2, 3, 5, 7 \sqsubset m, 8 \sqsubset 2, 9 \sqsubset 3, 10 \sqsubset 5\} \xrightarrow{(1)}$

6   $\{2, 3, 5, 7, 8 \sqsubset m, 8 \sqsubset 2, 9 \sqsubset 3, 10 \sqsubset 5, 14 \sqsubset 7\}$

This system is essentially a multi-agent version of Eratosthenes' sieve. It is very simple and efficient compared with the analogous example of the previous section, and with the analogous symbolic devices, based on Post systems, and Chomsky grammars [12] (p. 236-237), [16] (p.14). See [10] for a proof of its correctness and completeness.

References

[1] J. Barwise, ed., *Handbook of Mathematical Logic*, North-Holland, Amsterdam, 1977.

[2] J. L. Bell, M. Machover, *A Course in Mathematical Logic*, North-Holland, Amsterdam, 1977.

[3] N. Chomsky, Three models for the description of language, *IRE Transactions on Information Theory*, IT-2 (1956), 113 – 124.

[4] E. Csuhaj-Varju, J. Dassow, J. Kelemen, Gh. Păun, *Grammar Systems: A Grammatical Approach to Distribution and Cooperation*, Gordon and Breach, London, 1994.

[5] E. Csuhaj-Varju, V. Manca, *Grammar Systems and Metagrammatical Representations*, VI Tarragona Seminar on Formal Syntax and Semantics, GRLMC Report, Universitat Rovira y Virgili, Tarragona, Spain, 1997.

[6] J. Dassow, Gh. Păun, *Regulated Rewriting in Formal Language Theory*, Springer-Verlag, Berlin, 1989.

[7] T. Head, Formal language theory and DNA: An analysis of the generative capacity of recombinant behaviors, *Bulletin of Mathematical Biology*, 49 (1987), 737 – 759.

[8] W. Hodges, Elements of Classical Logic, in *Handbook of Philosophical Logic* (F. Guenthner, ed.), Vol. I, Reidel Publishing Comp., Dordrecht, 1983.

[9] A. Lindenmayer, Mathematical models for cellular interaction in development, I and II, *J. Theoret. Biol.*, 18, (1968), 280 – 315.

[10] V. Manca, D. Martino, From string rewriting to logical metabolic systems, in *Grammatical Models of Multi-agent Systems* (Gh. Păun, A. Salomaa, eds.), Gordon and Breach, London, 1998.

[11] S. Marcus, Contextual Grammars, *Rev. Roum. Math. Pures Appl.*, 14 (1969), 1525 – 1534.

[12] M. L. Minsky, *Computation: Finite and Infinite Machines*, Prentice-Hall, Inc., Englewood Cliffs, N.J., 1967.

[13] Gh. Păun, ed., *Artificial Life: Grammatical Models*, Black Sea University Press, Bucharest, 1995.

[14] Gh. Păun, G. Rozenberg, A. Salomaa, *DNA Computing. New Computing Paradigms*, Springer-Verlag, Berlin, 1998.

[15] G. Rozenberg, A. Salomaa, eds., *Handbook of Language Theory*, 3 Vol., Springer-Verlag, Berlin, 1997.

[16] A. Salomaa, *Formal Languages*, Academic Press, New York, 1973.

[17] C. Smoryński, *Logical Number Theory*, Springer-Verlag, Berlin, 1991.

[18] R. M. Smullyan, *Theory of Formal Systems*, Princeton Univ. Press, Princeton, New Jersey, 1961.

A Molecular Abstract Machine

Gabriel CIOBANU

Al. I. Cuza University, Faculty of Computer Science
Str. General Berthelot 16, 6600 Iaşi, Romania
gabriel@infoiasi.ro

Abstract. This paper describes a "molecular" abstract approach to the DNA computing, looking for theoretical developments related to a suitable abstract machine for this idea of computing using molecules, and to the computational power of the molecular machine – which has at least the same power as the Turing Machines. In order to define the abstract machine and to prove its computational power, we use notions and results of concurrent communicating systems (CCS) theory, π-calculus, and encodings of λ-calculus into π-calculus).

1 Introduction

DNA computing was born few years ago, when an instance of an NP-complete problem was encoded by using strands of DNA, and then solved by molecular biology techniques. In this way Leonard Adleman demonstrated the feasibility of carrying out computation at the molecular level; he described a biological solution of the direct Hamiltonian path problem[1]. Molecular computation brings hope of massive parallelism in computation. The first ligation step of Adleman's biological computer can solve much more operations/second than a supercomputer. Additionally, the energy efficiency of a biological system is far better than that of an electronic machine. One joule is sufficient for around 2×10^{19} operations in the ligation step of Adleman's experiment, whereas existing supercomputers execute less than 10^9 operations per joule. However, Adleman detected physical and logical difficulties in order to create a practical molecular computer. Most of the results of this new field (i.e., DNA computing) are and remain theoretical. This paper is on this theoretical stream, suggesting an abstract model of a molecular computer. The paper is also a survey of some results of the theory of concurrent and interactive systems which are relevant, in the author's opinion, to DNA computing. Some notions, results, and remarks of the paper can be found in various papers. In order to make them useful to the DNA computing community, we just adapt and put them together.

2 Power of Interaction

The formal language approach is based on Turing Machines. Turing Machines are widely used in the classical theory of sequential computation, and they are, together with Random Access Machines, primary tools within the theories of recursive functions and computational complexity. However Turing Machines are not the most powerful computing abstract machine. The view that Turing Machines are not the most powerful computing mechanisms has a distinguished pedigree: it was accepted by Turing, who showed in 1939 that Turing Machines with oracles were more powerful than simple Turing Machines. Milner [15] noticed in 1975 that concurrent processes cannot be expressed by sequential algorithms, while Manna and Pnueli [14] showed that non-terminating reactive processes like operating systems cannot be modeled by algorithms. The intuition that computing corresponds to formal computability by Turing Machines (i.e., the Church-Turing thesis) breaks down when the notion of what is computable is broadened to include interaction. Interaction extends computing to computable non-functions over histories, rather over strings. For instance, airline reservation systems and other reactive systems provide interactive services over time that cannot be specified by functions. In principle, driving without traffic can be modeled by off-line rules for closed systems. However, driving in traffic cannot be reduced to an algorithm even in principle: it depends on complex unpredictable on-line events that are not algorithmically or sequentially presented. The idea that behavior of processes is determined by external interaction histories rather than by inner state transitions is central to models of interactive computing. Algorithms are metaphorically blind because they cannot adapt interactively while they compute. In contrast, interactive systems model an external reality more demanding and expressive than inner algorithmic transformation rules. Extending algorithms with interaction transforms algorithms into concurrent processes. Therefore, Church's thesis remains valid in the narrow sense that Turing

Machines express the behavior of algorithms; however the broader assertion that algorithms precisely capture what can be computed is not valid. The new thesis is that Turing Machines alone do not capture the new intuitive notion of computing, since they cannot express interactive computing.

Concerning the expressive power of the interactive processes, more statements and opinions can be found in [24, 25].

3 Process Algebras and Calculi

Milner anticipated that functions cannot express meanings of concurrent processes and that they are not enough to specify the semantics of concurrency. However, starting from λ-calculus, substitution of an argument for a variable in λ-calculus reduction rule is considered as an interaction:

$$\lambda(x).M \, applied \, to \, N \; \Rightarrow \; substitute \, N \, for \, x \, in \, M$$
$$\Rightarrow send \, N \, to \, receive(x)M.$$

In the Calculus of Communicating Systems (CCS), and its later refinement the π-calculus, the *receive* command binds values emitted by a *send* command. The connection between a receiving process P and a sending process Q, indicated in the λ-calculus by textual proximity, is specified by a channel name, say n. The interactive reduction rule becomes:

$$n - receive(x)M.P \mid n - send(N).Q$$
$$\Longrightarrow (substitute \, N \, for \, x \, in \, M). \, P \mid Q.$$

Channels are established by a nondeterministic matching which dynamically binds senders to eligible receivers. Even if there are many pairs which can satisfy the matching condition, only a single receiver gets the commitment of the sender. CCS models computing by algorithmic reduction and interactive matching. From a semantic viewpoint, this kind of matching uses the "." prefix operator to symmetrically triggered input actions and complementary output actions. Both the sender and receiver offer their availability for communication, symmetrically narrowing their choice from a set of offered alternatives to a specific commitment. Similar mechanisms work in biology, chemistry, and in computation. Thus it models biological binding in DNA, where affinity between complementary pairs of nucleotides A-T and C-G determines binding of DNA sequences, as well as chemical binding of positive and negative ions realized by availability to all ions of opposite polarity. Considering biological and other domains mechanisms in terms of their interaction patterns rather than state-transition rules can provide useful qualitative insights, and new forms of abstraction for computational models. The π-calculus generalizes CCS by transmitting names rather than values across send-receive channels, while preserving matching as the control structure for communication. Since variables may be channel names, computation can change the channel topology and process mobility is supported. The π-calculus for names is an interactive analog of the CCS calculus for values. Milner emphasized the importance of identifying the "elements of interaction", and his π-calculus extends the Church-Turing model to interaction by extending the λ-calculus. Both CCS and the π-calculus extend the algebraic elegance of lambda reduction rules to interaction between a sender and a receiver. Interaction among processes is modeled by nonalgebraic protocols that express the anonymous binding of positive to negative ions of a chemical abstract machine, as well as the binding of complementary DNA sequences. Anonymous interaction among processes is useful in many contexts, reflecting the fact that chemical coupling is anonymous, and may be contrasted to the sending of messages to named receivers that characterizes the Actor model [2].

This paper deals also with the Chemical Abstract Machine (CHAM) and the chemical metaphor introduced by Banâtre and Le Métayer, then used by Berry and Boudol as a framework for parallel interactive computations.

4   CHAM as a Molecular Abstract Machine

The chemical abstract machine CHAM is intended to provide a framework for describing the operational semantics of parallel molecular computations. Idea is to provide a model where the concurrent components are freely moving in the system, and have an interaction when they come in contact. A state of a system is a *chemical solution* in which floating *molecules* can interact with each other according to some *reaction rules*. A magical mechanism stirs the solution allowing for possible contacts between molecules – this mechanism is assumed to be given, i.e., it is provided by the implementation (in chemistry this is the result of Brownian motion). The solution transformations process is truly parallel: any number or reactions can be performed in parallel.

We can see a chemical solution as a finite multiset of elements (molecules) $S = \{m_1, \ldots, m_k\}$. This approach accounts for the magical stirring mechanism – since the elements are unordered, and so assumed to be in contact. Comma represents the parallel composition of molecules:

it is not difficult to see that this composition is associative and commutative. In other words, the molecules are not linked up together into a rigid structure. A reaction rule is given by a pair (*matching condition, action*). Execution is given by replacing the elements of the multiset satisfying the matching condition by the results of the corresponding action. The reactions can be expressed by multiset rewritings. Final result is obtained when a stable state is reached, namely when no more reaction can proceed.

Definition 4.1 *A multiset is a set-like collection in which elements may be duplicated. A multiset M over a set X can be also be considered as a function $M : X \to Nat$ which maps each element to its multiplicity.*

Given two multisets M and N, we write $M \uplus N$ for multiset union, and

$$M \uplus N (x) = M(x) + N(x).$$

It is easy to see that multiset union is associative and commutative.

The multiset is the data structure that fits to the notion of concurrent interaction; a multiset can be viewed as the parallel composition of all its elements, and the communication between these elements can he modeled by multiset replacement; a multiset is viewed as a "chemical soup" of molecules. (It is worth remembering here that working with multisets increases the power of finite H systems, see [12], [13].)

Our molecular model consists of :

– a set of molecules,

– solutions (states) which are finite multisets of molecules,

– a transition relation on the set of states.

Another possible construction, called "membrane", transforms a solution into a molecule; this means that any solution is also a molecule.

At the abstract level, the molecules can be terms, or strings, or words, etc. built according to some given syntax. Then we can use the notion of context – which is a term (string, word...) with holes in it. We denote by $C[t]$ the term (string, word...) we get by filling the holes in the context C by the term (string, word...) t. The transition relation $S \to S'$ is given by means of rules. There are few general rules:

$$\frac{S \to S'}{S \uplus R \to S' \uplus R}$$

and

$$\frac{S \to S'}{C[S] \to C[S']}.$$

The first rule is the formal description of the "locality" principle which says that reactions can be performed freely within any solution; the other one asserts that solutions can evolve freely in any molecule context.

Any rule may be used by selecting a submultiset satisfying a required condition, usually by pattern matching, and operating on it. A general form is

$$cond(m_1, \ldots, m_k) \Rightarrow R(m_1, \ldots, m_k).$$

According to the locality property, if the condition *cond* holds for several disjoint subsets, the reactions can be carried out independently and simultaneously.

Example. Let assume the solution is originally made of all integers from 2 to n, and the reaction rule is that any integer destroys its multiples. The solution will end up containing the prime numbers between 2 and n. We can express this by:

$$prime(\{2, \ldots, n\}) : \quad cond(x, y) : multiple(x, y) \Rightarrow R(x, y) : x.$$

We have two kinds of transition relation; the transition relation $\to$ is the union of two binary relations over multisets denoted by $\rightleftharpoons$ and $\mapsto$. The transition $S \to S'$ represents reactions. The other kind of transition, $S \rightleftharpoons S'$, represents structural transformations. The transition relation $\rightleftharpoons$ is also decomposed in two relations: $\rightharpoonup$ and $\leftharpoondown$. The first one is called heating transformation, and intuitively it is intended to prepare the solution for reaction, breaking down a complex molecule into smaller ones. The other transformation is called cooling transformation, and it rebuilds complex molecules from simpler ones. We shall only use pairs of inverse heating and cooling rules, which are written as $m \rightleftharpoons m_1, \ldots, m_k$. We call *structural equivalence* the equivalence generated by reversible structural transformations $\stackrel{*}{\rightleftharpoons}$, i.e., the reflexive and transitive closure of $\rightleftharpoons$. To have a molecular reaction in a given CHAM, one starts with the initial solution, then heats it to get a solution where reactions can occur. This is formalized by saying that a sequence of transformations of the form $S \stackrel{*}{\rightleftharpoons} \mapsto \stackrel{*}{\rightleftharpoons} S'$ is a reduction step, or an evaluation step. A solution is stable if it cannot perform any reduction.

5   CCS - Calculus of Communicating Systems

To give a simple illustrative example, we define a CHAM for a fragment of Milner's process calculus CCS called TCCS [10].

Let $N = \{a, b, \ldots\}$ be a set of names, and $\mathcal{L} = \{a, \overline{a} \mid a \in N\}$ be the set of labels, or complementary polarities built on N. The fragment of CCS we consider is built from the constant *nil*, and then using prefixing operator $a.P$ and parallel composition $(P|Q)$.

Definition 5.1 *The processes are defined over the set N of names by the following syntactical rules:*

$$P ::= nil \mid a.P \mid (P|Q)$$

where a is any label from $\mathcal{L}$.

Note that this calculus is without a special label $\tau \in T$ which represents the internal communications, and which is claimed to be invisible for an observer.

In order to give an operational semantics for CCS, it is convenient to use labelled transitions, representing the possible interactions.

Following De Nicola and Hennessy's [10], we use two kinds of transitions, namely the interaction reductions $P \to Q$, corresponding to the the the transition $P \xrightarrow{\tau} Q$ of CCS, and the labelled transitions $P \xrightarrow{a} Q$. These transitions are given by the following rules:

- a subterm $a.P$ may exhibit its interaction capability:

$$a.P \xrightarrow{a} P;$$

- the recorded interaction capabilities are carried along the structure:

$$\frac{P \xrightarrow{a} Q}{(P|R) \xrightarrow{a} (Q|R) \ \& \ (R|P) \xrightarrow{a} (R|Q)};$$

- finally they may contribute to an interaction, i.e., to a "reaction":

$$\frac{P \xrightarrow{a} P' , \ Q \xrightarrow{\overline{a}} Q'}{(P|Q) \to (P'|Q')}.$$

As a consequence of the last two rules, an interaction may occur everywhere in a term:

$$\frac{P \to Q}{(P|R) \to (Q|R) \ \& \ (R|P) \to (R|Q)}.$$

Now we can define a molecular abstract machine for TCCS by the following steps: The molecules and the solutions are the terms of the calculus. Then

$$(P|Q) \rightleftharpoons P, Q;$$
$$a.P, \bar{a}.Q \mapsto P, Q.$$

An equivalent formulation of such a molecular abstract machine can be given by defining a structural equivalence $P \equiv Q$ on terms, which is the least equivalence compatible with parallel composition for which this operator is associative and commutative, given by the following rules:

1) $\dfrac{P \rightarrow Q}{(P|R) \rightarrow (Q|R)}$;

2) $\dfrac{Q \equiv P, \; P \rightarrow P', \; P' \equiv Q'}{Q \rightarrow Q'}$;

3) $(a.P \,|\, \bar{a}.Q) \rightarrow (P \,|\, Q)$.

A difference between the molecular semantics and the usual operational semantics is the use of the structural equivalence. Technically, this is very similar to the notion of rewriting modulo an equational theory (namely the equational theory of multisets). The structural equivalence is a step forward for liberating from the rigidity of the syntax. Indeed, the molecules could very well be something else than terms of an algebra; they could be nodes, or parts of a graph, for instance.

CHAM-like abstract machine provides a description of the reductions (that is the internal computations), and of the interaction capabilities of molecules.

We can define and use various observational equivalences, like bisimulations. The idea of observational equivalence as bisimulation is due to Milner and Park. Two machines that simulate each other are observationally equivalent in the sense that they can make the same observational distinctions. Bisimulation has stronger observational power than set inclusion, and it is able to capture dynamic distinctions among processes specified by an interaction grammar, processes which cannot be sometimes distinguish as strings.

Definition 5.2 *Strong bisimulation, written $\sim$, is defined over the processes as the largest symmetrical relation such that $T \sim S$ implies (whenever $T \rightarrow T'$ then there exists S' s.t. $S \rightarrow S'$ and $T' \sim S'$).*

A new operator can be used to extract the interaction capability a out of a solution $(a.P|Q)$ while preserving bisimilarity [4]. This is a binary construct $(p \lhd q)$ over molecules, which puts the molecule p apart from a solution. The rules concerning this construct are:

$$p|q \rightleftharpoons p \lhd q;$$
$$(a.p) \lhd q \rightleftharpoons a.(p \lhd q).$$

We have now $(a.p|q) \stackrel{*}{\rightleftharpoons} a.(p \lhd q)$, therefore the capability described by a is extracted from $(a.p|q)$.

We can also extend our fragment of CCS with the restriction construct $P\backslash a$, where $a \in N$. This construct combines two features. One allows computing locally within a context, and it is expressed by the rule

$$\frac{P \rightarrow Q}{P\backslash a \rightarrow Q\backslash a}.$$

On the other hand, a restricted process $P\backslash a$ allows interaction on names other that a, if these exist such interaction capabilities of P. The corresponding rule is:

$$\frac{P \stackrel{b}{\rightarrow} Q}{P\backslash a \stackrel{b}{\rightarrow} Q\backslash a}, \quad b \notin \{a, \bar{a}\}.$$

Summarizing, we have the following molecular abstract machine for a fragment of TCCS. The syntax of the processes and molecules is:

$$P ::= nil \mid a.P \mid (P|P) \mid P\backslash a;$$
$$m ::= P \mid a.m \mid (m|m) \mid m\backslash a \mid m \lhd m.$$

We use $P, Q, R, \ldots$ to range over processes, and $m, p, q, \ldots$ to range over molecules. The reaction rules is:

$$a.p \, , \, \bar{a}.q \mapsto p \, , \, q$$

and the structural rules are:

$$(p|q) \rightleftharpoons p \, , \, q;$$
$$(a.p|q) \rightleftharpoons a.(p \lhd q);$$
$$(p \lhd q) \rightleftharpoons (p|q);$$
$$(b.m\backslash a) \rightleftharpoons b.(m\backslash a), \quad where \; b \notin \{a, \bar{a}\}.$$

Solutions are determined by the molecules of a term (process); they are multisets of these molecules. By the notation $\{P\}$ we understand the solution determined by P, i.e., a multiset of molecules determined by the term (process) P. The operational semantics over TCCS solutions are given by the following labelled transitions rules:

$$S \xrightarrow{\tau} S' \ if \ S \rightleftharpoons \mapsto \rightleftharpoons S',$$
$$S \xrightarrow{a} S' \ if \ S \overset{*}{\rightleftharpoons} \{a.m\} \ and \ S' \overset{*}{\rightleftharpoons} \{m\}.$$

The relationships between the operational semantics and the molecular semantics, as it was defined here, is that they are in fact strongly bisimilar. According to the definition of the molecular labelled transitions, $S \overset{*}{\rightleftharpoons} S'$ implies that $S \sim S'$.

Proposition 5.1 *For any TCCS term P, for any TCCS solution S, and for any $l \in \mathcal{L} \cup \{\tau\}$ we have:*

(i) *if $P \xrightarrow{l} Q$, then $\exists S \overset{*}{\rightleftharpoons} \{Q\}$ such that $\{P\} \xrightarrow{l} S$,*

(ii) *if $S \overset{*}{\rightleftharpoons} \{P\}$ and $S \xrightarrow{l} S'$,*
    then there exists Q such that $S' \overset{*}{\rightleftharpoons} \{Q\}$ and $P \xrightarrow{l} Q$.

As a consequence of these properties we have the following

Theorem 5.1 *The relation $\{(P,S) \mid \{P\} \overset{*}{\rightleftharpoons} S\}$ is a strong bisimulation.*

For more details about the general philosophy of the chemical abstract machine CHAM, as well as more results and remarks, see [4, 6].

6 π-calculus

A second example of defining a molecular abstract machine is for the π-calculus of Milner, Parrow, and Walker [22]. The π-calculus has been introduced as an extension of CCS to mobile concurrent processes (see [17, 19, 20]). As for CCS, the semantics of the π-calculus is given by a transition system the states of which are process terms. Concerning its expressiveness, it was showed that even our fragment of this calculus is enough to encode the λ-calculus [18].

The basic construct is given by the guarded processes $g.P$, where the guard g may be one of the following two kinds:

(i) an input guard $x(y)$.

In this case the guarded process $x(y).P$ waits for a name to be transmitted along the link(channel) named x. The name y is bound, and the process $x(y).P$ can receive any name z, to be substituted for y in P, yielding $P[z/y]$.

(ii) an output guard $\overline{x}z$.

In this case the process $\overline{x}z.P$ sends the name z along the link (channel) x, and then triggers P.

An interaction may occur between two concurrent processes when one sends a name on a particular link and the other is waiting for a name along the same link. Denoting by the parallel composition of two processes by $(P|Q)$, an interaction is actually defined by a "sender" $\overline{x}z.S$ and a "receiver" $x(y).R$, and can be represented by the transition:

$$(\overline{x}z.S \mid x(y).R) \rightarrow (S \mid R[z/y]).$$

This is a synchronous interaction, where the send operation is blocking: an output guard cannot be passed without the simultaneous occurrence of an input action. In the synchronous π-calculus we propose, the interaction agents are just the (channel) names. We shall consider the names themselves as elementary agents, freely available for other processes waiting for them. This fits in very well with the idea of a chemical abstract machine for describing the evaluation process in the π-calculus. In this model, the state of a system of concurrent processes is viewed as a "chemical solution" (formally: a multiset of processes) in which floating "molecules" can interact with each other according to some "reaction rules". In this setting, a particular molecule $\overline{x}z.S$ floating in the solution can interact with any molecule $x(y).R$ floating in the same solution. The result of this interaction is a solution containing S and $R[z/y]$.

First of all we introduce the formal π-calculus framework. Let N be a countable set of *names*. The elements of N are denoted by $x, y, \ldots$ The terms of this formalism are called *processes*. The set of processes is denoted by $\mathcal{P}$, and processes are denoted by $P, Q, R, \ldots$

Definition 6.1 *The processes are defined over the set N of names by the following syntactical rules:*

$$P ::= \overline{x}z.P \mid x(y).P \mid 0 \mid (P \mid Q) \mid !P \mid (\nu y)P.$$

The prefixes $x(y)$ and (νy) binds all free occurrences of y in P. We denote by $fn(P)$ the set of the names with free occurrences in P. We denote by $P[v/u]$ the result of simultaneous substitution in P of all free occurrences of the name u by the name v, using the α-conversion wherever necessary to avoid the name capture. 0 is the inactive process. A replicated process $!P$ denotes a process that allows to generate arbitrary instances of P in parallel. $(\nu y)P$ is the restriction of y to P, denoted $P\backslash y$ in CCS, and parallel composition of processes P and Q is denoted by $P \mid Q$, as usual in CCS.

Over the set of processes it is defined a *structural congruence* relation; this relation defines a syntactical equivalence over processes, providing somehow a non-explicit semantics of some formal constructions. This structural congruence captures the fact that two processes are structurally, i.e., statically, the same. Roughly speaking, this means that the processes can be decomposed into the same concurrent subprocesses.

Definition 6.2 *The relation* $\equiv \subseteq \mathcal{P} \times \mathcal{P}$ *is called structural congruence, and it is defined as the smallest congruence over processes which satisfies the following requirements:*

1. $P \equiv Q$, *whenever P is α-convertible to Q;*

2. $P \mid 0 \equiv P$, $P \mid Q \equiv Q \mid P$, $(P \mid Q) \mid R \equiv P \mid (Q \mid R)$, $!P \equiv P \mid !P$;

3. $(\nu x)(\nu y)P \equiv (\nu y)(\nu x)P$,
   $(\nu x)P \equiv P$, *if $x \notin fn(P)$, and*
   $(\nu x)(P \mid Q) \equiv (\nu x)P \mid Q$, *if $x \notin fn(Q)$.*

Two processes are α-convertible if they are the same modulo a renaming of some bound names.

The structural congruence deals with the aspects related to the structure of the processes, not to their interaction and mobility. Therefore, a transition system is given such that structurally congruent processes are given the same behaviour, by definition. This separation of *structure* and *behaviour* is intuitively clear, and it simplifies the transition system. For instance, the commutativity and associativity of parallel composition are handled on the structural level.

Definition 6.3 *The reaction relation over processes is defined as the smallest relation* $\rightarrow \subseteq \mathcal{P} \times \mathcal{P}$ *satisfying the following rules:*

$$(COM) \qquad x(y).P \mid \overline{x}z.Q \rightarrow P[z/y] \mid Q;$$

$$(PAR) \qquad \frac{P \to P'}{P \mid Q \to P' \mid Q};$$

$$(RES) \qquad \frac{P \to P'}{(\nu x)P \to (\nu x)P'};$$

$$(STRUCT) \qquad \frac{P \equiv P' \quad P' \to Q' \quad Q' \equiv Q}{P \to Q}.$$

The COM rule formalizes the synchronous interaction between two processes along a link x. The $STRUCT$ rule embodies the idea that structurally congruent processes have the same behaviour. The replication $!P$ of process P allows to consider as many concurrent instances of P as we need. Because $!P$ represents a possible countable infinite number of concurrent copies of P, we use multisets in which elements may occur countable many times.

Usually an unlabelled transition system $(Q, \to)$ is determined by a set Q of states and a binary relation $\to$ on Q which is called the transition relation. A multiset transition system is a transition system where Q is a set of multisets, and $\to$ is the smallest relation on Q which contains a set of basic transitions and is closed under multiset union. More precisely, if BT is a set of Basic Transitions, and S_1, S_2 are multisets, then the transition relation $\to$ is defined by the following axiom and rule

i) If $(S_1, S_2) \in BT$, then $S_1 \to S_2$.

ii) $\dfrac{S_1 \to S_2}{S_1 \uplus S \to S_2 \uplus S}.$

The axiom and the rule correspond to the laws of the molecular abstract machine CHAM. Intuitively, a basic transition of BT models a local interaction among the elements of S_1, and as a result they turn into the elements of S_2. These laws express the locality of such interactions: if there is a basic transition from S_1 to S_2, then that transition can take place in any context S. It is not difficult to prove the following result.

Proposition 6.1 *Let P, Q be two multisets. $P \to Q$ if and only if there exist S, S_1, S_2 such that $P = S_1 \uplus S$, $Q = S_2 \uplus S$, and $(S_1, S_2) \in BT$.*

Following [11], we can define now a specific multiset transition system called MAM, and then we define a semantic mapping that associates a state of MAM to each process of the π-calculus. According to this mapping, the meaning of a process is a multiset, namely the multiset of its concurrent subprocesses. Let us now recall that the state S of a molecular abstract machine is a "solution", that is a finite multiset of molecules. Taking over the chemical terminology, the states of MAM are called solutions, which are multisets of molecules, defined in a recursive way as follows:

1) A solution is a multiset over the set of molecules.

2) A molecule is a pair $g.S$, where g is a molecular guard, and S is a solution.

The recursion starts with the empty solution $\emptyset$. Note that the ordered pair (g, S) is written as $g.S$ in order to remain closer to the syntax of the π-calculus. A molecule $g.S$ is a molecule with a "subsolution" S. Formally, the sets *Solutions* and *Molecules* are the smallest sets such that *Solutions* $\in \mathcal{P}(Molecules)$, and *Molecules* $\in$ *Guards* $\times$ *Solutions*, where $\mathcal{P}(Molecules)$ is the set of all multisets over *Molecules*.

MAM can be described as the multiset transition system defined by the set *Molecules* of all molecules, and the set of the following basic transitions (that model interaction between molecules):

$$\{x(y).S \ , \ \overline{x}z.S'\} \rightarrow (S[z/y]) \uplus S'.$$

These transitions are actually the ones that will simulate those of the π-calculus.

Remark 6.1 *The left-hand side multiset of a basic transition is always a set of two molecules. Note that, by the rule ii) of a multiset transition system, the transition relation of MAM consists of all transitions*

$$\{x(y).S, \overline{x}z.S'\} \uplus S'' \rightarrow (S[z/y]) \uplus S' \uplus S''.$$

The effect of such a transition is that the solutions S and S' which are hidden by guards in their molecules are changed (e.g. S into $S[z/y]$), and then added to the soup S'' of molecules which represent the current state of MAM.

The next result connects substitutions to the behaviour of the solutions.

Proposition 6.2 *Let $S, R \in$ Solutions, and $y, z \in N \cup New$, where New is an infinite set of new names, disjoint of N. If $S \rightarrow R$, then $S[z/y] \rightarrow R[z/y]$.*

We describe the transformations of MAM corresponding to the π-calculus. We will define a relation $\rightsquigarrow$ which associates a solution of MAM to a process of π-calculus. By $new(S)$ we denote the set of new names of S; $new(S)$ is a proper subset of the set New of all new names.

Definition 6.4 *The semantic relation $\rightsquigarrow$ is defined as the smallest relation that satisfies the following compositional requirements:*

(i) $0 \rightsquigarrow \emptyset$

(ii) *If $P_1 \rightsquigarrow S_1$, and $P_2 \rightsquigarrow S_2$, then $P_1 | P_2 \rightsquigarrow S_1 \uplus S_2$, whenever $new(S_1) \cap new(S_2) = \emptyset$.*

(iii) *If $P \rightsquigarrow S$, then $(\nu x)P \rightsquigarrow S[z/x]$, whenever $z \in New - -new(S)$.*

(iv) *If $P \rightsquigarrow S$ and g is a guard, then $g.P \rightsquigarrow \{g.S\}$.*

(v) *If $P \rightsquigarrow S_i$ for all $i \in Nat$, then $!P \rightsquigarrow \uplus_{i \in Nat} S_i$, whenever $new(S_i) \cap new(S_j) = \emptyset$ for all $i \neq j$.*

Rule (i) translates the inactive process into the empty multiset. Rules (ii) and (v) translate parallel composition and replication into multiset union. By rule (iv) the guard $\overline{x}z.P$ is translated into $\{\overline{x}z.S\}$, and $x(y).P$ is translated into $\{x(y).S\}$. In (iii), the name x of $(\nu x)P$ with the scope restricted to P is replaced by a completely new name z with a global scope.

Definition 6.5 *Two process terms P and Q are multiset congruent, and we denote this by $P \cong Q$, if $\{S \mid P \rightsquigarrow S\} = \{S \mid Q \rightsquigarrow S\}$, i.e., if P and Q have the same multiset semantics.*

It is easy to prove from the definition of the multiset semantic $\rightsquigarrow$ that $\cong$ is a congruence indeed. The following results show that the semantic relation is, up to a bijective choice of the new names, a function.

Proposition 6.3

(i) *For every process P there exists multiset solutions S such that $P \rightsquigarrow S$.*

(ii) *If $P \rightsquigarrow S$, then $P \rightsquigarrow S'$ if and only if the only difference between S and S' is given by a bijective renaming function of new names.*

The bijective renaming function of new names determines an equivalence relation over solutions. The previous result (ii) shows that the semantic relation $\rightsquigarrow$ associates an equivalence class of solutions to each π-process (term). Intuitively, the solutions in one such equivalence class are structurally isomorphic and they differ each other only by the new names.

The next result shows that the semantic relation is compositional with respect to substitution, and this implies that the multiset congruence $\cong$ is a congruence with respect to substitution.

Proposition 6.4 *Let y, z be names in N.*

(i) *If $P \rightsquigarrow S$, then $P[z/y] \rightsquigarrow S[z/y]$.*

(ii) *If $P[z/a] \rightsquigarrow S'$, then there exists S such that $P \rightsquigarrow S$ and $S' = S[z/y]$.*

Therefore the compositional semantic mapping that associates a molecular state with each process of the π-calculus is well defined and makes sense. The intuitive meaning of a process is the multiset of all its guarded subprocesses. The molecular machine has only one type of transition, which corresponds to the basic action in the π-calculus, namely the interaction between two guarded subprocesses. Now we would like to know that this semantic mapping is a strong bisimulation between the transition system of the π-calculus and the multiset transition system of our molecular abstract machine. The following result show that the semantic relation $\rightsquigarrow$ is a strong bisimulation.

Proposition 6.5

(i) *If $P \rightsquigarrow S$ and $P \rightarrow P'$, then there exists S' such that $S \rightarrow S'$ and $P' \rightsquigarrow S'$;*

(ii) *If $P \rightsquigarrow S$ and $S \rightarrow S'$, then there exists P' such that $P \rightarrow P'$ and $P' \rightsquigarrow S'$.*

Moreover, if two processes of the π-calculus are structurally congruent, then they have the same semantics; this means that if two processes have the same structure, they have also the same multiset semantics.

Proposition 6.6 *If $P \equiv Q$, then $P \cong Q$.*

We can extend the structural congruence by adding some new rules for replication and restriction. In this extended calculus the structural congruence and the multiset congruence coincide, i.e., two processes are structurally congruent if and only if they are multiset congruent [11]. Moreover, it is shown in [11] that it is decidable whether two processes are structurally congruent (which means also that it is decidable whether they are multiset congruent).

7   Encoding of the λ-calculus into MAM

The paper [18] is an important paper on mobile processes, and it is invaluable for a good exposition of what is going on in the interaction between processes. Milner's recognition of the importance of identifying the "elements of interaction" and his pursuit of this goal by extension of the λ-calculus provide remarkable insights into the foundations of interaction.

There are at least two reasons why the translation of the λ-calculus into various formal models for concurrency and interaction is interesting. First reason is concerning the expressive power of the model, and the second is related to the new properties of the λ-calculus which could be obtained as a consequence of such a translation into a more general context (from the point of view of concurrency and interaction). From the first point of view, we recall that Turing Machine and λ-calculus have the same computational power. Extending the λ-calculus, the π-calculus extends the Church-Turing model of computation. In many papers on DNA computing we can find results about molecular computing systems capable of universal computation. These results are mainly based on simulations, implementation, or encodings of Turing Machines.

We show here another way of proving that a molecular system has at least the same computational power as Turing Machines, namely by encoding the λ-calculus into the multiset molecular system.

In order to do this, we use the translation of λ-calculus into the π-calculus (which was given first time by Milner in [18] and then studied and extended in many other papers – see [7]), and the translation of the π-calculus into the multiset molecular system. Adaptations of Milner's proof shows that these translations are adequate. In fact, the λ-calculus can be encoded into the molecular abstract machine in a more efficient way than by composing the translations.

8   Further Research

The multiset molecular systems offer a natural notion of parallel computation. It is not difficult to see that in MAM we have more freedom of interaction. However, in the π-calculus, interaction is restricted by guards, and the decomposition of parallel processes is not complete. Thus, a parallel computation naturally induces a partial order on the basic transitions that occur during the computation. If we want to formalize a massive parallelism, then the freedom of interaction should be unrestricted. We may adopt the "membrane law" of CHAM: a solution is encapsulated within a "membrane" and can itself be considered as a single molecule. The membrane law asserts that transformations of solutions may occur within molecules, namely at the level of subsolution. This fact allows internal interactions, and we need new transition rules related to this membrane law.

Another way is to completely decompose the processes (π-terms) into their subprocesses. This means that guards are also decomposed. This approach asks for a deep investigation of the guards. A difficult part of this approach is to find an adequate dynamics, i.e., an adequate transition system. These aspects are presented into the forthcoming paper [23].

References

[1] L. M. Adleman, Molecular computation of solutions to combinatorial problems, *Science*, 226 (1994), 1021 – 1024.

[2] G. Agha, *Actors: A Model of Concurrent Computation in Distributed Systems*, MIT Press, 1986.

[3] P. Banâtre, D. Le Metayer, Programming by multiset transformation, *Comm. of the ACM*, 36 (1993), 98 – 111 .

[4] G. Berry, G. Boudol, The chemical abstract machine, *Theoretical Computer Science*, 96 (1992), 217 – 248.

[5] G. Boudol, Towards a lambda calculus for concurrent and communicating systems, *TAPSOFT'89, Lecture Notes in Computer Science* 351, Springer-Verlag, 1989, 149 – 161.

[6] G. Boudol, Asynchrony and the π-calculus, *INRIA Res. Report* 1702, 1992.

[7] G. Ciobanu, M. Rotaru, A faithful graphical representation for π-calculus, *Proc. 11th Conference on Control Systems and Computers Science (CSCS11)*, vol. II, Bucharest, 1997, 148 – 152.

[8] G. Ciobanu, M. Rotaru, Faithful π-nets. From π-calculus to faithful π-nets and back, *The 4th Workshop on Expressiveness in Concurrency*, Italy, 1997.

[9] G. Ciobanu, F. Olariu, Interaction process structures, *Rosycs'98*, Iaşi, Romania, 1998.

[10] R. De Nicola, M. Hennessy, CCS without τ's, *TAPSOFT'87, Lecture Notes in Computer Science* 249, Springer-Verlag, 1997, 138 – 152.

[11] J. Engelfriet, A multiset semantics of the π-calculus with replication, *Proc. CONCUR'93, Lecture Notes in Computer Science* 715, Springer-Verlag, 1993, 7 – 21.

[12] R. Freund, L. Kari, Gh. Păun, DNA computing based on splicing: The existence of universal computers, *Theories of Computer Science*, in press.

[13] T. Head, Gh. Păun, D. Pixton, Language theory and molecular genetics, Chapter 7 in vol. 2 of *Handbook of Formal Languages* (G. Rozenberg, A. Salomaa, eds.), Springer-Verlag, 1997, 295 – 360.

[14] Z. Manna, A. Pnueli, *The Temporal Logic of Reactive and Concurrent Systems*, Springer-Verlag, 1992.

[15] R. Milner, Processes: A mathematical model of computing agents, *Logic Colloquim'73*, North Holland, 1975.

[16] R. Milner, *Communication and Concurrency*. Prentice-Hall, Englewood Cliffs, NJ, 1989.

[17] R. Milner, The polyadic π-calculus: A tutorial, *ECS-LFCS Report Series* 91 – 180, Laboratory for Foundations of Computer Science, University of Edinburgh, 1991.

[18] R. Milner, Functions as processes, *Journal of Mathematical Structures in Computer Science*, 2 (2) (1992), 119 – 141.

[19] R. Milner, Elements of interaction, *Comm. of the ACM*, 36 (1993), 78 – 89.

[20] R. Milner, Action structure for the π-calculus, *Proceedings of the NATO Summer School on Logic and Computation*, Springer-Verlag, 1993.

[21] R. Milner, π-nets: A graphical form of π-calculus, *ESOP'94, Lecture Notes in Computer Science* 788, Springer-Verlag, 1994, 26 – 42.

[22] R. Milner, J. Parrow, D. Walker, A calculus of mobile processes, *Journal of Information and Computation*, 100 (1992), 1 – 77.

[23] M. Rotaru, G. Ciobanu, Explicit input guards and interaction automata, submitted, 1998.

[24] P. Wegner, Why interaction is more powerful than algorithms. *Comm. of the ACM*, May 1997.

[25] P. Wegner, Interactive foundation of computing, *Theoretical Computer Science*, 1998.

Hamiltonian Paths and Double Stranded DNA

Tom HEAD

Mathematical Sciences Department
Binghamton University
Binghamton, NY 13902-6000, USA
tom@math.binghamton.edu

Abstract. It is suggested that computations using DNA molecules as computing units may sometimes be simpler to carry out and less error prone if double stranded DNA molecules are used rather than the single stranded molecules currently being used. A detailed illustration is given of a biomolecular algorithm that uses double stranded DNA for the solution of the directed Hamiltonian path problem.

1 Introduction

Following the method used by Adleman [1] in his DNA solution of a seven vertex instance of the directed Hamiltonian path (DHP) problem, researchers continue to use the annealing of single stranded DNA in their molecular computing efforts. Single stranded DNA molecules (ssDNAs) are capable of behaviors that obstruct the desired annealings. Through partial self annealing they may form undesired hairpin structures. Undesired partial annealings between distinct ssDNAs may result in undesired ligations. Single stranded DNA molecules are undesirably flexible. Researchers devote substantial effort to finding good encodings into ssDNAs in order to reduce these naturally occurring bad behaviors, [2], [3].

The point of this article is to suggest that it may be possible to avoid the ssDNA problems through the simple expedient of using *double stranded* DNA molecules (dsDNAs). Annealing and ligation can still be used, but the annealing will take place only between relatively short sticky ends extending from relatively stiff double stranded DNAs. In order to clarify the possible use of dsDNAs, an alternate proposal is given here for solving DHP problems using dsDNAs rather than ssDNAs.

The Adleman procedure consists of two phases that will be referred to here as (1) the *generation* of DNA molecules that encode candidate solutions and (2) the *elimination* of those DNA molecules that encode non-solutions. The procedure proposed here has these same two parts: generation and elimination. In Adleman's solution the generation phase is conceptually quite attractive because it is a one step process. However, it allows the production of molecules encoding paths that fail to initiate at the proper vertex and of producing molecules of length far too great to represent solutions. Adleman's elimination phase has a number of steps proportional to the number of vertices in the directed graph for which a Hamiltonian path is sought. In the solution proposed here the molecules representing candidate solutions are *grown* in a number of steps proportional to the number of vertices in the graph, but in such a way that every candidate initiates at the required vertex and no molecule grows to a length greater than the length appropriate for a solution. The proposed elimination phase, like Adleman's, has a number of steps proportional to the number of vertices in the graph. Thus each method has a total number of steps that is linear in the number of vertices in the directed graph.

In order to communicate rapidly and precisely the biomolecular algorithm that is proposed, it will be sufficient to spell out the solution of an instance of the DHP for a tiny directed graph having six vertices and nine directed edges. The principle for solving larger DHPs will be clear once this example is understood – just as the principle for solving larger DHPs by Adleman's algorithm is clear once one understands the seven vertex example solved in [1]. It remains to be seen whether a non-trivial instance of the DHP problem will ever actually be solved in a wet lab using any biomolecular algorithm whatever. The report of a laboratory solution based on a graph with 50 vertices and 100 edges would be very reassuring. This author regards all biomolecular computing research as fundamental scientific research of major significance. Where these investigations will pay off for humanity is not clear. The pay off may be in advancing biotechnology or biochemistry rather than in computing as we know it today.

We recommend the papers [4], [5] by L. Kari as excellent introductory and background reading concerning DNA computing. See [8] for the solution, using DNA, of the problem of finding a maximal clique in a six vertex graph. See [9] for an alternate and powerful conceptual scheme for enzymatic computing.

2   An Encoding of an Instance of the Directed Hamiltonian Path Problem

Consider the directed graph G having vertex set $V = \{0, 1, 2, 3, 4, 5\}$ and set of directed edges $E = \{(0, 1), (0, 2), (1, 2), (2, 3), (2, 4), (3, 4), (3, 5), (4, 2), (4, 5)\}$. Let 0 be the initial vertex and 5 the terminal vertex. The DHP in this instance is the question: Is there a path in $G = (V, E)$ that initiates at 0, does not re-enter 0, enters each of the other five vertices exactly once, and terminates at 5 ? Inspection of $G = (V, E)$ yields immediately the answer 'yes'. In fact for this G there is a unique path that meets the condition, namely: (0,1) (1,2) (2,3) (3,4) (4,5). Examples are easily produced for which there are no paths meeting the specified conditions. Other examples have several paths meeting these conditions. The DHP problem is simply to determine whether the number of paths satisfying the specified conditions is positive. It is the DHP as understood in this way for which we will now illustrate a biomolecular algorithm using the specific example $G = (V, E)$ presented above.

We begin by labeling three test tubes: Initials, Middles, Finals. But let us abbreviate these labels to: INI, MID, FIN. In INI we will place two dsDNAs that will be chosen to encode the edges (0,1) and (0,2). In FIN we will place two dsDNAs that will be chosen to encode the edges (3,5) and (4,5). In Mid we will place five dsDNAs that will be chosen to encode the edges (1,2), (2,3), (2,4), (3,4), and (4,2). Of course, as is typical in DNA computing, each of these three test tubes will contain on the order of a picomole of each of the chosen molecules, i.e., on the order of a trillion of each. Each of the nine dsDNAs we choose will not quite be entirely double stranded. Each will be blunt on one end with a three base long sticky end on the other. With each of the interior vertices, 1,2,3,4, we associate a word over the alphabet $\{A, C, G, T\}$. Distinct vertices must be assigned distinct words. We choose, largely arbitrarily, 5'-ACG-3', 5'-CAG-3', 5'-GCT-3', 5'- TGA-3'. Note that the complement triples that will base pair with the chosen triples are 3'-TGC-5', 3'-GTC-5', 3'-CGA-5', 3'-ACT-5'. Dropping redundance, we will denote the first four as ACG3', CAG3', GCT3', TGA3' and the second four as 3'TGC, 3'GTC, 3'CGA, 3'ACT.

With the above choices made we can indicate, in a schematic way, the molecules that will encode the edges of the graph. The notations will be explained below.

In INI we place the molecules that encode (0, 1) and (0, 2), respectively:

```
N--NO--ON--N--N1--1N--NACG3'     N--NO--ON--N--N2--2N--NCAG3'
N--NO--ON--N--N1--1N--N          N--NO--ON--N--N2--2N--N
            *                                *
```

In FIN we place the molecules that encode $(3, 5)$ and $(4, 5)$, respectively:

```
      N--N5--5N--N                    N--N5--5N--N
3'CGAN--N5--5N--N              3'ACTN--N5--5N--N
```

In MID we place molecules that encode $(1,2)$, $(4,2)$, $(2,3)$, $(2,4)$, and $(3,4)$:

```
                      ___
      AAAN--N2--2N--NCACCAGGTGN--N
3'TGCTTTN--N2--2N--NGTGGTCCACN--N

                      ___
      AAAN--N2--2N--NCACCAGGTGN--N
3'ACTTTTN--N2--2N--NGTGGTCCACN--N

                      ___
      AAAN--N3--3N--NCACGCTGTGN--N
3'GTCTTTN--N3--3N--NGTGCGTCACN--N

                      ___
      AAAN--N4--4N--NCACTGAGTGN--N
3'GTCTTTN--N4--4N--NGTGACTCACN--N

                      ___
      AAAN--N4--4N--NCACTGAGTGN--N
3'CGATTTN--N4--4N--NGTGACTCACN--N
```

The symbols A, C, G, T denote the four bases, adenine, cytosine, guanine, and thymine, as usual. The symbol N is used to indicate that any one of the four bases may be used. The symbol – is used to indicate an unspecified number of occurrences of the symbol that lies at each of its ends. Thus N–N indicates an unspecified number of 'don't care' bases. The length of each of these N–N intervals can be determined later to accommodate further considerations. Each of the intervals of digits 1–1, 2–2, 3–3, 4–4, 5–5 must be chosen to be an encoding in the A,C,G,T alphabet of a word that names the vertex at which the edge terminates. The interval 0–0 may be regarded as an encoding of 'start'. There are many useful ways of choosing the base sequences of the form digit–digit to encode desired information. How one chooses a particular encoding scheme will be decided after one chooses the method that will be used in the final (elimination) phase of the algorithm. All this will be clear after the elimination phase has been discussed below.

What remains to be explained at this point is the specific choices of the five nine-base segments that share the pattern CACXXXGTG in the upper level. When these patterns are understood, it will be possible to explain the five three-base segments that have AAA in the upper level. These requirements are made because we will employ the enzyme *Dra*III in the generation phase we recommend. This restriction enzyme, like a few hundred others, is readily purchased. *Dra*III acts on dsDNA molecules having an occurrence of one of the 64 possible nine base pair sites [C/G][A/T][C/G][N/N][N/N][N/N][G/C][T/A][G/C] to cut the molecule as illustrated:

```
N--NCACNNNGTGN--N     gives    N--NCACNNN3'   &        GTGN--N
N--NGTGNNNCACN--N              N--NGTG              3'NNNCACN--N.
```

At appropriate points in the generation phase, *Dra*III will be introduced to trim newly grown tips of DNA molecules in order to expose sticky ends as a preparation for the next growth step. When *Dra*III is introduced it must not do any cutting except at the tips. This explains the triples AAA. The growth step will be accomplished by ligating matching sticky ends. The role of the AAA segments is simply to assure that at this particular location there is NOT a GTG segment (which would allow an undesired cut by *Dra*III later). Thus it is not specifically necessary to have AAA triples, but triples different from GTG are certainly required. The overlines in the five molecules in MID simply mark the sticky ends that will occur when *Dra*III is used. Finally, the star appearing below the single occurrence of an N in each molecule in INI indicates that biotin is to be attached at that base. Biotin is used because it adheres firmly to certain materials, such as streptavidin, that are in common use in biotechnology. This allows the molecules in INI to be attached to a surface – which will allow separation processes to be carried out conveniently later.

At this point the reader may wish to review the molecules that have been chosen and confirm that all the information specifying the given instance of the DHP has been encoded in these molecules (even with some redundancy).

3   A Procedure for Generating Molecules that Encode Candidate Solutions

This procedure will be described in the form of commented pseudocode.
Procedure DHP:
(1) Combine the contents of INI with a biotin attracting surface.
Comment: The two molecular varieties in INI may now be visualized as billions of leaves of seaweed waving in water rooted by biotin to the attracting

surface. (The surface itself may actually be realized in the laboratory as the combined surfaces of many separate metal microspheres.)

(2) Add MID. Add a ligase. Allow time for ligation.

Comment: Continue the seaweed imagery. We now have three molecular varieties which we will denote in a more compressed notation in which we abbreviate each occurrence of N–N to N and, for each digit, we abbreviate each occurrence of digit–digit to digit. The starred N indicates the region of attachment to the surface through the biotin connection.

```
       NON1NACGAAAN2NCACCAGGTGN,
       NON1NTGCTTTN2NGTGGTCCACN
          *
```

which encodes the path 012, and the following two molecules which encode the paths 023 and 024:

```
  NON2NCAGAAAN3NCACGCTGTGN          NON2NCAGAAAN4NCACTGAGTGN
  NON2NGTCTTTN3NGTGCGTCACN          NON2NGTCTTTN4NGTGACTCACN.
     *                                 *
```

[Traces of molecules encoding paths 01 and 02 may remain.]

(3) Wash away excess MID and ligase.

(4) Add *Dra*III. Allow time for *Dra*III to act.

(5) Wash away *Dra*III.

Comment: The tips have been removed leaving the three molecular varieties:

```
       NON1NACGAAAN2NCACCAG3'
       NON1NTGCTTTN2NGTG
          *

       NON2NCAGAAAN3NCACGCT3'
       NON2NGTCTTTN3NGTG
          *

       NON2NCAGAAAN4NCACTGA3'
       NON2NGTCTTTN4NGTG.
          *
```

[Traces of molecules encoding paths 01 and 02 may still remain.]

Repeat (2), (3), (4), (5).

Comment: We now have the four molecular varieties which encode the paths 0123, 0124, 0234, and 0242:

```
NON1NACGAAAN2NCACCAGAAAN3NCACGCT3'
NON1NTGCTTTN2NGTGGTCTTTN3NGTG
    *

NON1NACGAAAN2NCACCAGAAAN4NCACTGA3'
NON1NTGCTTTN2NGTGGTCTTTN4NGTG
    *

NON2NCAGAAAN3NCACGCTAAAN4NCACTGA3'
NON2NGTCTTTN3NGTGCGATTTN4NGTG
    *

NON2NCAGAAAN4NCACTGAAAAN2NCACCAG3'
NON2NGTCTTTN4NGTGACTTTTN2NGTG
    *
```

[Traces of molecules encoding paths 01, 02, 012, 023, and 024 may remain.]

Repeat (2), (3), (4), (5).

Comment: We now have the five molecular varieties which encode the paths 01234, 01242, 02342, 02423, and 02423. The reader may now supply these five molecules.

(6) Add FIN. Add ligase. Allow time for ligase to act.

(7) Wash away excess FIN and ligase.

Comment: We now have the two molecules that represent the paths that begin at 0, terminate at 5, and contain the number of edges appropriate for a solution (five). These crucial molecules represent the paths 012345 and 024235. We continue to have a very substantial number of molecules, each with a sticky end, that represent the paths 01242, 02342, and 02423.

[Traces may remain of additional molecules that represent paths that contain too few edges to be solutions. Some of these potential trace molecules may even specify paths from 0 to 5: 0235, 0245, 01235, 01245, and 02345.]

The generation of molecules that represent candidate solutions is now complete. As will be seen, the molecules that represent paths containing too few edges will present no difficulty at all no matter which of the candidate elimination procedures to be discussed one chooses.

The challenge to the design of elimination procedures is to determine whether, among the molecules that represent paths containing the appropriate number of edges, there is one representing a path in which no vertex occurs twice (or, equivalently, in which every vertex occurs).

4 Procedures for Eliminating Molecules that Do Not Represent Solutions

The generation procedure given above can be combined with various procedures in order to provide a complete biomolecular algorithm for the DHP problem. The description of the system for encoding given in Section 2 allows flexibility which we can use in order to discuss three different ways to link the generation procedure with further procedures to complete the solution.

Those who have learned to carry out with adequate error control the elimination procedure used by Adleman in [1] may use this elimination procedure after the generation procedure has been carried out as in Section 3. Merely heat to separate the strands attached through biotin from their complements and wash the complements away. Proceed with Adleman's elimination procedure to obtain the solution of the DHP instance. If this procedure is chosen then in the encoding procedure there is no need for the interval 0–0. All edges may be represented by molecules of the same length and need only be sufficiently different to allow the elimination procedure. This choice of elimination procedures is not the one we propose here since it involves ssDNA. The two remaining methods avoid ssDNA completely.

Arthur Weinberger has suggested the following simple procedure which may be of value in testing the generation procedure for small instances of the DHP, such as the six vertex example used here. For this choice the interval 0–0 is not needed. Merely choose the intervals 1–1, 2–2, 3–3, 4–4, 5–5 to have as their lengths successive powers of two. Keep the lengths of the remaining portions of the molecules constant. The length of a molecule representing a solution can be calculated and, thanks to an easily confirmed mathematical property of the sequence of powers of two, any molecule generated that has this length must be a solution of the DHP. With this in mind, no elimination process is required. To complete the solution of the DHP merely make a gel separation and inspect the gel to see if there is a band at the appropriate length for a solution. The choice of the lowest power of two to use must be based on the separation sensitivity expected from the gel. This method is only appropriate for graphs having a small number of vertices since the lengths of the coding regions grow exponentially.

We come now to the method we recommend as the conclusion of our proposed biomolecular algorithm. The DNA in maintained in double stranded form throughout.

We begin with the DNA molecules in the form in which they were generated – still attached to the surface through the biotin connection. We will circularize the candidate molecules that encode paths extending from the initial node to the terminal node. In our example this would be the molecules encoding the paths 012345, 024235, 01235, 01245, 02345, 0235, and 0245. Anticipating this circularization, we will have previously chosen the encoding so that the regions 0–0 and 5–5 are identical and are the site of a previously chosen restriction enzyme. To be specific, let us assume that 0–0 = 5–5 = $[G/C][G/C][G/C][C/G][C/G][C/G]$, which is the site at which the restriction enzyme ApaI acts. At an occurrence of such a site in a dsDNA, ApaI cuts to produce the sticky ends GGCC3′ and 3′CCGG. We continue with pseudocode:

(8) Add chosen enzyme (ApaI). Allow time for total enzyme digestion.

(9) Wash away all DNA not attached to the surface and all enzyme.

Comment: At this point ALL the DNA molecules have a newly created 3′CCGG sticky end near their point of attachment to the surface. It is precisely those that encode paths that reach the terminal vertex that now also have a newly formed GGCC3′ sticky end at the end further from their point of attachment to the surface. Note that these sticky ends can match by means of hydrogen bonds.

(10) Add ligase. Allow time for ligase to act.

(11) Wash away ligase.

Comment: The sticky GGCC3′ and 3′CCGG ends will now have annealed and been ligated. Consequently circular dsDNA molecules will have been produced from these candidates – and remain bound to the surface. It is only these circular molecules that could possibly encode solutions to the DHP. We must expect additional non-circular molecules to remain attached to the surface. These molecules will have protruding sticky ends. They are the molecules that encode paths that do not terminate at the destination vertex.

Comment: For the present elimination procedure, each of the remaining digit–digit regions in the encoding molecules will have been filled with distinct restriction enzyme sites. We recommend that all molecules encoding edges be of the same length. For the present example we may choose to use the sites for the enzymes AflII, AgeI, ApaLI, and AscI in the segments 1–1, 2–2, 3–3, and 4–4, respectively. For the purpose of a pseudocode representation for the remainder of our biomolecular algorithm, let the four chosen enzymes be denoted Enzyme(i), $i = 1, 2, 3, 4$, respectively.

To the circular and linear molecules attached to the surface:

(12) For $i := 1$ to 4 Do

Add Enzyme(i). Allow time for total enzyme digestion.

Wash away all DNA not attached to the surface and all enzyme.

Add ligase. Allow time for ligation.

Wash away ligase.

Comment: At the ith pass through the For loop above: Molecules having no site for Enzyme(i) are unaffected. Circular molecules having exactly one site for Enzyme(i) will be cut at that site, but will be religated after being cut (the minority that aren't successfully religated cause no problem). Circular molecules that have two or more sites for Enzyme(i) will be shortened (those that are too short for religation will remain linear causing no problem). Linear molecules having one or more sites for Enzyme(i) are shortened. Thus, after the execution of the For loop, any DNA molecule that remains and has the length that is appropriate to encode a solution of the DHP MUST encode a solution. Consequently, the biomolecular algorithm is completed with the following step:

(13) Detach DNA molecules from the surface. Make a gel separation of the remaining DNA molecules. If a band appears on the gel at the length appropriate for a solution to the DHP, then the DHP has a solution; if not, there is no solution.

Remarks: The portions of the molecules that we have left unspecified can be treated in virtually any way an experimenter wishes. The only restriction is to be sure that no additional sites for the enzymes to be used are created. The number of base pairs in each of the dsDNAs should be equal (unless one is using the 'powers of two' encoding). The lengths must be fully adequate to insure that the molecules representing solutions are sufficiently flexible so that the desired circularization is very highly probable. (When shorter molecules fail to circularize it can only be considered fortunate.)

To avoid false positives one must assure that it is highly unlikely that two DNA molecules attached to the surface lie sufficiently close that they may ligate to form 'hybrids' rather than to circularize in the elimination phase of the algorithm. A final protection against false positive results is to sequence the molecules that appear to represent solutions. This also gives the paths that constitute the solutions – and these paths may be desired for other reasons.

5   Overcoming Some of the Limitations of the Algorithm as Presented

Suppose we wish to use the biomolecular algorithm illustrated here on instances of the DHP based on graphs having 50 or more vertices. *Dra*III cannot play the role for such problems that it plays in our example. Although *Dra*III acts at 64 distinct sites, 32 is the largest number of these sites that can be used to encode vertices in the way we have encoded the four (interior) vertices in our example. This results from the fact that the encoding of each vertex actually uses up two sticky triple base ends. This is clear when one considers that the choice of the triple ACG3′ to encode vertex one eliminates the possibility of using its conjugate triple 3′TGC as a possible code for a second vertex. Thus *Dra*III can only be used, as we have used it, for instances of the DHP having at most 34 vertices (i.e., 32 interior vertices). According to catalog descriptions, enzyme *Hga*I cuts dsDNA at sites which allow all 1024 possible five-base sticky ends to be generated. Consequently it is conceivable that up to 512 interior nodes could be encoded using *Hga*I in place of *Dra*III.

Recall that our algorithm actually encodes each (interior) vertex v not only by a special three base sticky end, but also by a special restriction enzyme site that is placed in each edge that terminates at v. No two vertices can be assigned enzymes that act at a common site. Although there are now more than 2,500 known restriction enzymes, many pairs act at common sites. The number currently available provide less than 200 distinct sites. This limitation is therefore more strict than the 512 figure of the previous paragraph.

Note that the algorithm as presented does not use PCR amplification. We suggest that the algorithm be tested on graphs that are small enough that PCR will not be required. If laboratory tests of the algorithm are successful on small graphs, then larger graphs can be attempted but will require the insertion of PCR amplification steps into the algorithm. The flexibility in the choice of the molecules encoding the edges allows a common left and a common right primer region to be included for each edge. Exonuclease steps can also be inserted into the algorithm to prevent the accumulation of useless short dsDNAs with sticky ends. If this is to be done the molecules encoding edges will have been provided with an appropriate attachment at the blunt non-biotinylated end that protects against exonuclease attack.

If the proposed surface attachment of the DNA through biotin proves inadequate to withstand the numerous washing steps, then more powerful means of attachment can be considered. We have suggested that the biotin be attached at a non-terminal base. This may require attachment through a non-standard base, perhaps uracil.

Arthur Weinberger and Fernando Guzman have observed that by making use of the restriction enzyme sites appearing in the molecules encoding edges and then following with gel separations, one can make a laboratory determination of each of the paths representing solutions to the DHP. This is particularly simple when there is only one solution path, but not difficult in any case. Discovering this process is left as an exercise here.

We are currently planning a new prototype biomolecular system that, with a preassigned bound B on the number of vertices, will, with minimal programming, solve any instance of the directed Hamiltonian path problem based on a graph having B or fewer vertices. Some of the concepts from the system sketched here will be used, but the new system will never form a molecule that has a repeated vertex. We hope this new system will allow the solution of non-trivial instances of the DHP. The suggestions for sequencing of solutions, as made by Weinberger and Guzman, may work in a natural and simple way in the new scheme.

Acknowledgments. The ideas for the biomolecular algorithm proposed here have arisen from extended conversations with several scientists from various nations. In fact the article is a synthesis of ideas and suggestions from others. The idea of growing the paths one edge at a time came from discussions with Akira Suyama, University of Tokyo, who works with a group that has already grown paths, but using ssDNA rather than the dsDNA as proposed here [7]. This article would not have been written if the discussions with Akira Suyama had not taken place. Hendrick van Hogeboom, University of Leiden, suggested the specific method used here for incorporating, for each molecule representing an edge, the designation of the vertex at which the edge terminates. This provided a major improvement over my own previous version. The idea of circularizing the DNA molecules was provided by Satoshi Kobayashi, University of Electro-Communications, Chofu, Tokyo. Elizabeth Laun and K. J. Reddy, Binghamton University, have carried out fundamental laboratory work [6]. Sharing in their experience has provided my closest contact with the relevant biochemical realities. Without Gheorghe Păun's kind invitation to speak at his *Workshop on Molecular Computing 1997*, I would not have been stimulated to produce this article. All naivete and foolishness included here is my personal property, as none of the people mentioned here has had the opportunity to read or correct this article.

Partial support for this research through NSF CCR-9509831 and DARPA/NSF CCR-9725021 is greatfully acknowledged.

References

[1] L. Adleman, Molecular computation of solutions to combinatorial problems, *Science*, 266 (1994), 1021 – 1024.

[2] R. Deaton, R. Murphy, M. Garzon, D. Franceschetti, S. Stevens, Good encodings for DNA-based solutions to combinatorial problems, presented at 2nd Workshop on DNA computing, *DIMACS Series in Discrete Math. and Theor. Computer Sci., Amer. Math. Soc.*, to appear.

[3] R. Deaton, R. Murphy, J. Rose, M. Garzon, D. Franceschetti, S. Stevens, A DNA based implementation of an evolutionary search for good encodings for DNA computation, *Proceedings of 1997 IEEE International Conference on Evolutionary Computation*, Indianapolis, 1997, 267 – 271.

[4] L. Kari, DNA computing: Arrival of biological mathematics, *The Mathematical Intelligencer*, 19 (1997), 9 – 22.

[5] L. Kari, From micro-soft to bio-soft: Computing with DNA, *Proceedings on Biocomputing and Emergent Computations*, Skovde, Sweden, 1997.

[6] E. Laun, K. J. Reddy, Wet splicing systems, presented at 3rd Workshop on DNA computing, *DIMACS Series in Discrete Math. and Theor. Computer Sci.*, Amer. Math. Soc., to appear.

[7] N. Morimoto, M. Arita, A. Suyama, Solid phase DNA solution to the Hamiltonian path problem, presented at 3rd Workshop on DNA computing, *DIMACS Series in Discrete Math. and Theor. Computer Sci.*, Amer. Math. Soc., to appear.

[8] Q. Ouyang, P. Kaplan, S. Liu, A. Libchaber, DNA solution of the maximal clique problem, *Science*, 278 (1997), 446 – 449.

[9] P. W. K. Rothemund, A DNA and restriction enzyme implementation of Turing machines, presented at 1st Workshop on DNA computing, *DIMACS Series in Discrete Math. and Theor. Computer Sci.*, 27, Amer. Math. Soc., 1996, 75 – 119.

Graph Structures in DNA Computing

Nataša JONOSKA, Department of Mathematics
Stephen A. KARL, Department of Biology
Masahico SAITO, Department of Mathematics
University of South Florida
Tampa, Florida 33620, USA
jonoska/saito@math.usf.edu

Abstract. We show that 3-dimensional graph structures can be used for solving computational problems with DNA molecules. Vertex building blocks consisting of k-armed ($k = 3$ or $k = 4$) branched junction molecules are used to form the graph. We present the solution to the 3-SAT problem and to the 3-vertex-colorability problem. Construction of one graph structure (in many copies) in both procedures is sufficient for determining the answer to the problem. In our proposed solutions, for the 3-SAT, the number of steps required for the algorithm is equal to the number of variables in the formula, and for the 3-vertex-colorability problem, the algorithm requires constant number of steps, regardless of the size of the graph. The vertex and edge building blocks are made of stable molecules.

1 Introduction

The field of practical DNA computing opened in 1994 with the Adleman's paper [1], in which a laboratory experiment involving DNA molecules was used to solve a small instance of the Hamiltonian Path problem. In the follow up paper by Lipton [9], it was demonstrated how a large class of NP-complete problems could be solved by encoding the problem in DNA molecules. In particular, Lipton showed how another famous NP-complete problem, the so-called "satisfiability" problem (SAT) and subsequently any other NP problem could be encoded and solved using DNA molecules.

Shortly after, several authors have suggested applications of DNA methodology for computational purposes (for example [10], [11]).

In many algorithms, the general approach has been to treat the DNA molecules as linear strings where much of the information content is encoded in the order of nucleotides that make up the DNA. Recently reported research [14] however, demonstrated that it is possible to form higher order, three dimensional (3D) structures with DNA molecules, and in [7] it was shown that 3D structures could be used to solve the Hamiltonian Path problem in constant number of steps regardless of the size of the graph. This showed that the use of 3D DNA structures could significantly reduce the time and steps needed to identify a solution. Such an approach also was suggested in [15], where 2-dimensional structures are used.

In this paper we propose solutions to two other famous NP-complete problems, the satisfiability (SAT) and the 3-vertex-colorability problems. In both cases we use k-armed branched junctions molecules to represent the k-degree vertices of the graph. These molecules are much simpler than the vertex building blocks used in [7]. Their construction is fairly well understood (see [14] and [15]), and consequently, 3D construction of a graph using these molecules as vertex building blocks is much more feasible. For both problems (SAT and 3-vertex-colorability) we give algorithms that use 3D graph structures. In the first case the number of steps required for the algorithm is equal to the number of variables in the formula, and in the second case, the algorithm requires constant number of steps, regardless of the size of the graph. The laboratory protocols for the computational steps suggested are well known in molecular biology. The vertex building blocks are also characterized [14] and are known as stable.

We start the paper with the description of the SAT problem and the Lipton's algorithm. Then we describe our approach and the algorithm for SAT. The third section contains the description of the 3-vertex-colorability problem and our algorithm for its solution. We end our paper with a discussion of the feasibility of the 3D approach in computing.

2 The Satisfiability (SAT) Problem

2.1 Definition of SAT. Let $A = \{a_1, a_2, \ldots\}$ be a set of boolean variables. A *clause* is a formula $C = b_1 \vee b_2 \vee \cdots \vee b_k$ where $\vee$ is the logical "or" and each b_i is a variable in A or a complement to a variable in A. The complement of $a \in A$ is denoted by $\bar{a}$. A *logical formula* is a formula

$$\alpha = C_1 \wedge C_2 \wedge \cdots \wedge C_r$$

where each C_i is a clause.

The *satisfiability problem* (SAT) asks whether for a given logical formula α there is an assignment of $\{0, 1\}$ to the variables in α that would give the formula α the value 1.

It is convenient to write a "+" instead of a "∨" and a "·" or a simple concatenation instead of a "∧". In that case the formula α becomes $\alpha = C_1 C_2 \cdots C_r$.

Consider the example

$$\alpha = (x + y)(\bar{x} + y + z)(\bar{x} + y + \bar{z}). \qquad (*)$$

This formula $(*)$ has value 1 for the assignments $(x, y, z) \in \{(1, 1, 1), (1, 1, 0), (0, 0, 1), (0, 0, 0)\}$. For any other assignment the value of the formula is 0.

The SAT problem is one of the standard examples of the so called NP-complete problems [5] and it is well known that these problems require (in the worst case) an exponential number of steps with the size of the problem for their solution. Identifying a solution for even a moderately simple problem therefore, can require a prohibitively large number of steps. In fact, the excitement over DNA as a computational tool rests in the ability to perform in parallel some of the computational steps. Thus, the number of steps to identify a solution does not increase exponentially with increases in the size of the graph.

2.2 Review of Lipton's Algorithm. Lipton's solution contains the following steps:

1. If a formula $\alpha = C_1 \cdots C_r$ contains k variables, then form all k-bit strings of DNA molecules. Keep these in a tube t_0.

2. For each clause $C_i = b_1 + \cdots + b_m$ $(m \le k, i = 1, \ldots, r)$ for $j = 1, \ldots, k$ from tube t_{i-1} extract the k-bits that have the b_j-th value 1 (i.e., if $b_j = a$ then the b_j-th value is 1, if $b_j = \bar{a}$ then the j-th value is 0). Combine all in tube t_i.

3. If there is any DNA in the last tube t_r then the answer is "yes".

The first step presumably generates the exponential number of k-bits among which the correct assignments for the formula are contained. The rest of the steps are used to extract those k-bits that are solution of the formula α. This extraction procedure is used as many times as there are

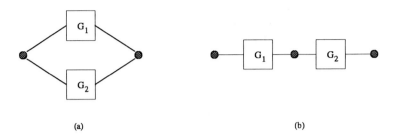

Figure 1: Contact network for (a) $\alpha = \beta + \gamma$ and (b) $\alpha = \beta\gamma$

operations in the formula, i.e., rk times (for r clauses and k variables). The laboratory procedure suggested for this step is the use of biotin label oligos that are complementary in DNA sequence to the bits that are of interest. The bit strings encoding a specific bit value are removed from the mix when hybridized to the biotin labeled oligo that is conjugated to paramagnetic beads. Other extraction techniques such as restriction endonuclease digestion have been suggested [3]. Since errors could accumulate, methods with fewer steps would be desirable.

2.3 Solving SAT by Constructing 3D Graphs with DNA. In [9], Lipton also shows the well established fact that SAT is equivalent to the "contact network" problem. The contact network problem is a graph with two distinguished vertices: the source s and the target t. Each edge in the graph is labeled with x or $\bar{x}$ where x is a variable. An edge is considered as connected if the value of the variable assigns value 1 to the edge (if the edge is labeled $\bar{x}$ then the value $x = 0$ keeps the edge connected). The SAT problem for contact network asks whether there is an assignment of the variables that would provide the source to be connected with the target.

Any logical formula can be transformed in a contact network problem. If $\alpha = \beta + \gamma$ where β and γ are simpler formulas with contact networks G_1 and G_2, then the contact network for α is obtained by placing G_1 and G_2 in parallel (Figure 1(a)). If $\alpha = \beta\gamma$, then the contact network for α is obtained by placing G_1 and G_2 in series (Figure 1(b)). Similarly, the contact network for the formula $(*)$ is presented in Figure 2.

It is a well known fact that the restricted SAT problem 3-SAT, when every clause C_i in the formula has at most 3 variables, is also an NP-complete problem [5] (the number of variables that appear in the formula is not necessarily 3). Hence we can concentrate on 3-SAT problems in which

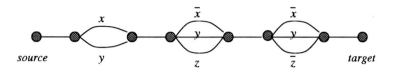

Figure 2: Contact network for formula (∗)

Source   *Target*

Figure 3: Molecules encoding the source and the target

case the corresponding contact network problem will contain vertices of degree at most 4.

We will refer to the formula (∗) and the contact network in Figure 2 as an example in our general discussion.

Now we construct contact networks by DNA to solve 3-SAT problems. First we list the molecular building blocks for the contact network graph.

- *Source and target.* The endpoints of the contact network (the source and the target) are presented with a hairpin structured molecule as depicted in Figure 3. Hydrogen bonds between the anti-parallel, complementary Watson-Crick bonds are depicted as dotted segments between the strands. Orientation of the strands of DNA are indicated with arrowheads being placed at the 3′ end. For simplicity, secondary structure (the double helix) of each of the molecules is not presented in any of the figures.

- *Vertex blocks.* Each vertex in the contact network has degree at most 4. One of the edges is a connection between the clauses and the other three edges are labeled with the three variables that are included in a clause as depicted in Figure 4.

  If a clause in the formula contains only two variables, then 3-armed branched molecules are used for the vertices. In the case when the clause has only one variable, then a single double stranded DNA molecule representing the variable is used.

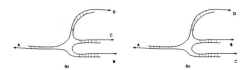

Figure 4: Molecules encoding the vertices for 3-SAT

- *Edges.* Edges of the contact network graph are represented by double stranded DNA molecules ligated to the vertices. The 3′ ends of the DNA strands in the edge molecule end with single stranded segments of 20 to 30 nucleotides length that are specific for the variable that is encoded. In particular, the single stranded segments of the edge encoding x and encoding $\bar{x}$ are distinct. The edge molecules have complementary sequence tails to the corresponding vertex blocks such that compatible vertex and edge blocks can form Watson-Crick paired double-stranded DNA segments and be joined together.

  The central part of an edge molecule encodes a restriction site that can be cleaved by a restriction enzyme. Different restriction enzymes are used for different variables and complements of the variables. Hence, if k variables appear in a formula, $2k$ restriction enzymes are used. For a variable x, the restriction enzymes that cleave the x-edge and $\bar{x}$-edge will be referred to as $(x = 0)$-enzyme and $(x = 1)$-enzyme respectively.

- *Caps.* Caps are used in the proposed solution to "cap off" the cleaved ends of edges. These caps are not used for building the given graph. This process will be explained in the next section. Caps have the same shape as sources and targets. Suppose a restriction site used in an edge consists of the sequence h and k. (The single stranded sequence is h and k, and they are complementary to each other making a double stranded segment.) Corresponding caps have end sequences h and k, respectively, for every restriction site used.

To form connections in a graph, vertex blocks are combined (together with the "source" and "target" blocks) with edge blocks, and their compatible ends are allowed to form double-stranded DNA. Once formed, the edges are locked together by sealing all open "nicks" in the DNA strands with DNA ligase. Figure 5 shows a formation of the contact network graph for

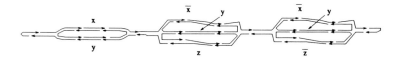

Figure 5: Molecules encoding the source and the target

the formula (∗). In every such graph structure, edges between two vertex blocks represent a clause.

2.4 Solving the 3-SAT Problem:

1. Combine the vertex blocks, edge blocks, the source and the target molecules in a single mix. Allow them to hybridize and then be ligated.

2. Remove the partially formed graphs that have open ends. This can be done with an exonuclease enzyme.

3. For each variable x in the formula repeat the following:

   (a) Separate the mix in two tubes.

   (b) In the first tube cleave with $(x = 0)$-enzyme and in the second cleave with $(x = 1)$-enzyme.

   (c) Add cap building blocks, allow them to hybridize and then be ligated. This step is depicted in Figure 6.

4. Using PCR primers complementary to the source and the target molecules (in both directions), separate molecules that start with "source" and end with "target". If any molecules are recovered, the connection from "source" to "target" is established and 3-SAT problem has a solution.

Proposition 2.1 *Let A be the set of molecules that contain a single stranded molecule connecting both the "source" and the "target" molecules present in the mix after steps 1 and 2 of the procedure are performed.*
Then $A \neq \emptyset$ if and only if there is a solution to the 3-SAT problem.

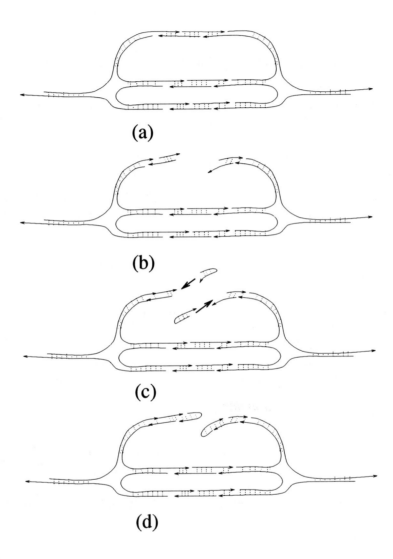

(a)

(b)

(c)

(d)

Figure 6: Capping off cleaved edges

Proof. Let α be a logical formula associated with a contact network. Suppose $A \neq \emptyset$. Then there is a single stranded molecule connecting the source and target molecule. This means that for every pair of vertices within a clause, there is an edge connecting them. Such an edge was not cleaved in the Steps 1 and 2, therefore there is an edge which received the value 1 (the variable $x \in \alpha$ which received 1 or the variable $\bar{x}$ which received 1). Furthermore, Step 2 ensures that the variable x receives the value 1 (not cleaved) if and only if $\bar{x}$ receives 0 (cleaved), and vice versa. Therefore the existence of such a molecule connecting the source and the target implies the existence of assignments of 0 and 1 to the variables that gives the value 1 for α. Thus the 3-SAT problem has a solution.

Conversely, if 3-SAT has a solution, then there are values for the variables that give each of the clauses value 1. This means that two adjacent vertices within a clause in the corresponding contact network problem are connected by an edge building block. Note that when an edge is not cleaved, then it connects the two vertices by a single stranded segment, because we capped off the cleaved edges by cap building blocks. The situation is depicted in Figure 6. It is shown that two vertices are connected by segments through the middle and the bottom edges in Figure 6(d) after the capping process Step 3 (e) (Figure 6(a) – (c)).

Therefore for the value 1, the corresponding edge block is not cleaved, and two vertices are connected by a single strand. Hence the target and the source molecules after Step 2 are connected by a single stranded DNA string. Thus $A \neq \emptyset$.

In fact, in the proof we assumed that only the original graph was constructed by DNA. By our construction, the DNA graphs we obtain in Step 1 are the original graph or its covering spaces. We will explain this point in Section 5.1.   □

It is important to note that none of the steps in the procedure depend on the number of vertices or edges in the graph. The procedure depends only on the number of the variables in the formula. This is an improvement from the Lipton's algorithm which depends on the size of the formula. In fact, from n variables there are $\binom{n}{k}$ choices of variables in a clause of k variables and there are 2^k clauses for each choice of k variables. Hence, for a 3-SAT problem with n variables, the length of the formula can be as long as $\frac{4}{3}n(n-1)(n-3)$, i.e., cubic polynomial of n.

In the Lipton's algorithm all possible paths from the source to the target vertices are formed and then the right answer is extracted. In the above

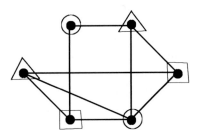

Figure 7: Coloring a graph

procedure, all possible paths from the source to the target are formed also, but not by obtaining exponentially more distinct molecules, but by constructing the graph itself which contains all possibilities. The connections from the "source" molecule to the "target" molecule that remain after Step 3, are solutions to the problem. However, the number of the graph structures is divided by 2 with each repetition of Step 3(a), so we need at least 2^k copies (k is the number of variables) of the graph.

3   3-Vertex-Colorability Problem

3.1 Definition of 3-Vertex-Colorability. Let G be a graph with vertices V and edges E. The graph G is *3-vertex-colorable* if there is a surjective (onto) function $f : V \to \{a, b, c\}$ such that if two vertices $v, w \in V$ are adjacent (connected with an edge) then $f(v) \neq f(w)$. The $n-$colorability is defined similarly.

This problem is also a well known NP-complete problem and has been addressed in the DNA computing literature as well (see [2], [3], for example). As with many other algorithms, the full power of the three dimensional feature of the problem is not used and explored. Here we show how a 3D graph structure of DNA can be used to solve the problem. In particular, we show that the whole graph can be constructed by DNA if and only if a solution to the problem exists.

In Figure 7, an example of a graph with a possible 3-vertex-coloring is given. In the figure, the three colors of the vertices are represented by small circles, triangles and squares surrounding the vertices.

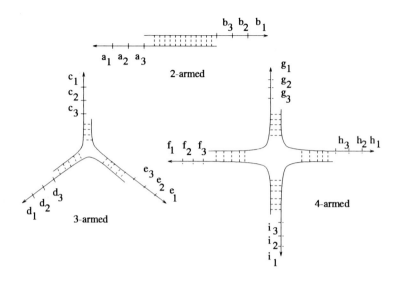

Figure 8: Building blocks for vertices

3.2 Construction of Graphs by DNA

- *Building blocks for the vertices.* As in the previous section, k-armed
  branched molecules are used as building blocks for the vertices. If
  a vertex has degree k, then a k-armed branched molecule is used
  for its building blocks. For the example presented in Figure 7, we
  will need one 2-armed branched molecule (which is just a regular
  double stranded molecule), four 3-armed branched molecules and one
  4-armed branched molecule (see Figure 8). The $3'$ ends of the k-
  armed branched molecules end with single stranded extensions. These
  extensions are 30 to 45 base pairs long and consist of three parts each
  10 to 15 base pairs long. The first and the third part (for example
  x_1 and x_3 where $x \in \{a, \dots, i\}$ in Figure 8) are specific encodings
  for the edge that is represented by the given "arm" of the molecule.
  The middle part of the encoding is the same for all "arms" of the
  vertex molecule and represents the color of the vertex. The sequence
  representing one of the colors is chosen to contain a restriction site for
  an enzyme which cuts the sequence at this site. For each vertex three
  blocks are needed (each corresponding to one of the three possible
  colors of the vertex).

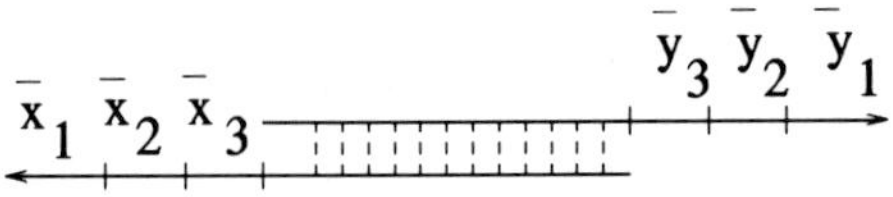

Figure 9: Edge building block

- *Edges.* Each edge is a regular double stranded molecule. The $3'$ ends of the molecule end with single stranded ends that are complementary to the corresponding "arm" of the vertex that is incident to the edge. Hence the first and the third part of the encoding at the $3'$ end is $\bar{x}_1$ and $\bar{x}_2$ (see Figure 9) which are Watson-Crick complementary to x_1 and x_2 ($x \in \{a, \ldots, i\}$) of the corresponding "arm" of the incident vertex molecule. The middle part of the encoding $\bar{x}_2$ is complementary to the color of the incident vertex, but here, the color sequence $\bar{x}_2$ at one $3'$ end is different from the color sequence $\bar{y}_2$ that is at the other $3'$ end of the edge molecule.

  For each edge we have exactly six double stranded molecules, each representing a pair of distinct colors at the endpoints of the edge molecule.

To form the graph, all edge molecules and all vertex building blocks are combined and their compatible ends are allowed to from double-stranded DNA. Once formed, the edges are locked together by sealing all open "nicks" in the DNA strands with DNA ligase. Three-dimensional DNA structures that do not contain open ends are referred to as *graph structures*.

Proposition 3.1 *For a given graph G, a graph structure can be formed by vertex building blocks and edge molecules if and only if G is 3-vertex-colorable or 2-vertex-colorable.*

Proof. The proof follows directly from the encodings of the single stranded $3'$ ends of the edge molecules and the vertex building blocks. If a graph structure is formed, then the endpoints of each edge molecule are "colored" distinctly. Since the colors at the $3'$ ends of the arms at a vertex building block are the same, only one color is associated with each vertex. The converse is equally straight forward. If G is 3-vertex-colorable, we can construct a graph structure from the building blocks as follows: if a vertex v has color a, we choose the vertex building block for v to have the color encodings at the arms *a*. If an edge *e* is incident to vertices *v* with color *a* and *w* with color *b*, we choose an edge molecule for *e* with *v*-end colored *a* and *w*-end colored *b*. These molecules can form a graph structure (with no open ends).

Again, we assumed that the graphs constructed are the copies of the given graph. We will discuss the problem of covering graphs in Section 5.1.

□

The procedure for solving the 3-vertex-colorability problem is as follows.

1. Combine all vertex building blocks with all edge molecules in a single tube and allow the complementary ends to hybridize and be ligated.

2. Remove partially formed 3D DNA structures with open ends that have not been matched. This could be done by using an exonuclease enzyme.

3. Remove the larger size of graphs formed in the above steps by gel electrophoresis. This step will be explained in Section 5.1.

   At this point, if there is no graph in the tube, then the graph is neither 2- nor 3-colorable.

4. Otherwise: we assume that if a graph is 2- or 3-colorable, then all coloring permutations of graphs are present in the tube. Add the restriction enzyme chosen for one of the colors and cleave the vertices that contain that color. There are three possibilities: (i) If the graph is only 2-colorable then graphs of the original size (colored with the remaining two colors) will be present together with vertex building blocks. (ii) If the graph is only 3-colorable, vertices are removed from each original graph, hence all remaining graphs are of smaller size than the original one. (iii) If the graph is both 2- and 3-colorable, then some graphs with the original size will be present (2-colorable ones), as well as smaller graphs (3-colorable) from which some of the vertices are removed. Gel electrophoresis is used to distinguish between the above cases.

The number of steps in this procedure again does not depend on the number of vertices (or edges) in the graph. Once the building blocks are formed, the procedure needs only four steps to perform.

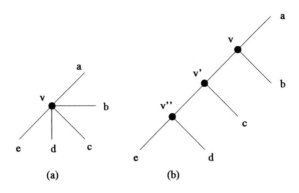

Figure 10: Perturbation of a vertex

4   Discussion

4.1 Complexity of the Building Blocks. The idea of using different building blocks to depict the entire graph structure was inspired by the work of N. Seeman and his research group [14] where they describe the construction of different 3D DNA structures. The building blocks proposed here are one of the simplest forms of the molecules described by these researchers (see references in [14] and also [15]). Furthermore, a plastic DNA model for a portion of vertex building blocks is built in [7], so that similar structures used in this paper are also feasible.

The algorithms presented assume that desired building blocks are already available. This is likely to be the most difficult and complex part of the procedures. However, the 3-armed and 4-armed branched molecules have already been obtained and it has bee reported that they are fairly stable [14]. It is certainly necessary to study the complexity and the stability of the k-armed branched molecules for k greater than 4. We are not aware of any experiments producing distinct k-armed branched molecules in large numbers. And a potentially large number of distinct such molecules are needed for carrying out both of the procedures described here.

Using k-armed branched molecules is certainly not necessary in either of the procedures described here: (1) the 3-SAT requires only 4-armed branched molecules and every SAT problem is equivalent to a 3-SAT problem by a simple renaming of the variables (see for example [5]); (2) There is a surjective homomorphism from a graph with vertices with degree at most three to any other graph. Statement (2) follows from a simple graph construction of substituting every vertex of degree 4 or higher with vertices of degree at most three. For example in Figure 10(a) a vertex v with degree 5 is depicted. In Figure 10(b), the same vertex is split into three vertices v, v', v'' each of degree 3. Two new edges connecting v with v' and v' with v'' are also introduced. The old 5 edges incident to v are split between v and the new vertices v' and v''. In checking the 3-vertex colorability using the new graph (having vertices with degree at most 3) we need to encode the edges connecting the new vertices (v with v' and v' with v'') to have only one color at the ends. Although k-armed junction DNA molecules have been successfully constructed, we are not aware of any experimental data on possibilities and efficiency of forming DNA graph structures proposed here.

5 Scaling up to Larger Graphs

In both procedures of this paper the complex molecules that are constructed at the end are the molecules that encode the solution of the problem. However, there are two difficulties. First, for constructing all the possible combinations that could encode the right answer to the problem, even in a fairly simple mixture, appropriate reaction conditions for annealing are important and difficult to determine [12]. The second difficulty is that constructing sufficient number of graph structures requires exponential increase in the DNA mass needed for the solution of the problem.

It is estimated that in a single test tube, there could be stored about 10^{15} DNA molecules. We could allow only 10^9 of these to be distinct. The building block of a 4 degree vertex can be formed with 200 nucleotides [14]. For the 3-SAT problem of a formula with r clauses, we need $2r$ vertex blocks $3r$ edge molecules $6r$ for capping off the cuts and 2 for source and target. So we need $11r + 2$ distinct DNA molecules. We need 2^k copies of DNA graphs, so that we get a bound $(11r + 2)2^k < 10^7$. This estimate might be very liberal, and it might be appropriate to count each 4-armed branched vertex building block as 4 distinct DNA molecules. In this case the bound of 10^7 does not get reduced by more than a factor of 2, i.e., $(11r + 2)2^k$ is bounded with 10^5. With this estimate k should be less than 15.

A very similar estimate could be made with the algorithm for the 3-vertex-colorability problem. For a graph with r vertices each of degree at most 3, we need $3r$ vertex blocks and $\frac{1}{2}6 \cdot 3r$ edge molecules, i.e. (counting each vertex block as three distinct molecules) we need $18r$ distinct DNA molecules in a tube. Hence, r is again bounded by 10^6. However, only a

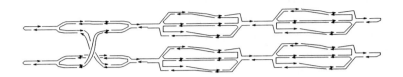

Figure 11: Covering graph.

small portion of the molecules used would actually form graph structures needed, and the solution might be difficult to detect.

We would like to point out that k-armed branched molecules as vertex building blocks can be also used for solving the Hamiltonian Path problem. This was hinted to us by E. Winfree [16]. Such building blocks simplify to a very large extent the 3D approach, with highly complex vertex and edge building blocks, suggested in [7].

5.1 Covering Graphs. In [7] it was shown that whenever a graph is constructed by vertex and edge DNA molecules as building blocks, we can potentially end with a covering space of the graph (regarding the graph as a 1-complex, see the next paragraph for explanations). Hence, here we are faced with that possibility too. Unlike the situation with the Hamiltonian path problem when formation of such graphs could lead to false conclusions, this is not true for the solution of SAT problem. For the 3-vertex-colorability problem, we use the same method as in [7].

First let us recall the covering graphs. The definition of a covering maps can be found in standard text books in Topology (see [13], for example.) Roughly, a covering $p : \tilde{X} \to X$ satisfies the condition that every point of X has a small neighborhood U in X such that $p^{-1}(U)$ consists of the same (homeomorphic) copies of U. Thus, when we construct a graph by DNA building blocks, we may get covering spaces of the given graph (regarded as 1-complex) that we wanted to construct, since the building blocks connect up themselves locally, and possibly build bigger pieces globally. Such a situation is depicted in Figure 11, where 2-sheeted covering space of the graph in Figure 5 is shown.

In SAT the construction of the contact network graph is oriented from the source to the target. If a covering graph is formed within the initial step of the computation, then this structure will be connected and in Step 3 (a), the whole structure will appear in a single tube. This means, either connections with variable x are removed from the whole structure or

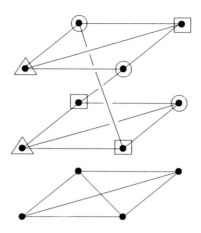

Figure 12: A covering graph of a complete graph

connections with $\bar{x}$ are removed from the whole structure. Hence, the connection from the source to the target will be established correctly even if the connection is part of a covering graph structure. (A lemma in the covering space theory (Lemma 4.1, p. 337, [13]) says that any path in a space lifts to a path in a covering space.) Therefore, if there is a solution to SAT, then the solution is realized in a covering graph as well, and conversely, if there is a path γ in a covering graph from a source to a target, then it gives a solution to SAT.

In the 3-vertex-colorability problem, it can happen that the original graph is not 3-colorable but a covering graph is. Such an example is shown in Figure 12. The square shaped graph in the bottom of the figure is the original graph given, which is easily seen to be non-3-colorable. The top of the figure is a covering graph, which is 3-colorable, as shown by circles, triangles and squares representing the colors of the vertices. Thus we need to exclude such possibilities from our solution. This is addressed in the Step 3 of the process we described. In this step, the graphs of larger size is eliminated. As the covering graphs have larger sizes than the original graph, such covering graphs will be eliminated.

We end this section with a note that, with the exception of forming the building blocks and graph structures, many of the laboratory techniques suggested are fairly standard and their detailed description can be found in [4]. Their application to complex DNA structures would, however, require extensive experimentation before they could be used as proposed.

References

[1] L. Adleman, Molecular computation of solutions of combinatorial problems, *Science*, 266 (1994), 1021 – 1024.

[2] L. Adleman, *On constructing a molecular computer*, 1995, preprint.

[3] M. Amos, A. Gibbons, D. Hodgson, Error-resistant implementation of DNA computations, in [10].

[4] F. M. Ausubel, R. Brent, R. E. Kingston, D. D. Moore, J. G. Seidman, J. A. Smith, K. Struhl, P. Wang-Iverson, S. G. Bonitz, *Current Protocols in Molecular Biology*, Greene Publishing Associates and Wiley-Interscience, New York, NY, 1993.

[5] M. Davis, R. Sigal, E. J. Weyuker, *Computability, Complexity, and Languages*, Academic Press, 1994 (sec. edition).

[6] J. Hartmanis, On the weight of computations, *Bulletin of the European Association for Theoretical Computer Science*, 55 (1995), 136 – 138.

[7] N. Jonoska, S. Karl, M. Saito, Creating 3-dimensional graph structures with DNA, in [11]

[8] J. Khodor, D. Gifford, The efficency of the sequence-specific separation of DNA mixtures for biological computation, in [11].

[9] R. Lipton, DNA solution of hard computational problems, *Science*, 268 (1995), 542 – 545.

[10] *Proceedings of the Second Annual Meeting on DNA Based Computers*, DIMACS Workshop, Princeton, NJ, June 10-12, 1996 (in press).

[11] *Proceedings of the Third Annual Meeting on DNA Based Computers*, DIMACS Workshop, University of Pennsylvania, June 23-25, 1997 (in press).

[12] N. Jonoska, S. A. Karl, Ligation experiments in DNA computations, *Proceedings of 1997 IEEE International Conference on Evolutionary Computation* (ICEC'97), April 13-16, 1997, 261 – 265.

[13] J. R. Munkres, *Topology, a First Course,* Prentice-Hall, 1975.

[14] N. C. Seeman et al., The perils of polynucleotides: The experimental gap between the design and assembly of unusual DNA structures, in [10].

[15] E. Winfree, X. Yang, N. C. Seeman, Universal computation via self-assembly of DNA: Some theory and experiments, in [10].

[16] E. Winfree, private communication.

Circuit Evaluation:

Thoughts on a Killer Application in DNA Computing

Mitsunori OGIHARA[1]

Department of Computer Science, University of Rochester
Rochester, NY 14627
ogihara@cs.rochester.edu

Animesh RAY[2]

Department of Biology, University of Rochester
Rochester, NY 14627, USA
ray@ar.biology.rochester.edu

1   Introduction

Biomolecular computing is the computing methodology in which biologically important molecules are used as memory. The 'aperiodic' nature of these polymeric molecules makes them suitable as memory units [16, 32]. Instead of monotonous repeating units of most synthetic polymers (such as polyethylene), certain biological polymers (such as DNA, RNA and protein) have sets of repeating (or information encoding) units, and these units can appear in any order. The use of atomic or molecular-scale particles for building machinery was first conceived by Feynman in 1959 [15], which gave rise to the recently emerging field of nanotechnology—the fabrication technology of molecular machines [13, 14]. The biomolecular computation we discuss here is somewhat different from what conventional nanotechnology pursues: here the definition of machines is obscure. These computational machines have specific geometrical forms only in terms of memory units; however, the architectures of the molecular computational processors are

---

[1]Supported in part by NSF Grant CCR-9701911.
[2]Supported in part by NSF Grants MCB-9630402 and IBN-9728239.

dynamic and transient, rather than static and permanent as in silicon-based microprocessors.

DNA has been chosen for molecular computation because of its role as a memory molecule in biological systems. Our discussions will be restricted to DNA computation, although alternatives have been proposed [29]. Biochemical processes that copy, edit, replace and amplify DNA sequences have evolved in natural systems, and are well-understood in molecular details [3]. The concept of molecular computation was born in 1994, when Adleman published his algorithm for solving Hamilton Path Problem (HPP) using recombinant DNA techniques, along with the report that he successfully solved a seven-node instance with the method [1]. Lipton [20] generalized Adleman's method and proposed an algorithm for SAT. Their algorithms both carry out parallel brute-force search in the entire search space as follows: assign a uniquely identifiable DNA strand to each solution; create the solution space as a DNA solution; perform biochemical operations to select correct solutions, if any, in parallel; and test whether any strand is separated at all.

This approach – trading nondeterminism with space – is possible because of two characteristics of DNA: massive parallelism, in the sense that billions (or trillions) of DNA strands can be processed concurrently, and large memory size (one liter of DNA can store approximately 10^{23} bits). The theoretical limit of information density in DNA is approximately 1 bit per 1 nm^3. Given the gigantic memory capacity of DNA, one may ask whether biomolecular computers will be a powerful tool to tackle NP-complete problems. However, the question is subtle. Both Adleman's and Lipton's algorithms can handle instance sizes of up to 70. Such instances are within the reach of silicon-based computers.

The urgent issue in DNA computing is to find "killer" applications where DNA computers outperform silicon-based computers. In this article, we summarize the chief questions in need of addressing. We emphasize our work on DNA-based methods for evaluating Boolean circuits, which indicates that these methods rival the efficiency of silicon-based computers.

2   The Background Information on DNA Chemistry

DNA is a charged copolymer of a repeating sugar-phosphate chain of indefinite length to which are attached one of four organic bases (A,T,G, and C) every 0.34 nm along the chain [32]. Each sugar-phosphate-base unit is a nucleotide. The chain is chemically and geometrically polar: one speaks of a 5′ end and a 3′ end. Maximum of 70-80 nucleotide sequences of DNA chain can be chemically synthesized without much error. A single polymer chain can have any sequence of bases from the 5′ end to the 3′ end. Each base offers one of four bonding surfaces. The bonding surface of A is complementary to that of T, and of G to that of C. Because of this complementarity rule, a DNA chain can pair with another chain only when their sequences of bases are mutually complementary.

There is one more restriction to this pairing operation: the chains must pair head-to-tail, i.e., antiparallel to each other. For example, a DNA of sequence 5′ACGGCTAACGCGTTGCC3′ can only pair with 5′GGCAACGCGTTAGCCGT3′. Occasional mispairing may occur over short stretches of mismatch within a long region of complementarity. Since the perfectly base-paired (double-stranded) state is energetically more favorable than a partially base-paired structure, mismatched pairing causes instability. The temperature at which 50% of the DNA strands of a double stranded molecule come apart (the melting temperature) is higher for a perfect duplex than for a mismatched duplex. This temperature also depends on the strand length, the ratio of $\frac{A+T}{G+C}$, the sequence itself, and on the presence of salt or other charged molecules.

When collections of complementary single strands are mixed, heated and then slowly cooled, the complementary strands find one another and double stranded DNA is formed (the process of annealing) by a second order reaction kinetics. This is DNA hybridization [31].

A DNA can be made to replicate in the test tube. To do this one adds a short primer (a short single stranded DNA) complementary to the 3′ end region of the DNA to the DNA to be replicated and anneal. The 3′ end of the primer now points inward along the long DNA. Adding four precursors of nucleotides, some magnesium, and an enzyme (DNA polymerase) will allow polymerization of nucleotide units growing on the 3′ end of the primer, with a sequence exactly complementary to the original long DNA strand. This step synthesizes the complement of the first strand, and is known as primer extension [31]. If another primer is available, whose sequence is identical to the 5′ end region of the first strand, these can be annealed next to the newly made second strand, prime the synthesis of a new first strand, and the reactions can be repeated. The number of DNA molecules will grow exponentially. This is the basic principle of Polymerase Chain Reaction (PCR) [31]. The progress of PCR or primer extension, conducted by automated instrumentation, can be monitored by either incorporating a radioactive atom or a fluorescent tag to the end of the primer molecules. The original HPP algorithm [1] used PCR for amplifying paths through the graph, and a method for bit addition [17] used primer extension for generating solutions.

Two double stranded DNA molecules can be covalently joined end to end by the action of the enzyme DNA ligase [31]. Very long DNA chains can be produced in this manner. The end ligation (joining) of two flushed-ended molecules is inefficient. The efficiency is many orders of magnitude higher if the two DNA molecules are partially single stranded and their ends are mutually complementary (DNA with 'sticky ends'), so that the two molecules can anneal by their sticky ends though weak bonding prior to covalent chain joining by ligase. Linear as well as circular DNA can be produced by ligation. Ligation has been used as an important method to generate memory molecules in virtually all algorithms.

DNA molecules can be separated on the basis of size by the process of electrophoresis [28]. This was important in the algorithms for HPP [1] and Boolean Circuit simulation [26, 27]. DNA can also be separated by virtue of their sequence: a mixture of single stranded DNA molecules of many different sequences is exposed to a short single stranded probe DNA of a specific sequence. The probe is labeled at one end with a chemical tag. The probe will anneal to its complementary sequence in the mixture. The mixture is then exposed to a reagent that binds strongly to the tag. The reagent is coupled to paramagnetic beads. The annealed DNA, therefore, binds to the bead, which can be trapped in a magnetic field. This is the method of sequence-specific affinity separation of DNA [28], an integral part of many DNA algorithms.

Adleman's HPP Algorithm. Suppose it is tested whether an n node directed graph G has a Hamiltonian path from 1 to n, and assume that 1 has no incoming edges and n has no outgoing edges. Choose 20-mer DNA patterns $p_1, \ldots, p_n$ to represent the nodes, where a directed path of G is the concatenation of the patterns corresponding to the nodes in the order that they are visited, and an edge (i, j) of is the antiparallel complementary pattern of the middle 20-mers of $v_i v_{i+1}$. First synthesize many copies of the node strands and edge strands, and let them anneal in a test tube. Ligation is triggered, and many double strands with a 10-mer overhang at each end are synthesized. Pull out from the test tube all those with p_1 by annealing the magnetically labeled antiparallel of the first half of p_1. Repeat this for p_n using the second half instead. The two separation steps complete the double strands. Remove the 'edge' parts by denaturing, then by discarding magnetically labeled strands. Next collect all the $20n$-mers. The collected strands are the length n paths from 1 to n. Now with $n - 2$ sequence-specific separation steps, collect those with all of $p_2, \ldots, p_{n-1}$. Finally, test whether any strands are collected at all.

3   Major Limitation of DNA Computing: Volume Complexity

In both Adleman's HPP algorithm and Lipton's SAT algorithm the total volume of DNA present in a test tube at any time of the computation grows exponentially as a function of input size. A practical limit to the maximum possible volume is the Avogadro number (6.02×10^{23}), and an exponential function reaches the limit very rapidly. This is the very reason that the largest input size that can be solved with those algorithms is only 60-70. Are there any algorithms for NP problems with smaller growth in the total volume of DNA molecules that are required? Or more generally, what are "killer" applications of DNA? In an attempt to answer the question, researchers have designed DNA algorithms for various problems. Proceedings of the first two DIMACS Workshops on DNA Based Computation [21, 22] are good sources of such algorithms.

A promising application is the attack of Data Encryption Standard (DES). In [9], an Adleman-Lipton type algorithm for finding a 56-bit key of DES is proposed. According to [9], the attack will take about three months. Note that an exhaustive search for the key could take $1,000$ years even at the rate of a million operations per second. In [2], the method is translated into a simpler one under the sticker-based model.

Another promising application is dynamic programming. With the method presented in [6], an n-item Knapsack instance with the total weight sum W and the total value sum V can be solved in time $O(n)$ with the volume $O(VW)$ of DNA.

The paper [5] presents a 3-Coloring algorithm, which uses the total volume $n1.89^n$ of DNA and assumes an operation for splitting test tube contents into weighted subsets.

The paper also presents a volume 1.67^n algorithm and a volume 1.51^n algorithm for Independent Set. The paper [24] shows that a practical SAT algorithm in [23] can be converted to an Adleman-Lipton type DNA algorithm, with the total DNA volume 1.62^n. These results have been improved in [8]. The authors observe that the 1.35^n time 3-Coloring algorithm in [7] can be converted to a nondeterministic algorithm with $(\log 1.35)n$ nondeterministic bits, thereby reducing an n-node 3-Coloring instance to a Circuit-SAT instance with $(\log 1.35)n$ inputs. Combined with the Circuit-SAT algorithm in [10], this yields a 3-Coloring algorithm with the total volume of 1.35^n. In a similar vein, [8] developed a DNA 3SAT algorithm with volume requirement 1.50^n, and a 1.23^n algorithm for Independent Set. In [25], an empirical result is shown that gradually extending partial assignments reduces the amount of DNA for 3SAT so that the space requirement becomes $2^{0.40n}$. However, even with this rate of DNA volume increase, practical instances of NP-complete problems are far beyond reach. Therefore, DNA based algorithms for NP-complete problems are unlikely to serve as killer applications. In view of the above limitation of DNA computing, it is important to identify other classes of problems in which DNA based computation has real advantage; Boolean circuit evaluation appears to be one such.

4   A Possible Killer Application

The goal of our research is to explore the possibility of using DNA to evaluate large parallel computations of Boolean circuits. Formally, a Boolean circuit is a node-labeled, directed graph without cycles. It takes a number of 0/1 values and outputs one 0/1-value (one can generalize the model so that the circuits have more than one value as the output). The number of input values is the *input-size* of the circuit. For a circuit of n inputs, there are $2n$ nodes with no incoming edges. They are called the *input gates* and have unique labels chosen from $x_1, \ldots, x_n, \overline{x}_1, \cdots, \overline{x}_n$. On an input $b_1 \cdots b_n$, for each $i, 1 \leq i \leq n$, the gate labeled x_i outputs b_i while the one labeled $\overline{x}_i$ outputs $1 - b_i$. Other nodes have labels from AND and OR. For such a node g, its *inputs* are those which have outgoing edges to g. If g is labeled AND (OR), its output is the conjunction (disjunction) of the outputs of g's inputs. It is customary that the gates of the circuit are assumed to be stratified so that each level consists of the same kind of gates, and edges can be drawn only from one level to the next level.

Classifications of the Boolean circuit models are in terms of the gate fan-ins (the number of maximum inputs that AND gates and OR gates can take). In *bounded-fan-in circuits*, both AND gates and OR gates may have at most two inputs. In *unbounded-fan-in circuits*, both AND gates and OR gates may have an arbitrary number of inputs. In *semi-unbounded-fan-in circuits*, OR gates can have an arbitrary number of inputs while AND gates have at most two inputs. There are two complexity measures for

Boolean circuits. The *size* of a circuit C is the number of gates in it and the *depth* of C is the length of the paths from input gates to the output gate. An alternative definition of the size of a circuit C is the number of wires (i.e., edges) in it. In the simulation methods we will describe below this alternative definition will make more sense. Also, the depth of a circuit is often viewed as the computation time of the circuit under an assumption that signal flow from one level to the next takes a unit time.

The results we have obtained so far are summarized as follows:

1. We showed that semi-unbounded-fan-in circuits can be simulated with a small overhead on computation time [26].

2. We improved the above simulation result and showed that bounded-fan-in circuits can be simulated with a constant run-time overhead, even with seemingly the smallest possible set of computational basis [27].

Below we will briefly describe these methods.

4.1   Simulating Semi-Unbounded-Fan-In Circuits

The basic idea is to assign unique single DNA strands of equal length to the gates of the semi-unbounded-fan-in circuit to be simulated, and concurrently simulate all the gates in the same level proceeding from the input level towards the output level. The simulation is performed so that for each level i, the simulation of the gates at level i produces the test tube T_i with the invariant property that for every gate g at the level i, the encoding of g is present in T_i if and only if g outputs 1. Strands encoding gates at level $< i$ may or may not present in T_i. There are three types of single-level simulation, the simulation of the input level, that of an OR level, and that of an AND level. In order to simulate the input level, we synthesize DNA strands corresponding to the input gates at that level so that the invariant property holds.

Simulation of an OR-level i is essentially the construction of T_i from T_{i-1}. For each gate g at level i, g outputs 1 if and only if at least one of the inputs to g outputs 1. Since the invariant property with respect to level $i - 1$ holds for T_{i-1}, g outputs 1 if and only if the encoding of at least one input gate of g is present in T_{i-1}. Based on this observation, we perform the following steps concurrently for all gates g at level i (see Figure 1):

- In a test tube U mix the contents of T_{i-1} and copies of the encoding of g. Also mix for each input h to g, copies of the "linker" strand of h

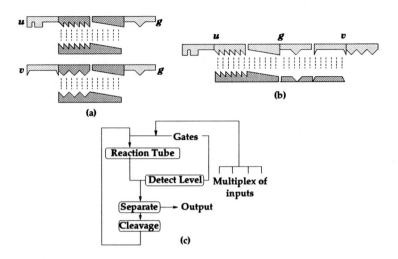

Figure 1: (a) simulation of an OR gate, (b) simulation of an AND gate, where g is the gate and u and v are inputs, (c) a diagram for automated experiments

and g, the complementary antiparallel X of the first half of g followed by that of the second half of h.

- Let the strands in U anneal in the presence of DNA ligase. Due to the invariant property with respect to $i-1$, if g outputs 1, then some strand h is ligated to the strand g; otherwise, no ligation occurs. In other words, g outputs 1 if and only if some strand and g have formed in U a double-length strand.

- Perform electrophoresis on the contents of U in a denaturing gel, blindly take out the "double-length" band and put it into a test tube V.

- Since only gates in two consecutive levels are connected, cutting all the strands in V in the middle completes generation of U (the strands from T_{i-1} will not affect the simulation of level $i+1$ at all). For that matter, design the encoding patterns so that they all have the same pattern A at the 3'-end and the same pattern unique pattern B at the 5'-end, and that the single-stranded AB is recognized by a restriction endonuclease (e.g., $RsaI$) whose restriction site is precisely between

> *A* and *B*. Now add the restriction enzyme into *V* and let it cut all
> the strands in *V*. This is the test tube T_i.

On the other hand, for an AND gate *g*, we need to leave the strand *g*
in the test tube T_i if and only if the two inputs to *g* are both present in
T_{i-1}. So, we produce triple-length strands instead of double-length strands
by connecting *g* after its first input and *g* before the strand of the second
input. Also, in the electrophoresis step, we take out the triple-length band.
One may choose a different order in which the three strands line up. If we
are to use a different order, then we must choose linker strands accordingly,
but this will not affect the overall efficiency of the evaluation.

In order to obtain the output of the circuit, we synthesize the strand for
the output gate with a fluorescent label. Upon production of the test tube
corresponding the topmost level of the circuit, which contains only the output gate, hybridize any fluorescent-labeled output strand to an immobilized
template (perhaps on an arrayed microgrid, [11]) remove any unhybridized
strands, and use a fluorescent-light detector to test whether the strand is
present or not, and thus, whether the circuit outputs 1 or 0.

A subtle issue in this logic-gate simulation is that the output from a
single gate may be shared by many other gates at the next level. This
implies that the amplitudes of the gates may show a great deal of de-
crease as simulation proceeds, and eventually, generation of double-length
or triple-length strands may fail due to the lack of enough single-length
strands. Therefore, we need to amplify the amplitude of strands in T_i. In
our method, we propose that this is done by repeating a pair of concurrent
primer-extension steps. In one test tube, we use the complement of *A* as
the primer and the gate strands as the template. Here the complement of
A is magnetically labeled. In the other test tube, we use *B* as the primer
and the gate-complements as the template. After each primer-extension
step, the strands are split into two groups: those with the magnetic labels,
which collectively form the second test tube in the next step, and those
without the labels, which collectively form the first test tube in the next
step.

As a result, the $\log_2$ of the amplification rate is incurred as an overhead
in the computation time. In order to simulate a semi-unbounded-fan-in
circuit of size *S* and depth *D* with maximum fan-out (the maximum number
of gates that share the same input) *M*, we would need at least $cD \log M$
computation steps and $c'SM$ many molecules of DNA, where *c* ranges from
15 min. to 3−4 hours, and c' is the base length of the gate encodings (see
Figure 1(b)).

An alternative simulation with the same order of efficiency can be derived by modifying the given circuit C to an equivalent one with fan-out bounded by 2, i.e., every node is supplying an input signal to at most two gates at the next level. Hoover, Klawe, and Pippenger [18] show a polynomial time method for converting a circuit to one with fan-out bounded by 2 with a reasonable increase in the size and the depth. Using their method, one can convert a size S, depth D circuit C with I inputs and with fan-in bounded by F to an equivalent circuit C' with I inputs, with fan-in bounded by F, and with fan-out bounded by 2, whose size is at most $FS + I - 1$ and whose depth is at most $(1 + \log_2 F)D + \log_2 I$. The circuit C' constructed by the method is not necessarily stratified. Stratifying the circuit may increase the size as well as the depth, but the increasing rates of both measures can be suppressed with in a multiplicative factor of 2. So, one can obtain from C an equivalent stratified, fan-out 2 circuit C'' of size $\leq 2(FS + I)$ and of depth $4\log_2 FD + 2\log_2 I$. Suppose we are to simulate this equivalent circuit C'' instead of C. By carefully choosing the amplitude of gate strands and linker strands that are newly added to the test tube T_i for simulation of the i-th level gates, we can guarantee that there is a constant $\varepsilon > 0$ such that for every $i > 0$ and a gate g at level i, the amplitude of g after level-i simulation is 0 if g outputs 0, and it is at least $1/(2 + \varepsilon)$ of the smallest of the non-zero amplitudes of the strands corresponding to an input gate to g. (The way one guarantees this will be discussed in the next section.) If ε is small enough, then one has only to run the amplification step twice per level. Thus, the total number of strands that are used will be $\leq 2(FS + I)$ and the number of biochemical steps required for the simulation will be cFS for some constant c.

An important issue in the study of DNA-based evaluation methods for Boolean circuits is the choice of a computational basis. The standard basis consists of the logical-AND, the logical-OR, and the negation. An alternate basis is the logical-NAND (the negation of logical-AND). It is a reasonable assumption that DNA-DNA hybridization is inevitably used for simulating logical units. However, DNA-DNA hybridization is an equilibrium process, thereby leaving some DNA strands unhybridized. The equilibrium nature of hybridization makes it very difficult to convert negative information to positive (or vice versa) at an operational step. This is because some strands that remain accidentally may amplify and deteriorate the signal. A novel scheme for simulating Boolean circuits with all NAND-gates presented in [4] suffers from this limitation. One way to overcome this problem is to use the deMorgan's law to push all the negation gates to the input level [26, 27].

This conversion only doubles the size of the circuit.

4.2 The Smallest Possible Set of Operations

Both the methods described above use several biochemical operations: DNA synthesis, mixing test tubes, electrophoresis, annealing, ligation (not used by [4]), cleavage, detection of labeled DNA, and affinity separation. Actually, even with a much smaller set of operations (DNA synthesis, electrophoresis, annealing, merger, and detect) efficient simulation of circuits is theoretically possible, and we presented a method for simulating bounded-fan-in circuits with a constant overhead in running time [27].

As in [26], simulation is level-wise in this method. A crucial difference is that each gate g is represented by a pair of single strands and the condition that the representatives are either both present or both absent in the test tube T's is maintained throughout the simulation. The use of representation by pairs enables us to amplify strands with repeated annealing steps and electrophoresis steps in a non-denaturing gel. An important observation is that with this representation, we can make the simulation "robust" in the sense that the simulation always outputs the correct answer. We are concerned with the following uncertain factors:

- Synthesis: We do not have a complete control of the amplitudes of strands that are synthesized. The actual amplitude of a synthesized strand fluctuates within a small error factor.

- Merge, Separation: A small fraction of strand will be lost during the merge operation.

- Anneal: A small fraction of strands may remain unhybridized, and so lost in the size-based separation that follows. Also a small fraction of strands may hybridize to a strand with incorrect match.

With this variety of uncertainty in mind we slice the given circuit into sub-circuits of some depth R. We convert the sub-circuits into trees and create edges for distributing the outputs of the sub-circuits of a slice among the input gates of the next slice. As long as the simulated circuit is a tree there is no contention for gate outputs, so there is no need to amplify. Thus, in this modified circuit, amplification steps are necessary only at the borders between slices. The distribution rate is at most S, the number of gates in the circuit. So, if we set R to $\log S$, then the total number of sequential operation steps becomes a constant factor of the depth. In order to make the simulation robust against the last error (namely, incorrect hybridization), we use S1 endonuclease. This enzyme chews cut DNA strands at the places where oligonucleotides remain unhybridized.

Therefore, with the use of S1 we can cut all incorrectly hybridized strands to those that are shorter than the double-length strands. As a result, the yield of annealing may be off by a small constant factor. Now we are left with the problem of recovering a small fraction of loss of strands at each operation step. In order to resolve the problem, we perform more amplification steps, but the number of these additional steps can be kept within a constant factor of R since the loss factors are small. Thus, a bounded-fan-in circuit of depth D and size S can be simulated in dD steps using $d'S$ molecules of DNA for some small constants d and d'.

Again, as in the previous simulation, an alternative method can be obtained by rewriting the given circuit into an equivalent stratified circuit with fan-out bounded by 2. The size increase factor is kept within 4 and the depth increasing factor is kept within 2.

The computational model with seemingly the smallest set of operations is called the Minimum Model [27]. What is the computational power of this model in computational complexity theory? To answer that question, consider language recognition problem, and suppose that a nonuniform program $\mathcal{C}$, consisting of Minimum Model algorithms $\{\mathcal{A}_n\}_{n \geq 1}$, is used to recognize a language, where for each $n \geq 0$, the n-th algorithm recognizes the length-n portion of the language.

Assume that the total number of DNA sequence patterns, single or double, can appear during computation for an input of size n is bounded by some fixed polynomial in n. We split computation of an algorithm under the Minimum Model into two parts: preparation and actual computation. In the first part, we synthesize DNA strands separately and merge them into test tubes, while in the second part, we use anneal, separate, and merge to construct the final test tube, and at the end of this part, the final test tube is examined by detect to obtain the computational result. The computational complexity of an Minimum Model algorithm is then majored by the number of sequential steps in the second part. In [27], we show for every $k \geq 1$, that (1) every language in NC^k is accepted by $O(\log^k n)$-time Minimum Model program, and that (2) every language accepted by $O(\log^k n)$-time Minimum Model program belongs to SAC^k. We can let the two bounds coincide with each other by further demanding that each pattern appearing in the computation has only a constant number of legitimate mates.

4.3 Is Circuit Evaluation a Killer Application?

Molecular circuit analysis described above may possibly be implemented for solving complex circuit evaluation problems on which digital computers are not so efficient. By a proof similar to that of the Gilbert-Varshamov Theorem (see [30]), one can show that there is a set of 1.6×10^{12} distinct 40 base oligonucleotide sequences such that (i) for any two sequences A and B, A disagrees with B and its complement at 10 positions and (ii) no sequence contains the pattern that is cleaved by the restriction enzyme *RSA*I. Artifactual annealing between short stretches of homology in otherwise noncomplementary oligonucleotides may be avoided by enzymatic annealing catalyzed by the DNA strand transfer protein RecA [19]. So, we can handle at least one trillion wires by encoding the gates as 40 base oligonucleotide sequences. On the other hand, one trillion wires are perhaps beyond the reach of digital computers. Then one might think that evaluation of such large circuits could be a "killer application" of DNA-based computation. However, one should note that synthesis of one trillion linker sequences of DNA has to be done sequentially unless the evaluated circuit has a highly regular gate-connection pattern so that much of gate-sequence synthesis process can be done concurrently. Therefore, one trillion wires are beyond the reach of DNA computer as well. However, we note that a technique for parallel synthesis of many DNA sequences currently exists [12].

Nevertheless, we still argue that our DNA-based evaluation method will be useful, in particular, when we need to evaluate one circuit on many distinct input sets. Suppose that we need to evaluate a Boolean circuit C of S wires, depth D, with I input gates on T different sets of input bits. If we are to conduct evaluation one input after another, then the total amount of time required is

$$O(T(S + I)) + O(TD),$$

where the first term is for preparation and the second for evaluation. Noting that the same set of linkers and gate strands are used for all the input sets, we can spare time by keeping the originals synthesized for the first input set and by using copies of the originals for the other input sets. Since copying process can be done level-wise, this modification reduces the total time to

$$O(T(D + I) + S) + O(TD),$$

thereby reducing the amortized cost to $O(D + S/T)$. For example, with $D = 10^4$, $I = 10^4$, $S = 10^8$, and $T = 10^4$, the total preparation time is only an order of 10^8 and the amortized running is an order of 10^4. On the other hand, the amount of time required for sequential digital processing is $TS = 10^{12}$.

In the same situation, we can even run the evaluation of all input sets concurrently in one test tube. To do this we employ a slightly different encoding scheme. We insert in the middle of each gate sequence a pattern that specifies a number $J, 1 \leq J \leq T$. This middle pattern is used to specify which of the input set the strand will be used for. We also make the linker strands specify their input names J by attaching the strand that is complementary antiparallel to the first half (the second half) of J at their $5'$-ends ($3'$-ends). If all the input names are assigned short common patterns at both ends, then we can generate the necessary strands for concurrent evaluation by controlling the volume of the strands and then by chemically connecting the components. Then the preparation cost is $O(T + S)$, and the total running time becomes $O(T + S + D)$. Again, the amount of time required for sequential digital processing is still $TS = 10^{12}$.

References

[1] L. Adleman, Molecular computation of solutions to combinatorial problems, *Science*, 266 (1994), 1021 – 1024.

[2] L. Adleman, P. Rothemund, S. Roweis, E. Winfree, On applying molecular computation to the Data Encryption Standard, *Proceedings of 2nd DIMACS Workshop on DNA Based Computers*, The American Mathematical Society, to appear.

[3] B. Alberts, D. Bray, J. Lewis, M. Raff, K. Roberts, J. Watson, *Molecular Biology of the Cell*, Garland Publishing Inc., New York, NY, 3rd edition, 1994.

[4] M. Amos, D. Hodgson, DNA simulation of boolean circuits, Research Report CTAG-97009, University of Liverpool, Liverpool, UK, December 1997.

[5] E. Bach, A. Condon, E. Glaser, C. Tanguay, DNA models and algorithms for NP-complete problems, *Proceedings of 11th Conference on Computational Complexity*, IEEE Computer Society Press, 1996, 290 – 299.

[6] E. Baum, D. Boneh, Running dynamic programming algorithms on a DNA computer, *Proceedings of 2nd DIMACS Workshop on DNA Based Computers*, The American Mathematical Society, to appear.

[7] R. Beigel, D. Eppstein, 3-coloring in time $O(1.3446^n)$: a no-MIS algorithm, *Proceedings of 36th Symposium on Foundations of Computer Science*, IEEE Computer Society Press, 1995, 444 – 453.

[8] R. Beigel, B. Fu, Molecular computing, bounded nondeterminism, and efficient recursion, Technical Report YALEU/DCS/TR-1116, Department of Computer Science, Yale University, November 1996.

[9] D. Boneh, C. Dunworth, R. Lipton, Breaking DES using a molecular computer, *Proceedings of 1st DIMACS Workshop on DNA Based Computers* (R. Lipton, E. Baum, eds.), The American Mathematical Society, 1996, 37 – 65.

[10] D. Boneh, C. Dunworth, R. Lipton, J. Sgall, On the computational power of DNA, *Discrete Applied Mathematics*, 71 (1996), 79 – 94.

[11] A. Caviani-Pease, D. Solas, E. Sullivan, M. Cronin, C. Holmes, S. Fodor, Light-generated oligonucleotide arrays for rapid DNA sequence analysis, *Proc. Nat. Acad. Sci. USA*, 91 (1994), 5022 – 5026.

[12] M. Chee, R. Yang, E. Hubbell, A. Berno, X. Huang, D. Stern, J. Winkler, D. Lockhart, M. Morris, S. Fodor, Accessing genetic information with high density DNA arrays, *Science*, 274 (1996), 610 – 614.

[13] B. C. Crandall, ed., *Nanotechnology*, MIT Press, Cambridge, MA, 1996.

[14] K. Drexler, *Nanosystems*, John Wiley & Sons, New York, NY, 1992.

[15] R. Feynman, There's plenty of room at the bottom, *Miniaturization* (D. Gilbert, ed.), Reingold, New York, 1961, 282 – 296.

[16] G. Gamow, A. Rich, M. Yčas, The problem of information transfer from the nucleic acids to proteins, *Adv. Biol. Med. Physics*, 4 (1956), 23 – 68.

[17] F. Guarnieri, M. Fliss, C. Bancroft, Making DNA add, *Science*, 273 (1995), 220 – 223.

[18] H. J. Hoover, M. M. Klawe, N. J. Pippenger, Bounding fan-out in logical networks, *Journal of the Association for Computing Machinery*, 31, 1 (1984), 13 – 18.

[19] S. C. Kowalczykowski, A. K. Eggleston, Homologous pairing and DNA strand-exchange proteins, *Ann. Rev. Biochem.*, 63 (1994), 991 – 1043.

[20] R. Lipton, DNA solutions of hard computational problems, *Science*, 268 (1995), 542 – 545.

[21] R. Lipton, E. Baum, eds., *Proceedings of 1st DIMACS Workshop on DNA Based Computers*, DIMACS Series volume 27, The American Mathematical Society, 1996.

[22] R. Lipton, E. Baum, eds., *Proceedings of 2nd DIMACS Workshop on DNA Based Computers*, The American Mathematical Society, to appear.

[23] B. Monien, E. Speckenmeyer, Solving satisfiability in less than 2^n steps, *Discrete Applied Mathematics*, 10 (1985), 287 – 295.

[24] M. Ogihara, Breadth first search 3SAT algorithms for DNA computers, Technical Report TR 629, Department of Computer Science, University of Rochester, Rochester, NY, July 1996.

[25] M. Ogihara, A. Ray, DNA-based parallel computation by counting, *Proceedings of 3rd DIMACS Workshop on DNA Based Computers*, 1997, 265 – 274.

[26] M. Ogihara, A. Ray, Simulating boolean circuits on DNA computers, *Proceedings of 1st International Conference on Computational Molecular Biology*, ACM Press, 1997, 326 – 331.

[27] M. Ogihara, A. Ray, The minimum DNA model and its computational power, *Unconventional Models of Computation* (C. C. Calude, J. Casti, M. J. Dinneed, eds.), Springer, Singapore, 1998, 309 – 322.

[28] J. Sambrook, E. F. Fritsch, T. Maniatis, *Molecular Cloning: a Laboratory Manual*, Cold Spring Harbor Press, NY, 2nd edition, 1989.

[29] W. Smith, DNA computers in vitro and vivo, *Proceedings of 1st DIMACS Workshop on DNA Based Computers* (R. Lipton, E. Baum, eds.), The American Mathematical Society, 1996, 121 – 185.

[30] J. van Lint, *Introduction to Coding Theory*, Springer-Verlag, 1991.

[31] J. Watson, M. Gilman, J. Witkowski, M. Zoller, *Recombinant DNA*, Scientific American Books, New York, NY, 2nd edition, 1992.

[32] J. Watson, N. Hopkins, J. Roberts, J. Steiz, A. Weiner, *Molecular Biology of the Gene*, Benjamin-Cummings, Menlo Part, CA, 4 edition, 1987.

Boolean Transitive Closure in DNA

Paul E. DUNNE, Martyn AMOS, Alan GIBBONS

Department of Computer Science, University of Liverpool
Liverpool L69 7ZF, England
martyn@csc.liv.ac.uk

Abstract. Existing models of DNA computation have been shown to be Turing- complete, but their practical significance is unclear. If DNA computation is to be competitive in the future we require a method of translating abstract algorithms into a sequence of physical operations on strands of DNA. In this paper we describe one such translation, that of *transitive closure*. We argue that this method demonstrates the feasibility of constructing a general framework for the translation of P-RAM algorithms into DNA.

1 Introduction

Since the publication of Adleman's seminal work [1], several models [9, 11] have been proposed which, in principle, establish the Turing-completeness of DNA computation. Although these simulations are of theoretical interest, their practical significance remains unclear. These models are often biologically infeasible, or have an unacceptable run-time. In addition, models are often constructed on an *ad hoc* basis, and fail to provide a general framework for the expression of a variety of algorithms. A realistic and useful model of DNA computation should require reasonable resources, whilst providing a general algorithmic framework. In this paper we describe initial work towards the development of such a model. We show how a particular computation – *transitive closure* – may be translated into DNA via Boolean circuits. In such a way we demonstrate the feasibility of converting a general algorithm into a sequence of molecular steps. We believe that this brings us closer to the development of a realistic and useful model of DNA computation. In a future paper we will describe a general model for the translation of P-RAM algorithms into DNA.

The rest of this paper is organized as follows. In Section 2 we provide the motivation behind the current work, placing it in the context of existing models. In Section 3 we describe background to the transitive closure problem. We then describe the structure of a $NAND$-gate Boolean circuit to compute the transitive closure of an $n \times n$ matrix. In Section 4 we briefly describe the implementation of the circuit in DNA, and we conclude in Section 5 with a brief discussion and suggestions for further work.

2   Motivation

Although several Turing-complete models of DNA computation have been proposed, they often require unrealistic resources (in terms of time or DNA volume). For instance, the model described by Reif [11] requires exponentially-sized volumes of DNA. It is clear that if such models of computation are to be realistic they should require, at most, a volume of DNA that is polynomial in the problem size. However, it is unlikely to be enough to simply have polynomial-volumed computations. We ought, at the same time, to ensure that the vast potential for molecular parallelism is utilised to obtain rapid computations. The view taken by the silicon-based parallel computing community [7] is that efficient parallel algorithms, within the so-called Parallel Random Access Machine (P-RAM) model of computation, should have polylogarithmic running time (and use a polynomial number of processors). Problems for which such solutions exist define the complexity class NC. If DNA computation is to compete within this domain, then we should clearly also look for polylogarithmic running times within polynomially-volumed computations.

In [9], Ogihara and Ray describe the simulation of Boolean circuits within a model of DNA computation. The authors claim claim real-time simulation of the class NC^1 [10] in time proportional to the depth of the circuit. Recall that NC^1 defines the class of problems of size n solved by bounded fan-in circuits of $O(\log n)$ depth and polynomial size. Ogihara and Ray claim that their simulation runs in time proportional to the circuit depth. In [5], we show that within a more realistic *strong* model of DNA computation, the simulation described in [9] actually runs in time proportional to the circuit *size*.

In [4], Amos and Dunne describe a DNA simulation of Boolean circuits that runs in time proportional to the *depth* of the circuit. In principle, this provides a real-time NC simulation. However, the model simulates circuits of 2-input $NAND$-gates, and, as such, allows only very low-level algorithmic specification. Classical P-RAM algorithms [7] are expressed in high-level pseudo-code. Such algorithms are difficult to translate into Boolean circuits. If DNA-based algorithms are to have a long-term future, we believe that it is imperative that a framework exists for the translation of high-level algorithms into the sequence of basic molecular operations required for their laboratory implementation. In this paper, we describe one such translation of a *transitive closure* algorithm. We show how transitive closure may be computed by a $NAND$-gate Boolean circuit, and describe the implementation of this circuit in DNA. In a forthcoming paper [3], we will describe a general framework for translation of P-RAM algorithms into operations on DNA.

3   The Transitive Closure Problem

The computation of the *transitive closure* of a directed graph is fundamental to the solution of several other problems, including shortest path and connected components. Several variants of the transitive closure problem exist; we concentrate on the *all-pairs transitive closure problem*. Here, we find all pairs of vertices in the input graph that are connected by a path.

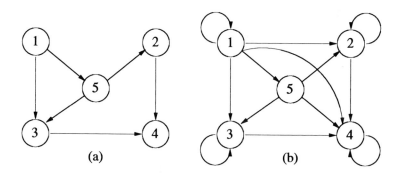

Figure 1: (a) Directed graph, G (b) The transitive closure of G

The transitive closure of a directed graph $G = (V, E)$ is the graph $G^* = (V, E^*)$, where E^* consists of all pairs (i, j), such that either $i = j$ or there exists a path from i to j. An example graph G is depicted in Figure 1(a) with its transitive closure G^* depicted in Figure 1(b).

We represent G by its incidence matrix, A. Let A^* be the incidence matrix of G^*. In [8], one shows how the computation of A^* may be reduced to computing a power of a Boolean matrix.

We now describe the structure of a $NAND$-gate Boolean circuit to compute the transitive closure of the $n \times n$ Boolean matrix, A. For ease of exposition we assume that $n = 2^p$.

The transitive closure, A^*, of A is equal to $(A + I)^n$, where I is the $n \times n$ identity matrix. This is computed by $p = \log_2 n$ levels. The ith level takes as input the matrix output by level $i - 1$ (with level 1 accepting the input matrix $A + I$) and squares it: thus level i outputs a matrix equal to $(A + I)^{2^i}$ (Figure 2).

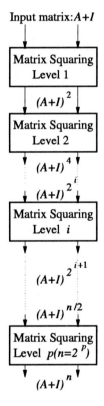

Figure 2: Transitive closure network

To compute A^2 given A, the n^2 Boolean values $A^2_{i,j}$ $(1 \le i, j \le n)$ are needed. These are given by the expression $A_{i,j}^2 = \overset{n}{\underset{k=1}{\vee}} A_{i,k} \wedge A_{k,j}$. First all the n^3 terms $A_{i,k} \wedge A_{k,j}$ (for each $1 \le i,j,k \le n$) are computed in two (parallel) steps using 2 $NAND$-gates for each $\wedge$-gate simulation. Using $NAND$-gates, we have $x \wedge y = NAND(NAND(x,y), NAND(x,y))$. The final stage is to compute all of the n^2 n-bit sums $\overset{n}{\underset{k=1}{\vee}} (A_{i,k} \wedge A_{k,j})$ which form the input to the next level (Figure 3).

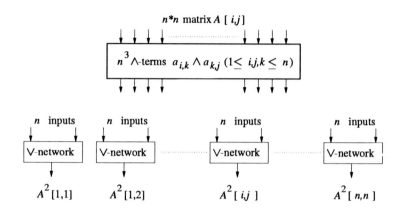

Figure 3: Matrix squaring network

The n-bit sum $\overset{n}{\underset{k=1}{\vee}} x_k$, can be computed by a $NAND$-circuit comprising p levels each of depth 2 (Figure 4). Let level 0 be the inputs $x_1, \ldots, x_n$; level i has 2^{p-i} outputs $y_1, \ldots, y_r$, and level $i+1$ computes $y_1 \vee y_2, y_3 \vee y_4, \ldots, y_{r-1} \vee y_r$. Each $\vee$ can be computed in two steps using 3 $NAND$-gates, since $x \vee y = NAND(NAND(x,x), NAND(y,y))$.

In total we use $2p^2$ parallel steps and a network of size $5p2^{3p} - p2^{2p}$ $NAND$-gates, i.e., $2(\log_2 n)^2$ depth and $5n^3 \log_2 n - 2n^2 \log_2 n$ size.

4   Implementation in DNA

In [4] we describe how a $NAND$-gate Boolean circuit, S, may be simulated in DNA. We now reproduce this here. We denote the depth of S by $D(S)$.

In what follows we use the term *tube* to denote a set of strings over some alphabet σ. We denote the jth gate at level k by g_k^j. We first create a tube, T_0, containing unique strings of length l, each of which corresponds to only those input gates that have the value 1. We then create, for each

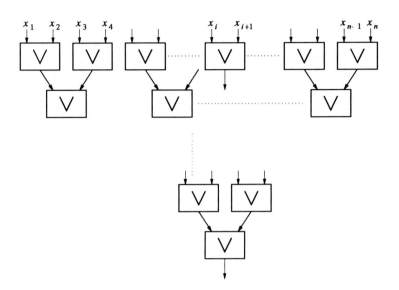

Figure 4: n-bit $\vee$-network $(n = 2^p)$

level $1 \leq k < D(S)$, a tube T_k containing unique strings of length $3l$ representing each gate at level k. We also create a tube U_k, containing strings corresponding to the complement of positions $2l - 4$ to $2l + 5$ for each g_k^j. We define the concept of complementarity in the next section, but for the moment we assume that if sequence x and its complement $\overline{x}$ are present in the same tube, the string containing sequence x is in some way "marked".

We then create tube $T_{D(S)}$, containing unique strings representing the output gates $< y_1, \ldots, y_m >$. These strings representing gates at level $1 \leq k < D(S)$ are of the form $x_k^j y_k^j z_k^j$. If gate g_k^j takes its input from gates g_{k-1}^m and g_{k-1}^n, then the sequence representing x_k^j is the complement of the sequence representing z_{k-1}^m, and y_k^j is the complement of the sequence representing z_{k-1}^n. The presence of z_k^j therefore signifies that g_k^j has an output value of 1.

The strings in tube $T_{D(S)}$ are similar, but the length of the sequence $z_{D(S)}^j$ is in some way proportional to j. Thus, the length of each string in $T_{D(S)}$ is linked to the index of the output gate it represents.

4.1   Level Simulation

We now describe how levels $1 \le k < D(S)$ are simulated. We create the set union of tubes T_{k-1} and T_k. Strings representing gates which take either of their inputs from a gate with an output value of 1 are "marked", due to their complementary nature. We then remove from T_k all strings that have been marked twice (i.e., those representing gates with both inputs equal to one). We then split the remaining strings after section y_k^j, retaining the sequences representing z_k^j. This subset then forms the input to tube T_{k+1}.

4.2   Final Read-out of Output Gates

At level D(S) we create the set union of tubes $T_{D(S)-1}$ and $T_{D(S)}$ as described above. We then, as before, remove from this set all strings that have been marked twice. By checking the length of each string in this set we are therefore able to say which output gate has the value 1, and which has the value zero by the presence or absence of a string representing the gate in question.

4.3   Physical Implementation

We now describe how the abstract model detailed in the previous section may be implemented in the laboratory using standard bio-molecular manipulation techniques. The implementation is similar to that of our *parallel filtering model*, described in [2, 6].

   We first describe the design of strands representing the input gates x_n. For each x_n that has the value 1 we synthesise a unique strand of length l. This population of strands constitutes T_0. We now describe the design of strands representing gates at level $1 \le k < D(S)$. We have already synthesised a unique strand to represent each g_k^j at the set-up stage. Each strand is comprised of three components of length l, representing the gate's inputs and output. Positions 0 to l represent the first input, positions $l+1$ to $2l$ represent the second input, and positions $2l+1$ to $3l$ represent the gate's output. Positions $l-2$ to $l+3$ and positions $2l-2$ to $2l+3$ correspond to the restriction site $CACGTG$. This site is recognized and cleaved exactly at its mid-point by the restriction enzyme PmlI, leaving blunt ends. Due to the inclusion of these restriction sites, positions 0 to 2, $l+1$ to $l+3$ and $2l+1$ to $2l+3$ correspond to the sequence GTG, and positions $l-2$ to l, $2l-2$ to $2l$ and $3l-2$ to $3l$ correspond to the sequence CAC. The design of the other sub-sequences is described in Section 4. A

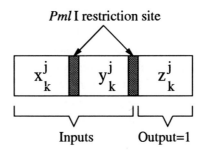

Figure 5: Structure of a gate strand

graphical depiction of the structure of each gate strand is shown in Figure 5.

The simulation proceeds as follows for levels $1 \leq k < D(S)$.

1. At k pour into T_k the strands in tube T_{k-1}. These anneal to the gate strands at the appropriate position. Allow sufficient time for complete annealing.

2. Add ligase to T_k in order to seal any "nicks".

3. Add to T_k the restriction enzyme $PmlI$. Allow sufficient time for complete digestion. Because of the strand design, the enzyme cleaves only those strands that have *both* input strands annealed to them. This is due to the fact that the *first* restriction site $CACGTG$ is only made fully double-stranded if both of these strands have annealed correctly. This process is depicted in Figure 6.

4. Denature the strands and run T_k through a gel, retaining only those strands of length $3l$. This may be achieved in a single step by using a denaturing gel [12]. Retrieve the product and place it in an (empty) tube T_k.

5. Add tube U_k to tube T_k. The strands in tube U_k anneal to the *second* restriction site embedded within each retained gate strand. Allow sufficient time for complete annealing.

6. Add enzyme $PmlI$ to tube T_k, which "snips" off the z_k^j section (i.e., the output section) of each strand representing a retained gate. Allow sufficient time for complete digestion.

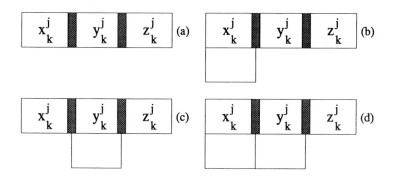

Figure 6: (a) Both inputs=0 (b) First input=1 (c) Second input=1 (d) Both inputs= 1

7. Denature and run T_k through another gel, this time retaining only strands of length l. Amplify the contents of T_k using a single PCR cycle. This tube, T_k of retained strands forms the input to T_{k+1}. We now proceed to the simulation of level $k + 1$.

At level $D(S)$ we carry out steps 1-7, as described above. However, at steps 4 and 7 we retain all strands of length $\geq 3l$, since these represent the output gates y_j. We are now then ready to implement the final read-out phase. This involves a simple interpretation of the final gel visualisation. Since we know the unique length u_j of each $z^j_{D(S)}$ section of the strand for each output gate y_j, the presence or absence of a strand of length $u_j + 2l$ in the gel signifies that y_j has the value one or zero respectively.

We now consider the biological resources required to execute such a simulation. Each gate is encoded as a single strand; in addition, at level $1 \leq k \leq D(S)$ we require an additional strand per gate for removal of output sections. The total number of strands required is therefore $n(-1 + 2\log n^2(-1+5kn))$ distinct strands, where k is a constant, representing the number of copies of a single strand required to give a reasonable guarantee of correct operation.

5   Conclusions and Further Work

In this paper we described how the computation of the transitive closure of an $n \times n$ matrix may be computed by a $NAND$-gate Boolean circuit.

Given that we have already demonstrated how such a circuit may be efficiently implemented in DNA, we have demonstrated the translation of this computation from a high-level description down to operations at a molecular level. Such results are crucial for the future of DNA computation, since they establish the potential for a general framework for translation of general algorithms into DNA operations. In a forthcoming paper, we will describe such a framework, yielding translations of P-RAM algorithms into DNA, via $NAND$-gate Boolean circuits.

Acknowledgements. Martyn Amos is partially supported by a Leverhulme Special Research Fellowship.

References

[1] L. M. Adleman, Molecular computation of solutions to combinatorial problems, *Science*, 266 (1994), 1021 – 1024.

[2] M. Amos, *DNA Computation*. PhD thesis, Department of Computer Science, University of Warwick, UK, September 1997. Available at http://www.csc.liv.ac.uk/~ctag/archive/th/amos-thesis.ab.html.

[3] M. Amos, P. E. Dunne, A. Gibbons, Efficient time and volume DNA simulation of CREW PRAM algorithms, in preparation.

[4] M. Amos, P. E. Dunne, A. Gibbons, DNA simulation of Boolean circuits, *Genetic Programming 1998: Proceedings of the Third Annual Conference* (J. R. Koza et al, eds.), Morgan Kaufmann, 1998.

[5] M. Amos, A. Gibbons, P. E. Dunne, The complexity and viability of DNA computations. *Proceedings of the First International Conference on Bio-Computing and Emergent Computation*, University of Skövde, Sweden, 1997, 165 – 173.

[6] M. Amos, S. Wilson, D. A. Hodgson, G. Owenson, A. Gibbons, Practical implementation of DNA computations, *Unconventional Models of Computation* (C. S. Calude, J. Casti, M. J. Dinneen, eds.), Springer-Verlag, Singapore, 1998, 1 – 18.

[7] A. Gibbons, W. Rytter, *Efficient Parallel Algorithms*, Cambridge University Press, 1988.

[8] J. JáJá, *An Introduction to Parallel Algorithms*, Addison-Wesley, 1992.

[9] M. Ogihara, A. Ray, *Simulating Boolean Circuits on a DNA Computer*, Technical Report 631, University of Rochester, August 1996.

[10] N. Pippenger, On simultaneous resource bounds, *20th Annual Symposium on Foundations of Computer Science*, Long Beach, Ca., USA, October 1979, IEEE Computer Society Press, 307 – 311.

[11] J. H. Reif, Parallel molecular computation: Models and simulations, *Proceedings of the Seventh Annual ACM Symposium on Parallel Algorithms and Architectures, Santa Barbara, June 1995,* Association for Computing Machinery, June 1995, 213 – 223.

[12] J. Sambrook, E. F. Fritsch, T. Maniatis, *Molecular Cloning: A Laboratory Manual,* Cold Spring Harbor Press, second edition, 1989.

Thermodynamic Constraints
on DNA-based Computing

Russell DEATON, Max GARZON

The Molecular Computing Group
The University of Memphis
Memphis, TN, 38152 USA
rjdeaton@memphis.edu, garzonm@hermes.msci.memphis.edu

Abstract. Computing with biological macromolecules, such as DNA, is fundamentally a physical/chemical process. The DNA chemistry introduces a level of complexity that makes reliable, efficient, and scalable computations a challenge. All the chemical and thermodynamic factors have to be analyzed and controlled in order for the molecular algorithm to produce the intended result. For instance, a computation based on DNA requires that the problem instance be encoded in single strands of DNA and that these strands react as planned, that molecular biology protocols, such as PCR or affinity separation, correctly extract the result, and that sufficient flexibility remains so that worthwhile computations can be done. In this paper, various thermodynamic and chemical constraints on DNA computing are enumerated. A similarity measure, based on Gibb's free energy of formation, is defined to judge the goodness of DNA encodings. Finally, the DNA computation problem for implementing molecular algorithms is defined, and it is likely that it is as difficult as the combinatorial optimization problems they are intended to solve.

1   Introduction

To solve hard combinatorial optimization (NP-complete [14]) problems, Adleman [1] introduced a method that utilized single-stranded DNA

138

molecules (oligonucleotides) and techniques from molecular biology [36]. The method, essentially, involves three steps: 1) ENCODING: an encoding that maps a problem instance onto a collection of oligonucleotides, 2) REACTION: template matching reactions, or hybridizations, between the oligonucleotides that do the massively parallel search for a solution, along with ligation, and 3) EXTRACTION: using basic biotech techniques [36] such as polymerase chain reaction (PCR), gel electrophoresis, and affinity separtation, to make the results known.

In Adleman's solution to the Hamiltonian Path Problem (HP) [1], the vertices and edges of the graph are encoded in oligonucleotides. DNA representations of paths through the graph are formed as the vertex oligonucleotides hybridize with the edge oligonucleotides. If the proper hybridizations occur, the DNA representation of the Hamiltonian path can be extracted with affinity separation and PCR. Since Adleman's pioneering work, numerous applications and algorithms have been proposed for DNA based computing [27, 8, 9, 17, 10]. In Adleman's algorithm, the fundamental reaction is hybridization between oligonucleotides that are Watson-Crick complements [1]. Hybridization to target strands on magnetic beads was used to extract the result. Subsequent proposals have continued to rely on the mechanism of hybridization to do the computation and extraction [27, 8, 9, 17, 10]. Other key chemical processes are ligation and PCR.

The complexity in Adleman's algorithm originates in the design of a set of oligonucleotides, both their sequence and concentrations, that will hybridize in preferred alignments, and a set of molecular biology protocols that will extract the desired result. The potential for mishybridizations between oligonucleotides necessitates that sequences be designed of prevent them [6], which in turn, bounds the size of the problem. Increasing the size of the problem requires exponentially increasing amounts of DNA which soon makes implementation impractical [18]. PCR can, also, be error-prone [3], and extraction, likewise, is difficult to do error-free [22]. Each of these steps involves the optimization of chemical and thermodynamic factors in order for the molecular algorithm to function as planned.

In general, molecular algorithms using DNA will use chemical steps which require some optimization in order for the computation to take place. For example, in splicing systems [19, 29], which have also been implemented in the lab [25], the fundamental reaction is enzymatic. Restriction enzymes cut double-stranded DNA molecules at locations of specific base sequence. The molecules are reassembled through hybridization and ligation of the pieces. It has been shown that these mechanisms are capable of producing regular languages over the set of DNA molecules. The chemical difficulty in splicing systems is in matching the optimal reactions conditions of different restriction enzymes, controlling the enzymatic reactions themselves, which are in general incomplete, and in extracting the result.

For any computation using biological macromolecules or enzymes, an issue is the feasibility of achieving the computation within the limits imposed by the thermodynamics and chemistry of the underlying molecular system. Examples of these physical constraints are:

1. Reaction Conditions (*e.g.*, time, temperature, solution)

2. Unwanted reactions (interactions) between oligonucleotides, enzymes, or other chemical constituents of the system.

3. Concentrations of the chemical constituents of the system.

4. Fidelity of the molecular biology protocols.

Taking all these into account, along with the original goal of producing a computational result, leads to what might be called the DNA Computation Problem. In what follows, the focus for discussion will be Adleman's molecular algorithm. Chemical and thermodynamic constraints on this algorithm will be discussed. A similarity measure based on the Gibb's free energy of hybrid formation is defined and suggested as an appropriate measure of the goodness of DNA encodings of problem instances. Finally, the DNA computation problem is defined and its computational complexity discussed.

2   Thermodynamic and Chemical Constraints

The pool of DNA oligonucleotides that represent the instance of the problem must meet several conditions. In a hybridization reaction, individual base pairs (bp) hydrogen bond in a highly specific way, with the purine base adenine (A) binding with the pyrimidine thymine (T), and the purine guanine (G) binding with the pyrimidine cytosine (C). These base pairs are the Watson-Crick complements of each other, and ideally, oligonucleotide hybridizations occur only between Watson-Crick complements. Depending upon several factors, however, base pairs that are not Watson-Crick complements can occur in hybridized DNA molecules [30]. The effect of the hybridization reaction conditions on the potential for non-Watson-Crick base pairs in a hybridized molecule is to introduce a possibility of error in both the reaction and extraction steps. In addition, oligonucleotides can shift out of their designed alignments and hybridize. Though these shifted alignments would not produce a false positive, they do use up oligonucleotide in wasteful reactions, and if enough occur, could produce a false negative.

Factors that influence the stringency of hybridization include the base sequences of the hybridizing oligonucleotides, the location of potential mismatches, the concentrations of the reactant oligos, the temperature of the reaction, the length of the oligonucleotides, and solvent concentrations [38]. These factors can be summarized in the melting temperature parameter, T_m. The melting temperature is the temperature at which 50% of the oligos are melted or single stranded. As reaction temperature increases, an increasing percentage of hybrids melt. For oligos in solution, the melting temperature is given by [4],

$$T_m = \frac{\Delta H^\circ}{\Delta S^\circ + R \ln([C_t]/4)}, \tag{1}$$

where ΔH° is the enthalpy, ΔS° is the entropy, R is the gas constant, and $[C_t]$ is the total molar strand concentration. Melting curves measured by UV absorbance techniques. Single-stranded oligos will absorb UV radiation, and therefore, as the temperature is increased, an increase in absorbance indicates melting. The width of the melting curve for equimolar complementary oligonucleotides is

$$\Delta T = \frac{6RT_m^2}{\Delta H^\circ}. \tag{2}$$

If the reaction temperature for the hybridization, T_r, is greater than $T_m + \Delta T/2$, then, the oligos involved in that reaction will not hybridize, and no errors from that mismatch are possible. Therefore, to prevent unwanted hybridizations, their $T_m + \Delta T/2$ should be less than the reaction temperature, T_r. This condition should hold for all possible binding configurations between oligos that are not wanted. In addition, we want the desired hybridizations to occur, and in sufficient number to enable an efficient and detectable computation. Therefore, for a desired hybridization, its T_m should be greater than the reaction temperature.

In addition, the oligonucleotide pool has to meet criteria associated with the reaction kinetics and thermodynamics of hybridization. As pointed out in [23], because of chemical effects, there is a possibility that certain graphs will be more difficult to solve with DNA than others. The essential reaction in a DNA computation is the hybridization reaction, as expressed by

$$a + b \rightleftharpoons c, \tag{3}$$

where a and b are oligonucleotides, c is the double-stranded hybridization product, and the $\rightleftharpoons$ indicates that the reactions are reversible. Assuming constant volume, the thermodynamic parameter that describes the state of the DNA computer at equilibrium is the Gibb's free energy, G [26]. The change in G for a small change in the chemical components is

$$dG = \sum_{i=1}^{k} \left(\frac{\partial G}{\partial n_i} \right)_{T,P,n_{j \neq i}} dn_i, \tag{4}$$

where n_i is the mole number of the ith of k components in the system. The chemical potential of component i is defined as

$$\mu_i \equiv \left(\frac{\partial G}{\partial n_i} \right)_{T,P,n_{j \neq i}}. \tag{5}$$

Reactions can be exothermic or endothermic. Exothermic reactions produce heat or the capacity for work, $dG < 0$. Endothermic reactions require heat or work input to proceed, and thus, $dG > 0$. The sign of dG determines whether the reaction can be spontaneous or not, and its direction (Table 1). Therefore, dG is the driving force for the reaction, and ultimately, will determined what reaction products are formed, and in what concentrations they occur. The reaction will be driven towards chemical equilibrium, where the rates to the left and right of $\rightleftharpoons$ in Eq. 3 are equal, and $dG = 0$. The condition for chemical equilibrium is

$$\sum_i \mu_i dn_i = 0. \tag{6}$$

For a general chemical reaction of species, A_i [26],

$$0 \rightleftharpoons \sum_i \nu_i A_i, \tag{7}$$

where the ν_i are the stoichiometric coefficients, which are negative for reactants and positive for products. The change in the amount of a species, A_i, in the reaction is proportional to the ν_i, $dn_i = \nu_i d\xi$, where the constant of proportionally is called the extent of the reaction, ξ. Substitution for dn_i into Eq. 6 produces another condition on chemical equilibrium,

$$\sum_i \nu_i \mu_i = 0. \tag{8}$$

The change in G with the extent of the reaction is

$$\frac{dG}{d\xi} = \sum_i \nu_i \mu_i, \tag{9}$$

which is zero at equilibrium.

ΔG	$A + B \rightleftharpoons AB$
< 0	Reaction proceeds to right
> 0	Reaction proceeds to left
$= 0$	Equilibrium

Table 1: Direction of reaction according to sign of change in free energy.

The free energy change can be written as [26]

$$\frac{dG}{d\xi} = \Delta G^\circ + RT \log Q, \tag{10}$$

where ΔG° is the free energy change under ideally dilute standard conditions, and $RT \log Q$ is a correction term for nonstandard conditions, with T the temperature and R the gas constant. The correction term for the solution of reacting oligos is

$$Q = \prod_i a_i^{\nu_i}, \tag{11}$$

where a_i is the activity of component i. The activity is defined as

$$a_i = \exp\left[\frac{(\mu_i - \mu_i^\circ)}{RT}\right], \tag{12}$$

where μ_i° is the chemical potential in the standard state of an ideally dilute solution [26]. Next, an activity coefficient is defined that is the ratio of the activity to the mole fraction,

$$\gamma_i = a_i / \chi_i, \tag{13}$$

where the mole fraction is $\chi_i = (n_i)/(\sum_i n_i)$, and $\sum_i^P \chi_i = 1$ for P species. Through Eq. 10 and 13, the free energy change is related to the concentrations of the reaction components. At equilibrium, $dG/d\xi = 0$, and the equilibrium constant is

$$K^{eq} = \prod_i (a_{(i,eq)})^{\nu_i} = \exp(-\Delta G^\circ / RT), \tag{14}$$

where $a_{(i,eq)}$ are the equilibrium activities. For an ideally dilute solution, $\gamma_i = 1$, $a_i = \chi_i$, and the thermodynamic parameters are expressed directly in terms of the mole fractions. In what follows, an ideally dilute solution is assumed.

Returning to Eq. 3, the equilibrium constant in terms of the mole fractions of the reactant oligonucleotides and the hybridization product is

$$K^{eq} = \frac{[c]}{[a][b]} = \exp\left(\frac{-\Delta G^\circ}{RT}\right), \tag{15}$$

where [ ] indicates the mole fraction. Therefore, the free energy change for the hybridization of Eq. 3 will determine the concentrations of the product formed, whether it is a desired hybridization product, a mismatched hybridization, or a hybridization with a shifted alignment. For a proper encoding of a problem instance, the free energy change for desired hybridizations should be maximized (more negative), and the free energy change for undesired hybridizations minimized (more positive). In addition, since $\Delta G^\circ = \Delta S^\circ - T\Delta S^\circ$, the melting temperature is very closely related to the free energy, and therefore, the free energy seems a likely candidate for characterizing the strength of a hybridization attraction between two oligonucleotides.

Another factor which can affect the results of a DNA computation is coupling among a set of hybridization reactions. In Adleman's algorithm [1], paths through the graph are represented by DNA molecules that are formed by successive hybridization reactions. The equilibrium constant for a series of hybridizations is proportional to the product of the individual hybridization equilibrium constants. Therefore, unfavorable reactions can be driven forward by coupling to favorable reactions. For instance, let

$$a + b \rightleftharpoons c, \tag{16}$$
$$c + d \rightleftharpoons e, \tag{17}$$

be a pair of coupled hybridization reactions. Therefore, the total equilibrium constant for the final product e is,

$$K_e^{eq} \propto K_{ab}^{eq} \times K_{cd}^{eq}. \tag{18}$$

This means that mishybridizations can be coupled with favorable hybridization to produce unforseen hybridization products. In addition, in certain graphs, if the molecule representing the Hamiltonian path is less favorable energetically than other paths, then, they will be formed preferentially over the Hamiltonian path. In fact, their relative concentrations will be

$$\frac{[\text{other path}]}{[\text{HP}]} \propto \frac{K^{eq}_{\text{other path}}}{K^{eq}_{\text{HP}}}. \tag{19}$$

Therefore, the encoding must account for these type of effects as well. In particular, relative concentrations of reaction components are potentially important for extraction operations.

Therefore, the requirements on a "good" encoding are:

1. The encoding adequately represents the problem instance.

2. The encoding enables extraction of the result.

3. Designed hybridizations are energetically favorable, while unplanned hybridizations are energetically unfavorable.

4. The encoding takes into account factors related to chemical effects, such as free energy coupling and kinetics, that are produced by the specific problem instance.

A pool of oligonucleotides for DNA computing should fulfill these criteria. Some of these constraints, particularly those associated with the last item, would have to be accounted for by most DNA computing schemes.

3   A Similarity Measure for Encodings

Various criteria have been proposed for the prevention of unwanted hybridization errors by a DNA encoding. In [7, 6], the Hamming distance between oligonucleotides was proposed to preclude the possibility of errors. A new metric [15] which considered errors from shifting oligonucleotides relative to each other has been developed. The problem with these distances is that they fail to capture all the complexity and criteria associated with the DNA chemistry. Therefore, in what follows, it is proposed that the Gibb's free energy is the appropriate measure for the strength of a hybridization attraction. The free energy is the driving force for the reaction, and takes into account or is related to most of the chemical requirements on an encoding. These are the melting temperature, coupled reactions, and equilibrium concentrations of hybridization products.

Similarity measures have been applied to problems in molecular biology [32, 35]. The application that has the most relevance to the encoding problem is that of alignment of two sequences. An alignment is an insertion of spaces in arbitrary locations so that two sequences, which may be of different initial length, are the same final length [32]. For each possible alignment between two sequences, a score is computed based on whether each column in the alignment contains a match, a mismatch, or a space. The similarity between two sequences in the maximum score over all alignments. The number of alignments between two sequences is exponential [35]. Nevertheless, dynamic programming [35] can be used to compute the similarity with quadratic complexity [32]. This alignment problem from molecular biology has much in common with the problem of determining a good encoding for a DNA computation. For an oligonucleotide encoding of a problem, we want to check all possible alignments between the oligonucleotides for hybridization. Unlike the traditional alignment problem, however, to measure the strength of a hybridization potential, we have to measure similarity under Watson-Crick complementation, and incorporate into the similarity measure the thermodynamic cost associated with hybridization of a given alignment.

The similarity measure is defined more precisely as follows [32, 35]. u, t are two sequences over a given alphabet Σ. By inserting spaces in u and t, a pair of new sequences u' and t' is obtained. An alignment $\alpha(u', t')$ between u and t must satisfy the following properties:

- $|u'| = |t'|$,

- u is obtained by removing all the spaces from u',

- t is obtained by removing all the spaces from t',

- Spaces cannot be stacked atop one another.

Next, we define an additive scoring system for an alignment. Typically, the scoring system, (p, g), consists of a function $p : \Sigma \times \Sigma \longrightarrow \mathbf{R}$, which assigns a score to each pair of symbols in an alignment, and a space penalty, g (typically < 0) which penalizes spaces. The similarity, then, is

$$s(u, t) = \max_{\alpha \in \mathcal{A}(u,t)} \text{score}(\alpha), \tag{20}$$

where $\mathcal{A}(u, t)$ is the set of all possible alignments.

The primary energetic factor for hybridization is not the energy of the hydrogen bonding between nucleotide bases, but is the nearest neighbor base stacking energies [4]. These base stacking energies must be measured, and are not unique. Nevertheless, from an energetic point of view, they are the parameters of choice to determine the potential for hybridiztion between oligonucleotides. Recently measured values for these parameters are given in Table 2 [31].

Sequence	ΔH° (kcal/mol)	ΔS° (eu)	ΔG°_{37} (kcal/mol)
AA/TT	-8.4	-23.6	-1.02
AT/TA	-6.5	-18.8	-0.73
TA/AT	-6.3	-18.5	-0.60
CA/GT	-7.4	-19.3	-1.38
GT/CA	-8.6	-23.0	-1.43
CT/GA	-6.1	-16.1	-1.16
GA/CT	-7.7	-20.3	-1.46
CG/GC	-10.1	-25.5	-2.09
GC/CG	-11.1	-28.4	-2.28
GG/CC	-6.7	-15.6	-1.77

Table 2: Thermodynamic parameters for DNA helix initiation and propagation in 1 M NaCl. Sequences are given $5' \to 3'/3' \to 5'$. [31].

To define an appropriate similarity measure for hybridization, we take as our alphabet all possible nearest neighbor pairs of nucleotide bases. Then, the cost measure is associated, in the case of perfect Watson-Crick complements, with the thermodynamic parameters (ΔG°) of Table 2. Mismatched base pairs could also be assigned a thermodynamic cost [38, 5, 37]. Spaces at the end of strings would be associated with the thermodynamic parameters associated with dangling ends [38, 5]. Spaces in the middle of the string would be bulge loops, with an associated thermodynamic penalty [38, 5] All the factors are added to produce a total free energy of hybridization for a particular alignment. Therefore, with this similarity measure, each alignment would have a cost associated with it, this cost would be the Gibb's free energy of formation at a standard temperature, and the similarity would be the maximum free energy, and therefore, would represent the most energetically favorable alignment for hybridization.

Typically [35], a similarity measure has the following properties:

$$p(a, a) > 0 \forall\ a \in \Sigma \qquad (21)$$
$$p(a, b) < 0 \text{ for some } a, b \in \Sigma. \qquad (22)$$

The purpose is to reward similar symbols from the alphabet, and penalize different symbols from the alphabet. For the encoding, however, the purpose is to reward Watson-Crick complements and penalize non-Watson-Crick complement sequences. Therefore, the proposed cost operator between two symbols is

$$p(a, b^r) = \Delta G^{\circ}_{a/b^r}, \tag{23}$$

where b^r indicates the reverse Watson-Crick complement.

Another advantage of the similarity measure defined above is that it is related to a distance metric [32, 35]. The distance, $d(u, t)$, and similarity, $s(u, t)$ are related by

$$s(u, t) + d(u, t) = \frac{M}{2}(m + n), \tag{24}$$

where u and t are sequences of length m and n, respectively, and M is an arbitrary constant.

4   The DNA Computation Problem

The goal of DNA computing is to implement algorithms in reactions among biological molecules. In so doing, the hope is to tap the generative power that is evident in the biological machinery of life. This is a thread that runs from cellular automata [34, 33] to genetic algorithms [20] and artificial neural networks [21]. At the very least, one would want to utilize the massive number of molecules for quicker solution of very difficult problems. In a DNA computer, a molecular solution will be reached. The important question is if that solution *in vitro* corresponds to the desired algorithmic solution. A similar situation exists when algorithmic descriptions of certain natural phenomena are attempted. For instance, computation of the folding of proteins [11], or the minimum energy state of a spin glass [2] are known to be NP-complete. Nevertheless, the actual protein folds in a fraction of the time that an algorithm would require to compute it. Therefore, given the chemical and thermodynamic factors that influence and constrain a DNA computation, a question occurs about how difficult it really is to implement algorithms in molecular systems. So far, it seems that the molecular implementation has differed for each problem [8, 9, 10], and when one says "molecular implementation," the actual laboratory procedures that produced a valid solution are meant, not the theoretical molecular algorithms. Therefore, in this section, the complexity of implementing algorithms in biomolecular systems is discussed. The process of converting an algorithm into a biomolecular systems is called the DNA Computation Problem.

DNA Computation Problem (DCP)

Instance. A DNA computer $D = (S, M, f)$, where S and M are finite sets of DNA molecules, and $f : S \to M$ is a mapping that represents a set of biomolecular operations (enzymatic reactions, PCR, gel electrophoresis, *etc* ...), and an algorithm, A, problem Π, and problem instance I.

Question. Is there a D that implements the application of A to I?

According to Church's thesis, there are any number of equivalent approaches to the notion of an algorithm, or finite procedure, *i.e.*, Turing Machines, λ-calculus. For the current discussion, the model is restricted to finite automata. While not as powerful as Turing Machines *et al.*, DNA implementations of finite automaton have been based on both Adleman's algorithm and splicing systems [19, 12, 13, 16].

As part of the DNA implementation of the finite automata, it would have to be determined if the finite automata and its DNA implementation recognized the same languages. Determining the inequivalence of finite state automata is a problem that is known to be NP-hard [14]. Therefore, DCP contains as a subset NP-hard problems, and therefore, is itself NP-hard.

Other aspects of DCP could involve difficult problems. Most significantly, the set of molecules M represent the reaction products produced by the chemical steps represented by f. Many chemical and physical interactions are possible between the chemical constituents of a DNA computer, as outlined above. Because of this, a DNA computer is similar to a frustrated physical system, like a spin glass [28]. Implementation of an algorithm might require that the minimum energy state be computed for the collection of chemical constituents, and if the DNA computer is a frustrated system, this problem would more than likely be NP-complete, as it is for spin glasses [2] and protein folding [11].

5 Conclusion

Given the difficulty of implementing DNA algorithms, what are the prospects for DNA based computing. Three possibilities present themselves: 1) Applications that do not require encodings have been suggested [24]. These applications might involve solution of problems in molecular biology and biotechnology with DNA computations. 2) DNA computing can continue down the track it has followed till now, and rely upon less than optimum implementations. It could be that implementations that are good enough for specific applications could be found with less effort. 3) Applications, such as artificial immune systems, evolutionary programs, and associative memories could be implemented in a DNA computer. These applications take advantage of the fuzziness of DNA chemistry to produce variation, fault tolerance, pattern recognition, and associative capabilities into the computation.

Adleman [1] developed a DNA based technique for the solution of a NP-complete problem. The massive parallelism of DNA hybridizations evoked a hope that these hard problems would be tractable in a DNA based computer. Implementing DNA based algorithms, however, has turned out to be very difficult, and there is evidence that it is just as hard as the original problems that they were intended to solve. The difficulty in the implementation originates in checking and controlling the many constraints imposed by the DNA chemistry. Failure to do so can produce molecular algorithms whose results are not those intended.

References

[1] L. M. Adleman, Molecular computation of solutions to combinatorial problems, *Science*, 266 (1994), 1021 – 1024.

[2] F. Barahona, On the computational complexity of Ising spin glass models, *J. Phys. A*, 15 (1982), 3241 – 3253.

[3] D. Boneh, C. Dunworth, J. Sgall, R. J. Lipton, Making DNA computers error resistant, in [9], 102 – 110.

[4] P. N. Borer, B. Dengler, I. Tinoco, Jr., O. C. Uhlenbeck, Stability of ribonucleic acid double-stranded helices, *J. Mol. Biol.*, 86 (1974), 843 – 853.

[5] C. R. Cantor, P. R. Schimmel, *Biophysical Chemistry: Part III The Behavior of Biological Macromolecules.* W. H. Freeman and Company, New York, 1980.

[6] R. Deaton, M. Garzon, J. A. Rose, D. R. Franceschetti, R. C. Murphy, S. E. Stevens Jr., Reliability and efficiency of a DNA based computation, *Phys. Rev. Lett.*, 80 (1998), 417 – 420.

[7] R. Deaton, R. C. Murphy, M. Garzon, D. R. Franceschetti, S. E. Stevens Jr., Good encodings for DNA-based solutions to combinatorial problems, in [9], 159 – 171.

[8] DIMACS, *Proceedings of the First Annual Meeting on DNA Based Computers* (Providence, RI), American Mathematical Society, 1996, DIMACS Proc. Series No. 27.

[9] DIMACS, *Preliminary Proceedings of the Second Annual Meeting on DNA Based Computers* (Providence, RI), American Mathematical Society, 1997, DIMACS Proc. Series.

[10] DIMACS, *Preliminary Proceedings of the 3rd DIMACS Workshop on DNA Based Computers* (Providence, RI), American Mathematical Society, 1997, Philadelphia, PA., June 23-27.

[11] A. S. Fraenkel, Complexity of protein folding, *Bull. Math. Biology*, 55 (1993), 1199 – 1210.

[12] R. Freund, Gh. Păun, G. Rozenberg, A. Salomaa, Watson-Crick finite automata, in [10], 305 – 317.

[13] Y. Gao, M. Garzon, R. C. Murphy, J. A. Rose, R. Deaton, D. R. Franceschetti, S. E. Stevens Jr., DNA implementation of nondeterminism, in [10], 204 – 211.

[14] M. R. Garey, D. S. Johnson, *Computers and Intractability.* Freeman, New York, 1979.

[15] M. Garzon, R. Deaton, P. Neathery, R. C. Murphy, S. E. Stevens Jr., D. R. Franceschetti, A new metric for DNA computing, in *Genetic Programming 1997: Proceedings of the Second Annual Conference*, AAAI, Stanford University, July 13-16, 1997, 479 – 490.

[16] M. Garzon, Y. Gao, J. A. Rose, R. C. Murphy, R. Deaton, D. R. Franceschetti, S. E. Stevens Jr., In-vitro implementation of finite-state machines, in *Proc. 2nd Intl. Workshop on Implementing Automata WIA'97, Lecture Notes in Computer Science*, Springer-Verlag, Berlin, 1998.

[17] F. Guarnieri, M. Fliss, C. Bancroft, Making DNA add, *Science*, 273 (1996), 220 – 223.

[18] J. Hartmanis, On the weight of computations, *Bulletin of the European Association for Theoretical Computer Science*, 55 (1995), 136 – 138.

[19] T. Head, Formal language theory and DNA: An analysis of the generative capacity of specific recombination behaviors, *Bull. Math. Biology*, 49 (1987), 737 – 759.

[20] J. H. Holland, *Adaptation in Natural and Artificial Systems.* MIT Press, Cambridge, MA, 1992.

[21] J. J. Hopfield, Neural networks and physical systems with emergent collective computational abilities, *Proceedings of the National Academy of Sciences of the U.S.A.*, 89 (1982), 2554 – 2558.

[22] R. Karp, C. Kenyon, O. Waarts, Error-resilient DNA computation, in *Proceedings of the 7th ACM-SIAM symposium on discrete algorithms*, ACM Press/SIAM, 1996, 458 – 467.

[23] S. A. Kurtz, S. R. Mahaney, J. S. Royer, J. Simon, Active transport in biological computing, in [9], 111 – 122.

[24] L. Landweber, R. J. Lipton, DNA^{2}DNA computations: A potential Killer App, in [10], 59 – 68.

[25] E. Laun, K. J. Reddy, Wet splicing systems, in [10], 115 – 126.

[26] I. N. Levine, *Physical Chemistry*. McGraw – Hill Book Company, Inc., New York, fourth ed., 1995.

[27] R. J. Lipton, DNA solution of hard computational problems, *Science*, 268 (1995), 542 – 545.

[28] M. Mezard, G. Parisi, M. A. Virasoro, *Spin Glass Theory and Beyond*. World Scientific, Singapore, 1987.

[29] Gh. Păun, G. Rozenberg, A. Salomaa, Computing by splicing, *Theoretical Computer Science*, 168 (1996), 321 – 336.

[30] J. Sambrook, E. F. Fritsch, T. Maniatis, *Molecular Cloning: A Laboratory Manual*, Cold Spring Harbor Laboratory Press, second ed., 1989.

[31] J. SantaLucia, Jr., H. T. Allawi, P. A. Seneviratne, Improved nearest-neighbor parameters for predicting DNA duplex stability, *Biochemistry*, 35 (1996), 3555 – 3562.

[32] J. Setubal, J. Meidanis, *Introduction to Computational Molecular Biology*, PWS Publishing Co., Boston, 1997.

[33] S. Ulam, On some mathematical problems connected with patterns of growth of figures, in *Essays on cellular automata* (A. E. Burks, ed.), Univ. of Illinois Press, Chicago, 1972, 219 – 231.

[34] J. von Neumann, *Theory of Self-Reproducing Automata*, University of Illinois Press, Urbana, IL, third ed., 1966.

[35] M. S. Waterman, *Introduction to Computational Biology*, Chapman and Hall, New York, 1995.

[36] J. D. Watson, N. H. Hopkins, J. W. Roberts, J. A. Steitz, A. M. Weiner, *Molecular Biology of the Gene*, The Benjamin/Cummings Publishing Co., Inc, Menlo Park, CA, fourth ed., 1987.

[37] H. Werntges, G. Steger, D. Riesner, H. J. Fritz, Mismatches in DNA double strands: Thermodynamic parameters and their correlation to repair efficiencies, *Nucleic Acids Res.*, 14 (1986), 3773 – 3790.

[38] J. G. Wetmur, Physical chemistry of nucleic acid hybridization, in [10], 1 – 25.

A Note on DNA Representation of Binary Strings

Somenath BISWAS

Department of Computer Science and Engineering
Indian Institute of Technology
Kanpur 208016, India
sb@iitk.ernet.in

1 Lipton's Encoding, Advantages and Disadvantages

After the pioneering work of [1], the scope of molecular computing has been greatly widened by Lipton and his colleagues [5, 3, 2, 4]. A key component of their work is a certain DNA representation of binary strings, first proposed in [5], which we call here as the Lipton encoding. A binary string x of length n, $x = x_1 x_2 \ldots x_n$, $x_i \in \{0, 1\}$, is represented in Lipton encoding (using the notation[1] given in [3]) as:

$$\updownarrow S_0 B_1(x_1) S_1 B_2(x_2) S_2 \ldots S_{i-1} B_i S_i \ldots S_{n-1} B_n(x_n) S_n$$

where *all* S_i's, $B_i(0)$'s, $B_i(1)$'s are distinct DNA sequences. S_i's act as separators, the presence of $B_i(0)$ in a DNA encoding captures the fact that the ith bit of the encoded binary string is 0, and similarly, the presence of $B_i(1)$ captures the fact that the ith bit of the encoded string is 1. There are several advantages of using this encoding. A standard paradigm in

---

[1]Briefly, the notation is: Let x be a string over $\{A, C, T, G\}$. A single stranded 5' to 3' DNA string whose ith nucleotide is as per the ith symbol of x, is denoted as $\uparrow x$. The Watson-Crick complement of this DNA string is denoted as $\downarrow x$, that is, the latter string will be a 3' to 5' string of nucleotides, and its ith nucleotide will be the complement of the nucleotide occupying the ith position in $\uparrow x$. (A and T are complementary to each other, and so are C and G). $\updownarrow x$ denotes the double stranded DNA string formed when $\uparrow x$ and $\downarrow x$ anneal to each other.

molecular computing is: while looking for a solution which is n bit long, begin with a DNA soup that contains encodings of all 2^n different binary strings of length n, and then prune out all those DNA encodings that represent non-solutions. To carry out this pruning, a bio-step that is often used implements an operation on a set of binary strings which we call here as *Select*. Suppose A is a set of binary strings of length n, i is a bit position, i.e., $1 \leq i \leq n$, and $b \in \{0, 1\}$, then $Select(A, i, b)$ is defined to be B where $B = \{x \in A \mid i\text{th bit of } x \text{ is } b\}$. When using the Lipton encoding, a bio-step that implements *Select* can be readily defined, as what is needed is to extract out all DNA strings (from amongst the ones which encode the binary strings on the set A) with the pattern $B_i(b)$. Another advantage of this encoding is that the DNA representations of all 2^n different binary strings of length n can be obtained in a straightforward manner as explained in [3]. Lipton encoding has some disadvantages as well. As it is required that we use a distinct DNA string for each pair $\langle i, b \rangle$, i is a bit position, and b a bit value, this of course means that to represent binary strings of length n, we need to first fix $2n$ distinct DNA strings, which itself is somewhat cumbersome. (Actually, as the separators will also be distinct if the encodings are obtained through the initialization graphs as in, say, [3], we need to fix $(3n + 1)$ distinct DNA strings.) Further, if we fix these DNA strings by randomly selecting $2n$ strands of a certain length as has been suggested, there is the possibility that two of the DNA strings selected are same, which of course, can lead to error in the computation. More seriously, Lipton encoding rules out procedural abstraction: suppose the result of one procedure is a test tube T_1 containing DNA strings representing a set A of binary strings, and the result of another procedure is the test tube T_2 containing DNA representations of a set B of binary strings. Now suppose we want to concatenate A to B. Clearly, we cannot do so if T_1 and T_2 are obtained independently, we must look into the *implementations* of the two procedures to check that the same string has not been used to represent different $\langle i, b \rangle$ pairs. These disadvantages disppear if we could use the same DNA sequence for every occurrence of 0 in every binary string that we encode, and similarly, the same DNA sequence for every occurrence of 1. We call such an encoding a *simple encoding*. To be useful, we of course, require that with simple encoding as well, we can implement all the operations on sets of binary strings that can be done using the Lipton encoding. We show that it is indeed the case, and that too, making use of no different bio-steps from the ones described in [3].

2 Simple Encoding

Let x be a binary string of length n, and let x_i denote the ith bit of x. In the proposed encoding, the DNA string which will represent x is:

$$\updownarrow S_0 B(x_0) S B(x_1) S \ldots S B(x_i) S \ldots S B(x_n) S'$$

where

$$B(x_i) = \begin{cases} \text{the DNA sequence } \alpha \text{ if } x_i = 0 \\ \text{the DNA sequence } \beta \text{ if } x_i = 1 \end{cases}$$

For $1 \leq i \leq n$, and $S_0 = \text{AATTA}\delta\text{GAATTA}$, $S = \text{GAATTA}\delta\text{GAATTA}$, and $S' = \text{GAATTC}\delta\text{GAATTA}$, α, β, δ, are three distinct DNA strings which do not cotain any occurrence of GAATTA, or GAATTC. (S acts as the separator between codes of two bits, S_0, and S' are the left, and the right end-marker respectively.)

Generation of encodings of all binary strings of length n: We shall describe in the next section how we can obtain the encodings of the concatenation of two sets A and B of binary strings, given the encodings of A and B. Using concatenation, we can therefore, build recursively the encodings of all binary strings of length n.

3 Implementing Operations on Sets of Binary Strings

Concatenation. We recall that DNA strings which represent binary strings are of the kind (using a mixture of regular expression notation and the notation to describe DNA strings):

$$\updownarrow S_0((\alpha + \beta)S)^*(\alpha + \beta)S',$$

where α and β represent bits 0 and 1 respectively, $S_0 = \text{AATTA}\delta\text{GAATTA}$, $S = \text{GAATTA}\delta\text{GAATTA}$, $S' = \text{GAATTC}\delta\text{GAATTA}$; and further, GAATTA or GAATTC do not occur as a substring in α, β or in δ. Let the test tubes T_A and T_B contain encodings of two sets A, and B, two sets of binary strings. The bio-procedure described below produces T_C, a test-tube containing encodings of $A \cdot B$. (We essentially use the idea behind implementing *append* as described in [3].)

Procedure Concatenate:

Step 1. EcoRI restriction endonuclease is added to T_A. The effect of this enzyme is to make the DNA strings encoding binary strings now to have the form

$$\updownarrow S_0((\alpha + \beta)S)^*(\alpha + \beta)G\uparrow\text{AATT}$$

Step 2. From T_B, we obtain a test tube $T_{B'}$ that will contain $\downarrow$ strands of T_B. (This can be done in the way Step 1 of *set-extract* in [3] is carried out. All strands in $T_{B'}$ will begin with $\downarrow$AATTA.

Step 3. We pour contents of $T_{B'}$ into T_A. $\uparrow$AATT ends of strings originally in T_A will anneal to $\downarrow$AATTA ends of the strings that came from $T_{B'}$.

Step 4. A polymerase enzyme is used to turn the strands in T_A to fully double stranded ones.

{End of procedure}

It can be seen that at the end of the above procedure, T_A will contain the encodings of $A \cdot B$.

Set_Difference. Given two test tubes, T_1 and T_2, both containing DNA strands, the procedure *set-extract* as described in [3], will give all DNA strands of T_1 which do not occur in T_2. If T_1 contained the encodings of a set A of binary strings, and B contained those of a set of binary strings, this therefore, implements *Set_Difference* which will take as inputs T_1 and T_2 as above, and return a test tube containing the encodings of $A - B$, i.e., $\{x \mid x \in A \text{ and } x \notin B\}$.

Select. We show here how we can implement *Select* using concatenation and set difference. Let A be a set of binary strings of length n, and let T_A be a test tube that contains the encodings of the strings of A. The following procedure implements *Select(A,i,b)*, given T_A, i, and b, $2 \leq i \leq (n-1)$, $b \in \{0,1\}$. (The boundary cases with $i = 1$ or n can be handled more directly.)

Procedure Select:

Step 1. Denote the set of all binary strings of length $(i - 1)$ as B_L, and the set of all binary strings of length $(n - i)$ as B_R. Prepare test tubes T_L and T_R which contain encodings of B_L, and B_R respectively. We also prepare a test tube $T_{\bar{b}}$ which will contain many encodings of the length 1 binary string with the bit $\bar{b}$, which is the complement of b.

Step 2. Invoke the procedure *Concatenate* with T_L and $T_{\bar{b}}$ as inputs, to obtain a test tube $T_{L'}$ which will contain the encodings of the set $B_L \cdot \{\bar{b}\}$.

Step 3. Invoke the procedure *Concatenate* with $T_{L'}$ and T_R as inputs, to obtain a test tube T' containing the encodings of the set $B_L \cdot \{\bar{b}\} \cdot B_R$.

Step 4. Invoke the procedure *Set_Diffrence* on inputs T_A and T' to obtain a test tube T as result.
{End of procedure Select}

It can be seen that T will contain the encodings of *Select(A,i,b)*.

4   Concluding Remarks

When one goes through the work on molecular computing by Lipton and his colleagues, it may appear that the DNA encoding of binary strings that they use, which we call here as the Lipton encoding, is essential for their work. What we have shown here that it is not so; all their results go through using the encoding proposed in this note. Moreover, the latter, arguably a more natural encoding scheme, has some advantages as well over the Lipton encoding. However, as it has been pointed out [4] that any bio-procedure will have errors, therefore, which of the two encodings is in general preferable will depend on perhaps by finding out which one is less error-prone.

References

[1] L. M. Adleman, Molecular computation of solutions to combinatorial problems, *Science*, 266 (1994), 1021 – 1024.

[2] D. Boneh, Ch. Dunworth, R. J. Lipton, Breaking DES using a molecular computer, *Tech. Report*, CS Dept., Princeton Univ., 1995.

[3] D. Boneh, Ch. Dunworth, R. J. Lipton, J. Sgall, On the computational power of DNA, *Tech. Report*, CS Dept., Princeton Univ., 1995.

[4] D. Boneh, R. J. Lipton, Making DNA computers error resistant, *Tech. Report*, CS Dept., Princeton Univ., 1995.

[5] R. J. Lipton, DNA solution of hard combinatorial problems, *Science*, 268 (1995), 542 – 545.

Silicon or Molecules ?
What's the Best for Splicing ?

Gheorghe ŞTEFAN

Politechnical University of Bucharest
Department of Electronics
Bv. Iuliu Maniu 1-3, 77202 Bucureşti, Romania
stefan@agni.arh.pub.ro

Abstract. We present the main ideas concerning the implementation in the solid state circuits of some *molecular mechanisms*: the *splicing* operation and the *insert/delete* operation. The physical support for these operations is based on the *Connex Memory* concept first introduced in [14]. We promote this solution because a pure biological process is very hard to be interfaced with machines in nowadays technologies. At the same time we believe that the mechanisms emphasized in the molecular process of computation are very good suggestions for silicon based machines devoted to perform a fine grain parallelism. Using a Connex Memory the splicing operation or the insert/delete operation is performed in linear time related to the length of the rules; the time does not depend on the length of the processed strings. In order to perform in *parallel* all possible applications of a rule in a set of strings, the function of the Connex Memory is extended over a *cellular automaton*, thus defining the *Eco-Chip*.

1   Introduction

The molecular computing is a very exciting theoretical approach from the '70s. M. Conrad is a pioneer in this domain [3]. T. Head emphasizes a basic operation named *splicing* [5], [6] that can be used as a model for some aspects of molecular processes involved in computation. At the end of 1994, L. M. Adleman reported his successful experiment of solving the Hamiltonian problem in a graph by manipulating DNA sequences [1]. Many scientists believe that these approaches will generate, maybe in short time, "biological computing machines".

At the same time these results *suggest* us two new directions in computer science research.

The first is to investigate the possibility of using the main molecular operations as basic generative rules in the formal language theory and in the theory of computation. In the last few years, the formal language theory was enriched by new formal systems devoted to modeling the typical mechanisms of molecular computing.

The second is to implement efficiently these operations using an appropriate hardware support. The degree of parallelism is very high in a system that uses these new operations. We believe that an adequate "smart memory" device can give full value of using these strange "molecular rules" in computation.

Starting from the first direction, opened by papers like [12], [13], our approach is related with the second by the aim of putting together the operations suggested by the molecular computing and the Connex Memory (CM) concept or the Eco-Chip (EC) model.

The *first step* is to prove that a system based upon the CM circuit performs these operations in a time related with the size of the rules instead of the standard system with RAM that performs the same operations in a time related to the storage space used for all the strings involved.

The *second step* uses the CM concept to build the EC as a cellular automaton. On this new circuit the insert and delete functions of the CM can be performed in many points in each clock cycle, rather than on the CM circuit that allows only one insert or delete function per clock cycle. Using the EC circuit, the similarity between molecular computing and the computation performed on this silicon machine becomes apparent, because many operations can be performed in parallel.

2 DNA Based Computing Mechanisms

So far, three distinct basic operations on DNA (and RNA) that can be used for molecular computation were mainly investigated:

- the *matching*, the basic operation in **P-systems**, introduced in [7] starting from the famous Adleman's experiment,

- the *splicing* operation, defined by Tom Head, is the basic element of **H-systems**,

- the *insertion/deletion* operation [8] can be used to define generative mechanisms in **I-systems**.

All the three operations lead to universal computability models, which are equivalent with the Turing Machine. For our purposes, it is enough to present only the second and the third operations because, from the point of view of the CM concept applications, the first two operations are very similar.

2.1   The Splicing Mechanism and H-Systems

A formal definition of the splicing operation is to be found in [12]. Following this definition, we specify one of the basic mechanism from DNA computing that can be borrowed for a silicon based machines.

Definition 2.1 *Let V be a finite alphabet and $\#$ a special symbol not in V. A splicing rule over V is a string $r = u_1 \# u_2 \# u_3 \# u_4$. For $x, y, z \in V^*$ and $r = u_1 \# u_2 \# u_3 \# u_4$, we write $(x, y) \vdash_r z$ if $x = x_1 u_1 u_2 x_2$, $y = y_1 u_3 u_4 y_2$ and $z = x_1 u_1 u_4 y_2$, for some $x_i, y_i \in V^*$.*

Starting from two strings x and y a new string z is obtained. The sites of the splicing are the places defined by the substrings $u_1 u_2$ and $u_3 u_4$. A machine must find the sites of the splicing, cut the two strings x and y in the sites of the splicing and concatenate the left part of x with the right part of y.

Definition 2.2 *An H-system is a pair $\sigma = (V, R)$ where V is an alphabet and $R \subseteq V^* \# V^* \# V^* \# V^*$ is a set of splicing rules.*

For each rule r a solid-state machine can be imagined. At the same time each H-system σ has an associated machine.

Definition 2.3 *A splicing machine is a couple $SM = (V, a_r)$ associated to a splicing rule r which has a random accessed memory (RAM) containing a language $S \subset V^*$ and a finite automaton a_r that performs the rule r on S.*

Our line of thought follows the conclusion of [13]: The splicing operation, essentially different from other language-theoretic operations, turns out to be surprisingly powerful. Easy characterizations of recursively enumerable languages (exhibiting Turing machine competence) are obtained in this framework. This, on the one hand, proves again the complexity of the DNA structure and the power of the mechanisms manipulating it, on the other hand, suggests that universal "computers" can be constructed on this basis.

2.2 The Insert/Delete Mechanism and I-Systems

Another main molecular mechanism that has a simple mathematical model refers to the local *mutations* occurred in a DNA or RNA strings. A mutation can be assimilated to an insertion or a deletion. In some places defined by specific substrings, a new substring can be inserted or a substring can be deleted. Two distinct set of rules, one for *insert* and another for *delete*, lead to a class of systems called *I-systems* [8].

Definition 2.4 *An* I-system *is a construct:* $\gamma = (V, T, A, I, D)$, *where:* V *is a finite alphabet,* $T \subseteq V$ *is the terminal alphabet,* $A \subset V^*$ *is the finite set of the axioms,* $I, D \subset V^* \times V^* \times V^*$ *are finite subsets of the* insertion *rules and of the* deletion *rules having the form* (u, z, v) *and acting as follows:*

1. $x = x_1 u v x_2$ *becomes* $y = x_1 u z v x_2$, *for* $x_1, x_2 \in V^*$ *and* $(u, z, v) \in I$, *thus performing an* insertion,

2. $x = x_1 u z v x_2$ *becomes* $y = x_1 u v x_2$, *for* $x_1, x_2 \in V^*$ *and* $(u, z, v) \in D$, *thus performing a* deletion; *for such* $x, y \in V^*$ *we write* $x \Longrightarrow y$. *The reflexive and transitive closure of* $\Longrightarrow$ *is denoted by* $\Longrightarrow^*$.

Definition 2.5 *The language generated by the* I-system γ *is*

$$L(\gamma) = \{w \in T^* \mid x \Longrightarrow^* w, \text{ for } x \in A\}.$$

The basic action in an *I-system* is to find all the places where the substrings uv or uxv are located. After that, the substring x is inserted between the substrings u and v or is deleted from between the substrings u and v.

3 The Connex Memory

3.1 The Definition

The CM is a physical support for a string in which we can find any substring, identifying in this way any place for *reading, inserting,* or *deleting* a symbol or a substring. The CM is *a sort of CAM* (Content Addressable Memory), structured as a *bi-directional shift register* in which a significant point is *marked*, as a consequence of an *associative sequential mechanism* used to find a *name* in a number of steps equal to the length of the name.

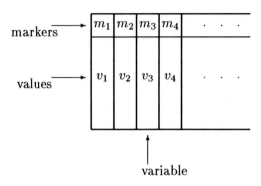

Figure 1: The content of the Connex Memory

Definition 3.1 *The* connex memory *CM is a physical support of a string of variables (see Figure 1) having* values *from a finite set of symbols and two* states: *non-marked or* marked, *over which we can apply the following set of functions (CM's commands):*
CMCOM = {RESET s, FIND s, CFIND s, INSERT s, READup, READ-down, READ, DELETE}, *where:*

- *RESET s: all the variables take the value s,*

- *FIND s: all the variables that follow a variable having the value s switch to the marked state and the rest switch to the non-marked state,*

- *CFIND s: (conditioned find) all the variables that follow a variable having the value s and being in the marked state switch in the marked state and the rest switch in the non-marked state,*

- *INSERT s: the value s is inserted before the first marked variable,*

- *READ up | down | − : the output has the value of the first marked variable and the marker moves one position to right (up) or to left (down) or remains unchanged (−),*

- *DELETE: the value stored in the first marked position is deleted, the position remains marked (the output has the value of the first marked variable) and the symbols from the right are moved one position left.*

All these functions are executed in time $O(1)$ (one clock cycle).

3.2 How Does Each Function Work ?

We shall answer to this question by giving a set of examples in which: $S(t)$ is the string stored in the memory in the current clock cycle when a certain function is applied, $OUTPUT(t)$ is the value of the output of the memory in the current cycle, $S(t+1)$ is the content of the memory as a result of the function applied in the previous clock cycle. The marked variables are bolded.

RESET p
$S(t) = $ roivndkgotrun...
$S(t+1) = $ ppppp....p...

FIND b
$S(t) = $ ...(bu**b**u(big brother's gun))...
$S(t+1) = $ ...(bu**b**u(big brother's gun))...

CFIND u
$S(t) = $ ...(bub**u** (big brother's gun))...
$S(t+1) = $ ...(bub**u**(big brother's gun))...

INSERT c
$S(t) = $ ...(bu**b**u(big brother's gun))...
$S(t+1) = $ ...(bu**c**bu(big brothe's gun))...

READ
$S(t) = $ ...(bu**b**u(big brother's gun))...
$OUTPUT(t) = $ b
$S(t+1) = $ ...(bu**b**u(big brother's gun))...

READ up
$S(t) = $ ...(bu**b**u(big brother's gun))...
$OUTPUT(t) = $ b
$S(t+1) = $ ...(bub**u**(big brother's gun))...

READ down
$S(t) = $ ...(bu**b**u(big brother's gun))...
$OUTPUT(t) = $ b
$S(t+1) = $ ...(b**u**bu(big brother's gun))...

DELETE
$S(t) = $ ...(bu**b**u(big brother's gun))...

OUTPUT(t) = b

S(t) = ...(bu**u**(big brother's gun))...

3.3 The Application Domains of the Connex Memory

The main domain of applications for CM based architecture is the *string oriented symbolic processing*. Some of them were presented in other papers and some represent works in progress.

1. The paper [17] is an exercise of using CM for implementing a *Lisp Oriented Machine*.

2. In [18] and in the present paper is offered a solution for a silicon-based machine that performs efficiently the *splicing mechanism* emphasized in molecular computing.

3. In the present paper we describe a very efficient system for the *insert/delete* mechanism, also characteristic for molecular computing.

4. The work [16] presents an application of the CM concept in implementing *eco-grammar systems* [2].

5. Another domain is the implementation of the unification mechanism in the Prolog language.

6. By expanding the CM functions over the cells of a cellular automaton the Lindenmayer grammars [9] gain a very good physical support for their parallel-executed substitutions.

7. We have in progress a work in which we will present applications of CM in implementing *Markov rewriting systems* [10].

4 Molecular Mechanisms on Connex Memory

In order to explain how CM works in finding strings of symbols having different length it is useful to expand the function $FIND\ s$ to the macro-function $SFIND\ S$ (string find), where: $S = s_1 s_2 \ldots s_n$ is a string of symbols. The sequence of commands that emphasizes the end of all occurrences of the S string, marking the variables that immediately follow the last symbol s_n, is:

$$SFIND\ S = FIND\ s_1, CFIND\ s_2, CFIND\ s_3, \ldots, CFIND\ s_n.$$

Thus, all occurrences of the string S in CM are found and marked in time $T_{SFIND\ S} \in O(n)$, i.e., *the searching space dimension does not matter*. In order to find the string S, having the length $l(S) = n$, *it is enough to waste only the time nedeed to utter it.*

4.1   Implementing Splicing with CM

The aim of this section is to prove the efficiency of the CM in performing the splicing operation.

If the splicing rule is $r = u_1 \# u_2 \# u_3 \# u_4$ and the initial content of the CM is

$$\ldots \$x_1 u_1 u_2 x_2 \$ \ldots \$y_1 u_3 u_4 y_2 \$ \ldots$$

(\$ is used for delimiting the strings in the memory), then an intermediate form is

$$\ldots \$x_1 u_1 \& u_2 x_2 \$ \ldots \$y_1 u_3 u_4 y_2 \$ \ldots$$

after identifying the first cutting point (emphasized with the special symbol & inserted between the substrings u_1 and u_2) and the final content of the CM is

$$\ldots \$x_1 u_1 u_4 y_2 \$u_2 x_2 \$ \ldots \$y_1 u_3 \$ \ldots$$

The next procedure describes this mechanism.

Procedure SPLICING
    if the string $u_1 u_2$ is found [**step 1**]
        then back before u_2 [**step 2**]
            insert & before u_2 [**step 3**]
            if the string $u_3 u_4$ is found [**step 4**]
                then back before u_4 [**step 5**]
                    move $u_4 y_2$ before & [**step 6**]
                    substitute & with \$ [**step 7**]
              else delete &
            endif
        endif
    end SPLICING

Definition 4.1 *A splicing machine with CM is a pair $SMCM = (V, a)$, see Figure 2, where CM contains a set of strings $S \subset V^*$, FA is a finite automaton that executes the procedure* **SPLICING**. *On the* Input tape *is stored the splicing rule. The current symbols accessed on the input tape and/or from the CM together with the current state of the automaton generate, in each clock cycle, the next state of the automaton. In each state the automaton can move, if needed, left or right the input tape head, or commands CM using one of their functions.*

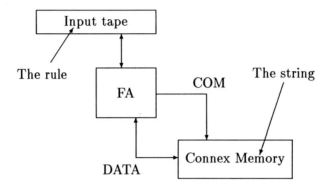

Figure 2: The splicing machine with CM

Indeed, executing the macro-function *SFIND S* the first marked variable in the CM is the variable after the first occurrence of the string S. The time for finding the string S in CM is in $O(l(S))$. More precisely, $l(S)$ is the number of clock cycles needed to find S. Using this "macro" the behavior of the finite automaton associated with the procedure SPLICING becomes clear and the execution time for each main step is:

step1 : $T \in O(l(u_1) + l(u_2))$,
step2 : $T \in O(l(u_2))$,
step3 : $T \in O(1)$,
step4 : $T \in O(l(u_3) + l(u_4))$,
step5 : $T \in O(l(u_2))$,
step6 : $T \in O(l(u_4) + l(y_2))$,
step7 : $T \in O(1)$.

Comparing SM and the $SMCM$ we observe that the *size* for both is in the same order: $O(\Sigma l(S_i))$ where $\Sigma l(S_i)$ is the total length of the strings $S_i \in S$. But, there is a significant difference between the execution time for the splicing operations in the two approaches. In the first machine, which uses a standard RAM memory, we have $T_{SM} \in O(f(\Sigma l(S_i)))$ but the second solution has a better performance: $T_{SMCM} \in O(max(l(S_i)))$.

Although the actual size of CM is bigger than the size of the RAM memory (around 20 times), the gain in the speed justifies using CM because the time is proportional only with the biggest string S_i stored in the memory instead of the sum of the length of all strings from the memory.

4.2   Implementing Insert/Delete Operations with CM

The *Insert/Delete* operations are described by the following procedures.

> **Procedure** INSERT
> > **if** the string uv is found (applying $SFIND\ uv$) [**step 1**]
> > > **then** back before v [**step 2**]
> > > > insert the string z [**step 3**]
> > **endif**
> **end** INSERT

> **Procedure** DELETE
> > **if** the string uzv is found (applying $SFIND\ uzv$) [**step 1**]
> > > **then** back before z [**step 2**]
> > > > delete the string z [**step 3**]
> > **endif**
> **end** DELETE

The execution time for both procedures in a system build with CM is

$$T_{I/D} \in O(l(uzv))$$

because:

step 1 is executed in $T_{SFIND\ uv/uzv} \in O(l(uv))/O(l(uzv))$,

step 2 is executed in $T_{BACK\ v/zv} \in O(l(v))/O(l(zv))$ or in $O(1)$ if a simple trick is used (inserting a special symbol after the string u),

step 3 is executed in $T_{i/d} \in O(l(z))$.

It is obvious that the machine that executes *insert/delete* operations has the same structure as the machine for the splicing mechanism (see Figure 2).

4.3   Limits to Be Removed

In the previous two approaches, for the *splicing* operation and for *insert/delete* operations, a rule is applied with a partial efficiency only, because the operation can be completed in only one place. Indeed, the places where the rule must be applied are all identified in parallel, but the effective concatenation or effective insertion/deletion is performed in only one place because of the structure of CM that consists in a sort of a shift register. For example, an insert in two distinct points in the string stored in the CM implies that a part of the internal register of CM is shifted one position and another part is shifted with two positions.

5   A Cellular Automaton as a Support for Connex Memory Functions

5.1   Definition

In order to add the possibility *to access all marked points* of the stored string, we adopt a two-dimensional support for CM. Instead of the one-dimensional structure of a shift register we use a *two-dimensional CA*. The string length can be now modified synchronously in many points by insertion or deletion. This new feature is enabled by the liberty of adding more new symbols in any places on the two-dimensional area of a CA.

Each symbol of the string is stored in the state of a cell and the link is done by the adjacency in the CA area. Because in our CA each cell has eight neighbors, it is very easy to add a new adjacency in the string.

The CA consists of *active cells* and of *inactive cells*. The first contain the string and are linked by three-bit pointers to the eight cells neighborhood. In each cell two processes are performed:

- the process of executing operations implied by the CM function:

  - the subset that does not modify the length of the string, only in the active cells,

  - the subset that modifies the length of the string,

- the self-organizing process of cells towards the state of each active cell to be surrounded by a maximum number (i.e., 6 if possible) of inactive cells.

The self-organizing process generates the conditions to have enough space to insert in many places, synchronously, the same symbol. The same process removes, step by step in a sequential way, the inactivated cells.

5.2   An Algorithm for the Multiple Access CM

Specific for the Multiple Access CM (MACM) are three types of actions:

1. the insert function from the function set of CM,

2. the delete function from the function set of CM,

3. the self-organizing process offering space for new insert actions.

In this paper, only the case of dispersed marked points is studied.

5.2.1 *INSERTs* in **MACM**

There are two typical situations in which a new symbol is inserted in a stored string. In Figure 3a, the marked cell (containing the arrow) has as the next cell the pointed cell (containing a circle). The second situation is presented in Figure 3b. These two configurations can be rotated three times for obtaining all the cases.

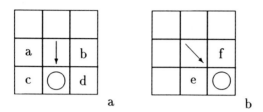

Figure 3: Insert configurations

In the first case the next marked positions can be the cells labeled by a, b, c and d. The algorithm for selecting the new active cell is presented in Table 1.

In the second case the next marked position can be selected between the positions e and f. The corresponding algorithm is presented in Table 2.

Deadlock situations are solved as a consequence of the self-organizing process described below.

5.2.2 *DELETE* in **MACM**

The CM's function DELETE is performed in two steps:

1. inactivate the content of the marked cell but maintain the cell connected to the string as an *empty cell*,

2. eliminate the empty cell by shifting them:

   - until the string form allows to *point over*, inactivating the empty cell,

   - or until the end of the string is reached.

Figure 4: The reference cells for DELETE

a b c d	The cell used for expanding the string
0 0 0 0	randomly *a* or *b*
0 0 0 1	*a*
0 0 1 0	*b*
0 0 1 1	**if** *c* and *d* do not point to the marked symbol (the arrow) **then** randomly *a* or *b* *deadlock*; **if** *c* points the marked symbol **then** *b*; **if** *d* points the marked symbol **then** *a*; **if** *c* and *d* point the marked symbol **then** *deadlock*
0 1 0 0	*a*
0 1 0 1	*a*
0 1 1 0	**if** *c* does not point the marked symbol **then** *a*, **else** *d*
0 1 1 1	**if** *c* does not point the marked symbol **then** *a*, **else** *deadlock*
1 0 0 0	*b*
1 0 0 1	**if** *d* does not point the marked symbol **then** *b*, **else** *c*
1 0 1 0	*b*
1 0 1 1	**if** *d* does not point the marked symbol **then** *b*, **else** *deadlock*
1 1 0 0	**if** *a* and *b* are not connected with the target of the marked symbol (the circle), **then** randomly *c* or *d*; **if** *a* and *b* are connected with "circle", **then** *deadlock*; **if** *a* is not connected with "circle", **then** *c*; **if** *b* is not connected with "circle", **then** *d*
1 1 0 1	**if** *a* is not connected with "circle", **then** *c*, **else** *deadlock*
1 1 1 0	**if** *b* is not connected with "circle", **then** *d*, **else** *deadlock*
1 1 1 1	*deadlock*

Table 1: The algorithm for finding the new active cell when the marked cell points horizontally or vertically

The *point over algorithm* is the main problem. In Figure 4 we present the typical situations (all the rest is reducible to rotating this representation). The symbol (◇) represents the empty cell.

1. $a \rightarrow \diamond \rightarrow d, e$, no points over

2. $b \rightarrow \diamond \rightarrow d, e$, no points over,

3. $c \rightarrow \diamond \rightarrow d \Rightarrow c \rightarrow d$, points over,

4. $c \rightarrow \diamond \rightarrow e \Rightarrow c \rightarrow e$, points over.

5.2.3   Self-Organizing in MACM

Many inserts in the same marked place generate *deadlocks* that can be solved only by reorganizing the string over the CA's cells. Our aim is to emphasize a set of *local* rules that solve this *global* problem. An example of deadlock is presented in Figure 5a, where in the cell 14 is stored the marked symbol. Any insertion is impossible in this configuration. The string must be expanded on the CA's surface, so allowing new insertions.

e f	Action
0 0	random e or f
0 1	e
1 0	f
1 1	*deadlock*

Table 2: The algorithm for finding the new active cell when the marked cell points diagonal.

The self-organizing algorithm consists in applying two rules:

1. moving the active cells in new positions so as to minimize the number of neighbors,

2. generating "fluctuations" of some cells in equivalent positions.

The effects of the moving cells are presented in Figure 5b-d and Figure 6. In Figure 6h the process of spacing the string can not be continued without applying the second rule, generating a "fluctuation" of the 12th cell in a new position in which the number of neighbors is the same. But in this new configuration the first rule can be used in two steps again (Figure 7j and Figure 7k). New "fluctuations" are needed in Figure 7l. The dispersing process ends when each cell has no more than two neighbors.

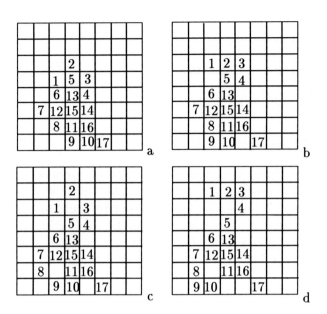

Figure 5: Self-organizing process

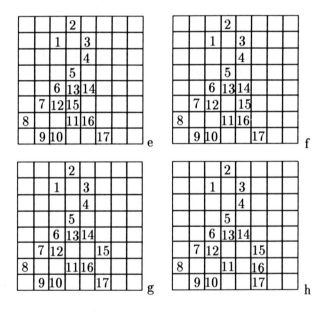

Figure 6: Self-organizing process (cont.)

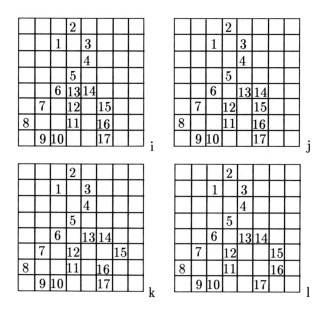

Figure 7: Self-organizing process (cont.)

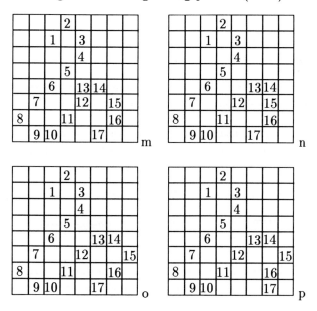

Figure 8: Self-organizing process (cont.)

6 The Eco-Chip: a Cellular Automaton with CM's Features

6.1 The Definition

In order to perform the two molecular mechanisms previously described on the EC, the CM function set must be adapted redefining a few simple functions. The main new features introduced by these new functions are related to the possibility to act in all the marked places on the string or on the strings stored in the cellular automaton.

Definition 6.1 *The* Eco-Chip *(EC) is a bi-dimensional cellular automaton each cell having two states:*

- *the* inactive *state in which the cell is unconnected with any other cell,*

- *the* active *state in which the cell is connected in a string with other cells and stores a string of variables, each having a value and two states: marked and non marked.*

Over the strings stored in the active cells the following functions can be executed:

- *RESET s: all the cells from the first diagonal become active, take the value s and switch to the state non-marked,*

- *FIND s: all the variables that follow a variable having the value s switch to the marked state and the rest switch to the non-marked state,*

- *CFIND s: (conditioned find) all the variables that follow a variable having the value s and being to the marked state switch to the marked state and the rest switch in the non-marked state,*

- *INSERTA s: the value s is inserted before* all *marked variables,*

- *LEFT | RIGHT: all the markers move one position to the left or to the right,*

- *DELETEA: the values stored in the* all *marked position are deleted, the positions remain marked and the symbols from the right are moved one position left,*

- *CUT: in all marked points the strings are divided (the link is removed) in two independent substrings,*

- *PASTE s, t: all the strings move over the area of the cellular automaton and when an end of a string having the value s meets another string having the end with the value t, the two strings are concatenated and the symbols s and t are deleted.*

Excepting the last function, all others are executed in a single clock cycle. The function $PASTE\,s, t$ depends on a random process and the end can be tested waiting for the disappearance of all the symbols s or of all the symbols t. A very important parameter in this process is the string's "mobility". The work is in progress for a simulator for experimenting algorithms that offer a lot of "mobility" to the strings stored over the cellular automaton. Many algorithms can be used, but the efficiency in performing the PASTE function must be measured only in some formal experiments using a simulator.

The previous definition describes only a theoretical model. For an actual circuit some simple input-output functions must be added.

6.2 The Full Parallel Insert/Delete Operations on the EC

The insert/delete operations described in Subsection 2.2 by the procedures INSERT and DELETE can be performed using the EC function set. The input data consists of many strings stored in the active cellular automata's cells. All the occurrences of the substrings uv or uxv are marked at the same time as for the CM applications, but now the insert or the delete of the string x is parallel performed in **all** marked points, rather than in the previous case when the operation was performed only in the first marked point.

The execution time for all insert or delete operations that can be performed over the content of EC depends only on the length of the $(u, x, v) \in I \,|\, D$, thus:

$$T_{allI/D} \in O(l(uxv)).$$

The size of the structure that performs these operations is in $O(n)$, where n is the number of cellular automata's cells.

6.3 The Full Parallel Splicing Operation on the EC

The algorithm for performing splicing operations over strings stored in EC starts similarly as in the case of the system using CM, but is completed in a different way, because all the splicing operations that can be performed will be performed in parallel. The main steps are the following:

1. all the occurrences of the substring u_1u_2 (see Definition 2.1) are found and the special symbol α is inserted between u_1 and u_2,

2. all the occurrences of the substring u_3u_4 (see Definition 2.1) are found and the special symbol β is inserted between u_3 and u_4,

3. perform the sequence: $FIND\,\alpha, CUT$, thus generating a set of strings having the form $x_1u_1\alpha$ (and another set of strings having the form u_2x_2),

4. perform the sequence: $FIND\,\beta, LEFT, LEFT, CUT$, thus generating a set of strings having the form βu_4y_2 (and another set of strings having the form y_1u_3),

5. perform $PASTE\,\alpha,\beta$ until all αs or all βs disappear.

The first four steps are performed in a time related with the length of the rule r, $T \in O(l(r))$. The execution time in the last step depends on the "mobility" of the strings stored in EC.

7   Conclusions

In this paper we have presented two solutions for performing molecular operations: the Connex Memory and its extension over a cellular automaton: the Eco-Chip. Both can be used to perform the splicing operation, the insert/delete operations or the matching operations (the last one was ignored in this approach). All the three operations have two main steps:

1. identifying the places where the operations can be applied,

2. performing the proper operation.

The first step is performed in parallel in CM or in EC, but the second one can be executed in parallel only with EC.

The performances of the system built around the CM circuit are the following:

1. The execution time for *each* splicing operation is

$$T_S \in O(max.\ length\ of\ a\ string)$$

for any number of strings stored in the Connex Memory. For the insert/delete operations the execution time for **one** insert or delete operation is $T_{I/D} \in O(l(uzv))$.

2. The size of the structure of Connex Memory is: $S_{CM} \in O(n)$, where n is the number of cells.

3. Therefore, using an $O(n)$ sized data structure (containing many strings) we can perform on it only one splicing operation or insert/delete operation in $O(1)$.

4. Instead of the von Neumann architecture:

Processor - Channel - RAM

we propose a new one:

Processor - Channel - CM

in which a *very fine grain parallelism*, performed by the CM, avoids the main effect of the channel's bottleneck.

5. Substituting the RAM with the Connex Memory the area on the silicon is multiplied only by a constant and the time decreases from $O(n)$ to $O(1)$ for a system executing one splicing, deletion or insertion at a time.

The performances of the system built around the EC circuit are the following:

1. The execution time for *any* number of insert or delete operations associated to one applications of the operations is in $O(l(uzv))$. For the splicing operations the time depends on the "mobility" of the strings that wind over the cellular automaton.

2. The size of EC is proportional with n, the number of cells.

3. An EC can be used as a performant co-processor for specific applications.

At the end of this paper we make a suggestion. In order to improve the "mobility" of the snakes of symbols that wind over the cellular automaton surface, a *multi-level cellular automaton* is proposed. Thus, an additional level of cells can be used to propagate the "smell" of the arguments of the function PASTE. Each cell on the basic level is connected with a correspondent cell in the second layer, used to propagate the "smells". The snakes of symbols will be oriented after the "smells" received from the second layer. We hope that using this model the strings $x_1 u_1 \alpha$ will meet as soon as possible the strings $\beta u_4 y_2$. The experiments made with a simulator will decide on the opportunity to use "smells" to accelerate the splicing operation.

We believe that implementing in silicon the molecular mechanisms is a real challenge for the implementation in molecules:

1. The interface with a silicon machine is simpler.

2. The size of the silicon machine has the smallest dimension.

3. The time for the insert/delete operation is the smallest possible: it is the time to express the rule.

4. The time for the splicing operation is in the same order on the silicon and on the molecules, being related to the "mobility" of strings/molecules on the physical support.

5. But the molecular computing has a definitive advantage: it is an amazing suggestion for silicon based computing.

References

[1] L. Adleman, Molecular computation of solutions to combinatorial problems, *Science*, 226 (Nov. 1994), 1021 – 1024.

[2] E. Csuhaj-Varú, J. Kelemen, A. Kelemenová, Gh. Păun, Eco-grammar systems: A grammatical framework for studying-like interactions, *Artificial Life*, 3, 1 (1997), 1 – 28.

[3] M. Conrad, On design principles for a molecular computer, *Communications of the ACM*, 28 (May 1985), 464 – 480.

[4] Z. Hascsi, Gh. Ştefan, The connex content addressable memory (C^2AM), *ESSCIRC'95 Twenty-first European Solid-State Circuits Conference*, Lille-France, 19-21 Sept., 1995.

[5] T. Head, Formal language theory and DNA: An analysis of the generative capacity of specific recombinant behaviours, *Bull. Math. Biology*, 49 (1987), 737 – 759.

[6] T. Head, Splicing schemes and DNA, in *Lindenmayer Systems: Impacts on Theoretical Computer Science and Developmental Biology* (G. Rozenberg, A. Salomaa, eds.), Springer-Verlag, Berlin, 1992, 371 – 383.

[7] L. Kari, Gh. Păun, A. Salomaa, S. Yu, DNA computing, sticker systems, and universality, *Acta Informatica*, to appear.

[8] L. Kari, Gh. Păun, G. Thierrin, S. Yu, At the crossroads of DNA computing and formal languages: Characterizing RE using insertion-deletion systems, *Proc. 3rd DIMACS Workshop on DNA Based Computing*, Philadelphia, 1997, 318 – 333.

[9] A. Lindenmayer, Mathematical models of cellular interactions in development I, II, *Journal of Theor. Biology*, 18, 1968, 280 – 325.

[10] A. A. Markov, *The Theory of Algorithms*, Trudy Matem. Instituta im V. A. Steklova, 42 (1954) (Translated from Russian by J. J. Schorr-kon, U. S. Dept. of Commerce, Office of Technical Services, no. OTS 60-51085, 1954).

[11] Gh. Păun, ed., *Artificial Life. Grammatical Models*, Black Sea University Press, Bucureşti, 1995.

[12] Gh. Păun, On the power of the splicing operation, *Intern. J. Computer Math.*, 59 (1995), 27 – 35.

[13] Gh. Păun, A. Salomaa, DNA computing based on the splicing operation, *Math. Japonica*, 43, 3 (1996), 607 – 632.

[14] Gh. Ştefan, V. Bistriceanu, A. Păun, Toward a natural mode of the Lisp implementation (in Romanian), Comm. to the Second National Symposium on Artificial Intelligence, Romanian Academy, Sept. 1985; published in *Systems for Artificial Intelligence*, Romanian Academy Pub. House, Bucharest, 1991.

[15] Gh. Ştefan, Memoria conexă, in *CNETAC '86*, 1986 (in Romanian)

[16] Gh. Ştefan, M. Malitza, The Eco-chip: A physical support for Artificial Life systems, in [11], 260 – 275.

[17] Gh. Ştefan, M. Malitza, Chaitin's toy-Lisp on Connex Memory machine, *Journal of Universal Computer Science*, 2, 5 (1996), 410 – 426.

[18] Gh. Ştefan, M. Malitza, The splicing mechanism and the Connex Memory, in *Proc. of the 1997 IEEE Int. Conf. on Evolutionary Computation*, Indianapolis, April 13-16, 1997, 225 – 229.

APPENDIX

The Structure of the Connex Memory

The whole structure of the CM is represented in Figure 9, where:

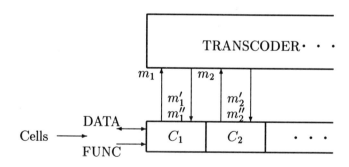

Figure 9: The structure of CM

The cells C_i, each storing a variable, are serially connected. All are externally connected through the bi-directional DATA bus and by the FUNC bus. In addition, each cell generates the marker m_i toward TRANSCODER and receives signals that classify them.

- C_i, for $i = 1, 2, \ldots, n$, represents the i-th cell

- Transcoder is a combinational circuit that receives from each cell the markers
$$m_0 m_1 \ldots m_n = 00 \ldots 01XX \ldots X,$$
$(X \in \{0, 1\})$ and generates:
$$m_0' m_1' \ldots m_n' = 00 \ldots 011 \ldots 1,$$
substituting with 1 all the symbols after the first occurrence of 1 and
$$m_0'' m_1'' \ldots m_n'' = 00 \ldots 010 \ldots 0,$$
emphasizing the first occurrence of a marked symbol.

In consequence, the cells of the CM can be divided in three classes:

- the class of cells before the first cell having the marker $(m_i = 1)$,

- the first marked cell,

- the class of cells after the first marked cell.

Using the signals m_i', m_i'', E_{i-1} and m_{i-1}, according to the current command (COM), each cell has enough information to switch into the next state. The size of this structure is $O(n)$ for the string of the cells and is $O(n \ log \ n)$ for Transcoder. In order to reduce the size of CM to $O(n)$ we must use a bi-dimensional solution for the cell array. It follows that the transcoder is substituted by two transcoders, each having the size $O(\sqrt{n} \ log \ n)$. This second solution allows us to define a CM with $O(n)$ complexity.

Beside the connections with the transcoder, each cell is connected with the previous and the next cell. In addition, each cell is connected to the two buses: the bi-directional DATA bus and the FUNC bus. The cell structure is presented in Figure 10, where:

- R is a p-bits register that stores the value of the variable v_i,

- D is a D (delay) flip-flop that stores the value m_i of the marker,

- MUX_0 is a multiplexer that selects in each clock cycle, according to the bits c_1 and c_2, the value to be stored in the register R,

- MUX_1 is a multiplexer, selected by c_3 and c_4, which allows storing the current value of the marker in D, in each clock cycle

- CP is a comparator that shows by the output E_i if v_i is equal to the value applied to the input DIN,

- CLC is a combinational circuit that generates the control bits $c_0, c_1, \ldots, c_5$, according to: the command received from COMP, the bits E_{i-1}, m_{i-1}, received from the previous cell, and the bits m_i', m_i'', generated by the Transcoder (see Figure 9),

- c_0 is the value of the locally generated marker,

- c_5 is the enable input for the tristate circuit that drives the bi-directional bus DATA.

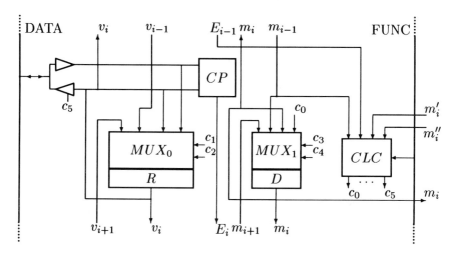

Figure 10: The organization of a CM cell

Bidirectional Sticker Systems

and Representations of RE Languages
by Copy Languages

Rudolf FREUND

Department of Computer Science
Vienna Univeersity of Technology
Karlsplatz 13, A-1040 Wien, Austria
rudi@logic.at

Abstract. We prove how any recursively enumerable language $L \subseteq T^*$ can be represented as $h_T(h_1(L_0) \cap h_2(L_0))$, i.e., as the projection of the intersection of two morphic images of a specific language L_0, e.g., L_0 can be chosen as the copy language $cp(V^+) = \{w\bar{w} \mid w \in V^+\}$ or the reversed copy language $rcp(V^+) = \{w\bar{w}^R \mid w \in V^+\}$ for some alphabet V. Moreover, we point out how these results arise from bidirectional sticker systems, a model of DNA computing introduced recently.

1 Introduction

In [4], the authors prove the representation of an arbitrary recursively enumerable language L (by using the idea of equality sets) as $h_T(h_1(E(h_1, h_2)) \cap R)$, where h_1, h_2 are two non-erasing morphisms, the equality set of h_1 and h_2 is defined as $E(h_1, h_2) = \{w \in \Sigma_2^* \mid h_1(w) = h_2(w)\}$, R is a regular language, and h_T is a projection. Based on this result, in [3] a basic result for bidirectional sticker systems is obtained, from which it is easy to show that L can be represented as the projection h of the intersection of two minimal linear languages L_1, L_2, i.e., $L = h(L_1 \cap L_2)$. (A *minimal* linear language is generated by a linear grammar with only one non-terminal symbol S; if, moreover, the only terminal rule is $S \to \lambda$, then the linear grammar and the generated linear language are called *strictly minimal*.) With a small refinement of this proof given in [3], in this paper we show that a strictly minimal language L' can be written as $h'(L_0)$, where L_0 is the reversed copy language $rcp(V^+) = \left\{w\bar{w}^R \mid w \in V^+\right\}$ for some alphabet V. ($\bar{V} = \{\bar{a} \mid a \in V\}$ contains the barred versions of the letters in V.) From these representations we immediately infer another representation of L that is based on one single generator language L_0 as

$$L = h_T(g_1(L_0) \cap g_2(L_0)),$$

where g_1, g_2 are two deterministic generalized sequential machines and $L_0 = rcp\left(\{0,1\}^+\right)$.

Another representation of a recursively enumerable language L can be obtained by using (regular) matrix grammars (with a string axiom instead of a start symbol). Based on this result characterizing L as

$$L = h_T(L(G_1) \cap L(G_2))$$

for two regular matrix grammars with axioms, G_1 and G_2, we are able to obtain a similar representation as above, but with L_0 being a specific copy language, i.e.,

$$L = h_T(g_1(L_0) \cap g_2(L_0)),$$

where g_1, g_2 are two deterministic generalized sequential machines and $L_0 = cp\left(\{0,1\}^+\right)$.

Some of the problems addressed above are described in [5] and, in some extent, were discussed at the Workshop on Molecular Computing, held in Mangalia, Romania, at the Black Sea University in the summer of the year 1997 from August 17th to August 24th. It was somewhat surprising to realize that the solutions to the problems considered above go back to a fundamental result obtained for bidirectional sticker systems, which were introduced in [3] and present an interesting model of computation in the field of DNA computing. Hence, part of this paper is devoted to reconsider the main features of bidirectional sticker systems and the characterization of recursively enumerable languages in this model of DNA computing.

Sticker systems as introduced in [4] are language generating devices based on the *sticker operation*, which, in turn, is a formal model for the techniques used by L. Adleman in his successful experiment of deciding the existence of a Hamiltonian path in a graph by using DNA as described in [1]. DNA sequences are double stranded (helicoidal) structures composed of four nucleotides, A (adenine), C (cytosine), G (guanine), and T (thymine); the pairs $\{A, T\}$ and $\{C, G\}$ are called Watson-Crick pairs according to their complementary occurrence in double-stranded DNA molecules. If we have two DNA molecules with sticky ends of single-stranded sequences of A, C, G, T nucleotides that are matching according to the Watson-Crick complementarity, the two sequences may be glued together (by hydrogen bonds). This matching of complementary nucleotides now is the biological mechanism we have in mind when formally defining the sticker operation for (bidirectional) sticker systems, where we prolong to the left and to the right a given domino of (single or double) symbols by using given single stranded strings or even more complex dominoes with sticky ends, gluing these ends together with the sticky ends of the current sequence according to a complementarity relation.

In the second section we specify some notions and notations from formal language theory. The model of bidirectional sticker systems is described in the third section, where we also recall the characterization of a recursively enumerable language by such a system as shown in [3]. The main results of the paper, i.e., the characterizations of a recursively enumerable language by a projection of the intersection of two morphic images or two deterministic generalized sequential machine mappings of specific reversed copy languages or of specific copy languages, are exhibited in the fourth section.

2   Prerequisites

We first specify a few notions and notations from formal language theory ([2], [6]). For an alphabet V, by V^* we denote the free monoid generated by V under the operation of concatenation; the empty string is denoted by λ. Moreover, we denote $V^+ = V^* \setminus \{\lambda\}$. By $[m...n]$ we denote the set of natural numbers between m and n, i.e., $[m...n] = \{i \mid m \leq i \leq n\}$.

A linear grammar $G = (N, T, P, S)$ is called *minimal,* if G contains only one non-terminal symbol, i.e., $N = \{S\}$. If, moreover, any production in P either is of the form $S \to uSv$ with $uv \in T^*$ or equal to $S \to \lambda$, then we will call G *strictly minimal.*

A *deterministic finite automaton* (dfa) will be presented in the form $M = (Q, V, s_0, F, \delta)$, where Q is the set of states, V is the alphabet, s_0 is the initial state, F is the set of final (accepting) states, and $\delta : (Q \times V) \to Q$ is the transition mapping.

A *deterministic generalized sequential machine* (dgsm) g is a deterministic finite automaton with outputs: $g = (Q, V_I, V_O, s_0, F, \delta)$, where Q is the set of states, V_I is the input alphabet, V_O is the output alphabet, s_0 is the initial state, F is the set of final states, and $\delta : (Q \times V_I) \to (Q \times V_O^*)$; $\delta(s, a) = (s', x)$ has the following meaning: reading the symbol a in state s, g writes the string x and goes to state s' (for $a \in V_I$, $x \in V_O^*$, $s, s' \in Q$); for arbitrary $w \in V_I^*$, $z \in V_O^*$, we then write $zsaw \Rightarrow_g zxs'w$, thus defining the relation $\Rightarrow_g$. For $w \in V_I^*$, the result of the translation of w by g is defined by

$$g(w) = y \text{ if and only if } s_0 w \Rightarrow_g^* y s_f \text{ for some } s_f \in F,$$

where $\Rightarrow_g^*$ is the reflexive and transitive closure of the relation $\Rightarrow_g$. Observe that $g(w)$ – if defined – is the uniquely determined output of g for the input w. The mapping g is extended in the natural way to languages by defining $g(L) = \{g(x) \mid x \in L \text{ and } g(x) \text{ is not undefined}\}$, for any $L \subseteq V_I^*$.

A *matrix grammar with axioms* G is a construct (V_N, V_T, M, A), where V_N is the set of non-terminal symbols, V_T is the alphabet of terminal symbols, M is a finite set of matrices (a matrix is a finite sequence of productions over $V_N \cup V_T$) and A is a finite set of axioms (words over $V_N \cup V_T$). For $w, v \in (V_N \cup V_T)^*$, we write $w \Rightarrow_G v$ if and only if is the result of applying a matrix $[p_1, \ldots, p_k] \in M$ to w, i.e.,

$$w = w_0 \Rightarrow_{p_1} w_1 \Rightarrow_{p_2} \ldots \Rightarrow_{p_k} w_k = v,$$

where $w_{i-1} \Rightarrow_{p_i} w_i$ $(1 \leq i \leq k)$ denotes the usual derivation of w_i from w_{i-1} by using the production p_i. The language generated by G is defined by

$$L(G) = \{v \in V_T^* \mid w \Rightarrow_G^* v \text{ for some } w \in A\},$$

where $\Rightarrow_G^*$ is the reflexive and transitive closure of $\Rightarrow_G$. G is called regular (of type X), if every production in every matrix in M is regular (of type X, respectively).

The main results shown in this paper, i.e., the representation of recursively enumerable languages as the projection of the intersection of two languages being the morphic images or the dgsm images of specific generator languages, are based on the following characterization of recursively enumerable languages shown in [4]:

Proposition 2.1 *For each recursively enumerable language L, $L \subseteq T^*$, there exist two non-erasing morphisms $h_1, h_2 : \Sigma_2^* \to \Sigma_1^*$, a regular language $R \subseteq \Sigma_1^*$, and a projection $h_T : \Sigma_1^* \to T^*$ defined by*

$$h_T(a) = \begin{cases} a, & \text{if } a \in T, \\ \lambda, & \text{if } a \in \Sigma_1 \setminus T, \end{cases} \quad \text{such that } L = h_T(h_1(E(h_1, h_2)) \cap R), \text{ where}$$

$$E(h_1, h_2) = \{w \in \Sigma_2^* \mid h_1(w) = h_2(w)\}.$$

3 Bidirectional Sticker Systems

In this section we recall the formal definitions for bidirectional sticker systems from [3] as well as the characterization of any recursively enumerable language as the morphic image of a language generated by a bidirectional sticker system.

Take an alphabet V (a finite set of abstract symbols) endowed with a relation ρ (of *complementarity*), $\rho \subseteq V \times V$; moreover, consider a special symbol, #, not in V, denoting an empty space (the *blank* symbol). Using the elements of $V \cup \{\#\}$ we construct the following sets of *composite* symbols:

$$\binom{V}{V}_\rho = \left\{ \binom{a}{b} \mid a, b \in V, (a, b) \in \rho \right\},$$

$$\binom{\#}{V} = \left\{ \binom{\#}{a} \mid a \in V \right\}, \quad \binom{V}{\#} = \left\{ \binom{a}{\#} \mid a \in V \right\}.$$

Moreover, we denote $W_\rho(V) = W_\rho^s(V) \cup S(V)$ (the set of *well-formed sequences* or *dominoes*) and $W_\rho^s(V) = S(V) \binom{V}{V}_\rho^+ S(V)$ (the set of *well-started sequences* or *dominoes*), where $S(V) = \binom{\#}{V}^* \cup \binom{V}{\#}^*$. Stated otherwise, the elements of $W_\rho^s(V)$ in the middle have pairs of symbols in V, as selected by the complementarity relation, and at the left and at the right they end either by pairs $\binom{\#}{a}$ or by pairs $\binom{b}{\#}$, for $a, b \in V$ (the symbols $\binom{\#}{a}$, $\binom{b}{\#}$ are not mixed). The elements of $S(V)$ at the left and at the right end of a domino in $W_\rho^s(V)$ are called *sticky ends*; if such an element is λ, then this end is called a *blunt end*.

The *sticker operation*, denoted by μ, is a partially defined mapping from $W_\rho(V) \times W_\rho(V)$ to $W_\rho^s(V)$; for $x, y \in W_\rho(V)$ with $\{x, y\} \cap W_\rho^s(V) \neq \emptyset$ and $z \in W_\rho^s(V)$ we write $\mu(x, y) = z$ if and only if one of the following cases holds:

1. $x = w \binom{a_1}{\#} \cdots \binom{a_r}{\#} \binom{a_{r+1}}{\#} \cdots \binom{a_{r+p}}{\#}, w \in S(V) \binom{V}{V}_\rho^+,$

$$y = \binom{\#}{c_1} \cdots \binom{\#}{c_r}, \ z = w \binom{a_1}{c_1} \cdots \binom{a_r}{c_r} \binom{a_{r+1}}{\#} \cdots \binom{a_{r+p}}{\#}, \text{ or}$$

$$y = \binom{a_{r+p}}{\#} \cdots \binom{a_{r+1}}{\#} \binom{a_r}{\#} \cdots \binom{a_1}{\#} w, \ w \in \binom{V}{V}_\rho^+ S(V),$$

$$x = \binom{\#}{c_r} \cdots \binom{\#}{c_1}, \ z = \binom{a_{r+p}}{\#} \cdots \binom{a_{r+1}}{\#} \binom{a_r}{c_r} \cdots \binom{a_1}{c_1} w, \text{ for}$$
$$r \geq 0,$$

$$p \geq 0, \ a_j \in V, 1 \leq j \leq r+p, \ c_i \in V, 1 \leq i \leq r, \ (a_i, c_i) \in \rho, 1 \leq i \leq r;$$

2. $$x = w \binom{\#}{c_1} \cdots \binom{\#}{c_r} \binom{\#}{c_{r+1}} \cdots \binom{\#}{c_{r+p}}, \ w \in S(V) \binom{V}{V}_\rho^+,$$

$$y = \binom{a_1}{\#} \cdots \binom{a_r}{\#}, \ z = w \binom{a_1}{c_1} \cdots \binom{a_r}{c_r} \binom{\#}{c_{r+1}} \cdots \binom{\#}{c_{r+p}}, \text{ or}$$

$$y = \binom{\#}{c_{r+p}} \cdots \binom{\#}{c_{r+1}} \binom{\#}{c_r} \cdots \binom{\#}{c_1} w, \ w \in \binom{V}{V}_\rho^+ S(V),$$

$$x = \binom{a_r}{\#} \cdots \binom{a_1}{\#}, \ z = \binom{\#}{c_{r+p}} \cdots \binom{\#}{c_{r+1}} \binom{a_r}{c_r} \cdots \binom{a_1}{c_1} w, \text{ for}$$
$$r \geq 0,$$

$$p \geq 0, a_i \in V, 1 \leq i \leq r, \ c_j \in V, 1 \leq j \leq r+p, \ (a_i, c_i) \in \rho, 1 \leq i \leq r;$$

3. $$x = w \binom{a_1}{\#} \cdots \binom{a_r}{\#}, \ w \in S(V) \binom{V}{V}_\rho^+,$$

$$y = \binom{\#}{c_1} \cdots \binom{\#}{c_r} \cdots \binom{\#}{c_{r+p}},$$

$$z = w \binom{a_1}{c_1} \cdots \binom{a_r}{c_r} \binom{\#}{c_{r+1}} \cdots \binom{\#}{c_{r+p}}, \text{ or}$$

$$y = \binom{a_r}{\#} \cdots \binom{a_1}{\#} w, \ w \in \binom{V}{V}_\rho^+ S(V),$$

$$x = \binom{\#}{c_{r+p}} \cdots \binom{\#}{c_r} \cdots \binom{\#}{c_1},$$

$$z = \binom{\#}{c_{r+p}} \cdots \binom{\#}{c_{r+1}} \binom{a_r}{c_r} \cdots \binom{a_1}{c_1} w, \text{ for } r \geq 0, p \geq 0, \ r+p \geq 1,$$

$$a_i \in V, 1 \leq i \leq r, \ c_j \in V, 1 \leq j \leq r+p, \ (a_i, c_i) \in \rho, 1 \leq i \leq r;$$

4. $x = w \begin{pmatrix} \# \\ c_1 \end{pmatrix} \dots \begin{pmatrix} \# \\ c_r \end{pmatrix}, \; w \in S(V) \begin{pmatrix} V \\ V \end{pmatrix}_\rho^+,$

$y = \begin{pmatrix} a_1 \\ \# \end{pmatrix} \dots \begin{pmatrix} a_r \\ \# \end{pmatrix} \dots \begin{pmatrix} a_{r+p} \\ \# \end{pmatrix},$

$z = w \begin{pmatrix} a_1 \\ c_1 \end{pmatrix} \dots \begin{pmatrix} a_r \\ c_r \end{pmatrix} \begin{pmatrix} a_{r+1} \\ \# \end{pmatrix} \dots \begin{pmatrix} a_{r+p} \\ \# \end{pmatrix}, \text{ for}$

$y = \begin{pmatrix} \# \\ c_r \end{pmatrix} \dots \begin{pmatrix} \# \\ c_1 \end{pmatrix} w, \; w \in \begin{pmatrix} V \\ V \end{pmatrix}_\rho^+ S(V),$

$x = \begin{pmatrix} a_{r+p} \\ \# \end{pmatrix} \dots \begin{pmatrix} a_r \\ \# \end{pmatrix} \dots \begin{pmatrix} a_1 \\ \# \end{pmatrix},$

$z = \begin{pmatrix} a_{r+p} \\ \# \end{pmatrix} \dots \begin{pmatrix} a_{r+1} \\ \# \end{pmatrix} \begin{pmatrix} a_r \\ c_r \end{pmatrix} \dots \begin{pmatrix} a_1 \\ c_1 \end{pmatrix} w, \text{ for } r \geq 0, p \geq 0, r+p \geq 1,$

$a_j \in V, 1 \leq j \leq r+p, \; c_i \in V, 1 \leq i \leq r, \; (a_i, c_i) \in \rho, 1 \leq i \leq r;$

5. $x = w \begin{pmatrix} a_1 \\ \# \end{pmatrix} \dots \begin{pmatrix} a_r \\ \# \end{pmatrix}, \; w \in S(V) \begin{pmatrix} V \\ V \end{pmatrix}_\rho^+, \; y = \begin{pmatrix} a_{r+1} \\ \# \end{pmatrix} \dots \begin{pmatrix} a_{r+p} \\ \# \end{pmatrix},$

$z = w \begin{pmatrix} a_1 \\ \# \end{pmatrix} \dots \begin{pmatrix} a_r \\ \# \end{pmatrix} \begin{pmatrix} a_{r+1} \\ \# \end{pmatrix} \dots \begin{pmatrix} a_{r+p} \\ \# \end{pmatrix}, \text{ or}$

$y = \begin{pmatrix} a_r \\ \# \end{pmatrix} \dots \begin{pmatrix} a_1 \\ \# \end{pmatrix} w, \; w \in \begin{pmatrix} V \\ V \end{pmatrix}_\rho^+ S(V), \; x = \begin{pmatrix} a_{r+p} \\ \# \end{pmatrix} \dots \begin{pmatrix} a_{r+1} \\ \# \end{pmatrix},$

$z = \begin{pmatrix} a_{r+p} \\ \# \end{pmatrix} \dots \begin{pmatrix} a_{r+1} \\ \# \end{pmatrix} \begin{pmatrix} a_r \\ \# \end{pmatrix} \dots \begin{pmatrix} a_1 \\ \# \end{pmatrix} w, \text{ for } r \geq 0, p \geq 0, r+p \geq 1,$

$a_i \in V, 1 \leq i \leq r+p;$

6. $x = w \begin{pmatrix} \# \\ c_1 \end{pmatrix} \dots \begin{pmatrix} \# \\ c_r \end{pmatrix}, \; w \in S(V) \begin{pmatrix} V \\ V \end{pmatrix}_\rho^+, \; y = \begin{pmatrix} \# \\ c_{r+1} \end{pmatrix} \dots \begin{pmatrix} \# \\ c_{r+p} \end{pmatrix},$

$z = w \begin{pmatrix} \# \\ c_1 \end{pmatrix} \dots \begin{pmatrix} \# \\ c_r \end{pmatrix} \begin{pmatrix} \# \\ c_{r+1} \end{pmatrix} \dots \begin{pmatrix} \# \\ c_{r+p} \end{pmatrix}, \text{ or}$

$y = \begin{pmatrix} \# \\ c_r \end{pmatrix} \dots \begin{pmatrix} \# \\ c_1 \end{pmatrix} w, \; w \in \begin{pmatrix} V \\ V \end{pmatrix}_\rho^+ S(V), \; x = \begin{pmatrix} \# \\ c_{r+p} \end{pmatrix} \dots \begin{pmatrix} \# \\ c_{r+1} \end{pmatrix},$

$z = \begin{pmatrix} \# \\ c_{r+p} \end{pmatrix} \dots \begin{pmatrix} \# \\ c_{r+1} \end{pmatrix} \begin{pmatrix} \# \\ c_r \end{pmatrix} \dots \begin{pmatrix} \# \\ c_1 \end{pmatrix} w, \text{ for } r \geq 0, p \geq 0, r+p \geq 1,$

$c_i \in V, 1 \leq i \leq r+p;$

7. $x = w \begin{pmatrix} a_1 \\ \# \end{pmatrix} \dots \begin{pmatrix} a_r \\ \# \end{pmatrix}, \; w \in S(V) \begin{pmatrix} V \\ V \end{pmatrix}_\rho^*, \; y = \begin{pmatrix} \# \\ c_1 \end{pmatrix} \dots \begin{pmatrix} \# \\ c_r \end{pmatrix} v,$

$$v \in \binom{V}{V}_{\rho}^{*} S(V), \ wv \in W_{\rho}^{s}(V), \ z = w \binom{a_1}{c_1} \ldots \binom{a_r}{c_r} v, \ \text{or}$$

$$x = w \binom{\#}{c_1} \ldots \binom{\#}{c_r}, \ w \in S(V) \binom{V}{V}_{\rho}^{*}, \ y = \binom{a_1}{\#} \ldots \binom{a_r}{\#} v,$$

$$v \in \binom{V}{V}_{\rho}^{*} S(V), \ wv \in W_{\rho}^{s}(V), \ z = w \binom{a_1}{c_1} \ldots \binom{a_r}{c_r} v, \ \text{for } r \geq 0,$$

$$a_i \in V, 1 \leq i \leq r, \ c_i \in V, 1 \leq i \leq r, \ (a_i, c_i) \in \rho, 1 \leq i \leq r.$$

In case 1 (case 2) we add complementary symbols on the lower (upper) level of x without completing all the blank spaces. In case 3 and case 4 we not only complete the blank spaces, but may even prolong the strand. In case 5 and case 6 the sticky end of x itself is prolonged. In case 7 we combine two dominoes whose sticky ends exactly fit together. Of course, for dominoes x, y which do not satisfy any of the previous conditions, $\mu(x, y)$ is not defined. Note that in all cases we allow the prolongation of "blunt" ends of strings in $W_{\rho}^{s}(V)$; moreover we have $\mu(\lambda, x) = \mu(x, \lambda) = x$ for every $x \in W_{\rho}^{s}(V)$; finally, observe that if all (intermediate) results are defined, then we have $\mu(\mu(x, y), z) = \mu(x, \mu(y, z))$, i.e., the sticker operation is associative.

A *bidirectional sticker system* is a construct $\gamma = ((V, \rho), A, P)$, where V is an alphabet, $\rho \subseteq V \times V$ is a relation on V, A is a finite subset of $W_{\rho}^{s}(V)$ (of axioms), and P is a finite set of pairs (D_l, D_r), where D_l is the domino to be adjoined at the left-hand side of the current object and D_r is the domino to be adjoined at the right-hand side of the current object, repectively; $D_l, D_r \in W_{\rho}(V)$.

For two objects $x, z \in W_{\rho}(V)$ we write

$$x \Longrightarrow_{\gamma} z \ \text{if and only if} \ z = \mu(\mu(D_l, x), D_r) \ \text{for some} \ (D_l, D_r) \in P.$$

By $\Longrightarrow_{\gamma}^{*}$ we denote the reflexive and transitive closure of the relation $\Longrightarrow_{\gamma}$.

A sequence $x_0 \Longrightarrow_{\gamma} x_1 \Longrightarrow_{\gamma} \ldots \Longrightarrow_{\gamma} x_k, x_0 \in A$, is called a *computation* in γ (of length k). A computation as above is *complete* when $x_k \in \binom{V}{V}_{\rho}^{+}$.

The language generated by γ, denoted by $L(\gamma)$, is defined by

$$L(\gamma) = \left\{ w \mid w \in \binom{V}{V}_{\rho}^{+}, \ x \Longrightarrow_{\gamma}^{*} w \ \text{for some} \ x \in A \right\}.$$

Therefore, only the complete computations are taken into account when defining $L(\gamma)$. Note that a complete computation can be continued, because we allow prolongations starting from blunt ends.

The following result uses the characterization of a recursively enumerable language L shown in Proposition 2.1 for proving how L can be represented as the morphic image of a bidirectional sticker system (for better understanding of the succeeding proofs, we shortly recall the proof of this result in a slightly modified version):

Theorem 3.1 *For any recursively enumerable language L, $L \subseteq T^*$, there exist a bidirectional sticker system γ and a morphism g such that $L = g\left(L\left(\gamma\right)\right)$.*

Proof. Let L be given according to Proposition 2.1 as $h_T(h_1(E(h_1, h_2)) \cap R)$, and R be given by the dfa $M = (Q, \Sigma_1, \delta, q_0, F)$. We now define a bidirectional sticker system $\gamma = ((V, \rho), A, P)$ with $L = g\left(L\left(\gamma\right)\right)$ by $V = \Sigma_1 \cup \tilde{\Sigma}_1 \cup Q \cup \{C, E, X, Z\}$, $\rho = \{(x, x) \mid x \in V\}$, $A = \left\{ \begin{pmatrix} q_0 \\ \# \end{pmatrix} \begin{pmatrix} X \\ X \end{pmatrix} \begin{pmatrix} Z \\ \# \end{pmatrix} \right\}$
and P containing the following pairs of adjoining rules:

For any $a \in \Sigma_2$, $h_1(a) = b_1 \ldots b_k$, $h_2(a) = c_1 \ldots c_m$, for any arbitrary states q_{i_j} from Q, $0 \leq j \leq m+1$, such that $\delta\left(q_{i_j}, c_j\right) = q_{i_{j+1}}, 0 \leq j \leq m$:

$$\left(\begin{matrix} q_{i_{m+1}} & \tilde{c}_m & q_{i_m} & \cdots & q_{i_2} & \tilde{c}_2 & q_{i_1} & q_{i_1} & \tilde{c}_1 & q_{i_0} \\ \# & \# & \# & \cdots & \# & \# & \# & \# & \# & \# \end{matrix} \right.,$$

$$\left. \begin{matrix} b_1 & C & Z & b_2 & C & \cdots & Z & b_k & C & Z \\ \# & \# & \# & \# & \# & \cdots & \# & \# & \# & \# \end{matrix} \right).$$

To the left we produce the reversed image of a word $w \in \Sigma_2^*$ under h_2 at the same time guessing a valid path through the finite automaton M, whereas to the right we produce the image of w under h_1 the symbols of which separated by the characters C, Z.

$$\left(\begin{pmatrix} \# \\ \tilde{b} \end{pmatrix}, \begin{pmatrix} \# \\ b \end{pmatrix} \begin{pmatrix} \# \\ C \end{pmatrix} \right) \text{ for all } b \in \Sigma_1;$$ by these rules, the correspondence (equality) of symbols on the left-hand side and on the right-hand side is checked.

$$\left(\begin{pmatrix} \# \\ q \end{pmatrix} \begin{pmatrix} \# \\ q \end{pmatrix}, \begin{pmatrix} \# \\ Z \end{pmatrix} \right) \text{ for all } q \in Q;$$ by these rules, the validity of the previously guessed paths through the finite automaton M is checked.

$$\left(\begin{pmatrix} E \\ \# \end{pmatrix} \begin{pmatrix} q_f \\ \# \end{pmatrix}, \begin{pmatrix} E \\ \# \end{pmatrix} \right) \text{ for all } q_f \in F; \left(\begin{pmatrix} \# \\ E \end{pmatrix}, \begin{pmatrix} \# \\ E \end{pmatrix} \right).$$

If the checks described above are all successful and M has reached a final state, the sticky ends can be completed, which for the first time yields a configuration with blunt ends on both sides. If such an object is prolonged, it never can yield another object with blunt ends at both sides any more.

Finally, we define $g : \begin{pmatrix} V \\ V \end{pmatrix}_\rho^* \to T^*$ by $g\left(\begin{pmatrix} x \\ x \end{pmatrix}\right) = \begin{cases} x, & \text{if } x \in T, \\ \lambda, & \text{if } x \in V \setminus T. \end{cases}$

$\square$

4 Copy Language Representations of Recursively Enumerable Languages

In this section we exhibit our main results, i.e., the representations of a recursively enumerable language as the projection of the intersection of two morphic images or two dgsm images of a single specific generator language L_0, e.g., a specific reversed copy language or a specific copy language.

4.1 Reversed Copy Languages

Based on the result exhibited in Theorem 3.1 above, in [3] the following representation of recursively enumerable languages was proved (as the construction is essential for the succeeding proofs, we shortly recall the proof of this result in a slightly improved version):

Theorem 4.1 *For any recursively enumerable language L, $L \subseteq T^*$, there exist two minimal linear languages $L_1, L_2 \subseteq \Gamma^*$ and a projection $h_T : \Gamma^* \to T^*$ such that $L = h_T (L_1 \cap L_2)$.*

Proof. Let L be given according to the characterization given above in Proposition 2.1 as $h_T (h_1 (E (h_1, h_2)) \cap R)$, and R be given by the dfa $M = (Q, \Sigma_1, \delta, q_0, F)$. We now define the following two minimal linear grammars G_1 and G_2:

$$G_i = (\{S\}, \Gamma, P_i, S), \ i \in \{1, 2\},$$
$$\Gamma = \Sigma_1 \cup \tilde{\Sigma}_1 \cup Q \cup \{C, E, X, Z\},$$
$$P_1 = \{S \to q_{i_{m+1}} \tilde{c}_m q_{i_m} \dots q_{i_2} \tilde{c}_2 q_{i_1} q_{i_1} \tilde{c}_1 q_{i_0} S b_1 C Z b_2 C \dots Z b_k C Z \mid a \in \Sigma_2,$$
$$h_1 (a) = b_1 \dots b_k, h_2 (a) = c_1 \dots c_m, \delta \left(q_{i_j}, c_j \right) = q_{i_{j+1}}, 0 \le j \le m \}$$
$$\cup \ \{S \to E q_f S E \mid q_f \in F\} \cup \{S \to q_0 X Z\},$$
$$P_2 = \{S \to q q S Z \mid q \in Q\} \cup \{S \to \tilde{b} S b C \mid b \in \Sigma_1\}$$
$$\cup \ \{S \to E S E, S \to X\}.$$

The projection $g_T : \Gamma^* \to T^*$ is defined by $h_T(a) = \begin{cases} a, & \text{if } a \in T, \\ \lambda, & \text{if } a \in \Gamma \setminus T. \end{cases}$ $\square$

The following theorem is a slight modification of the preceding theorem, where now the linear grammars are strictly minimal and not only minimal; yet we have to pay a price for that because we have to eliminate the empty word (e.g., by taking the intersection with Γ^+):

Theorem 4.2 *For any recursively enumerable language* L, $L \subseteq T^*$, *there exist two strictly minimal linear languages* $L_1, L_2 \subseteq \Gamma^*$ *and a projection* $h_T : \Gamma^* \to T^*$ *such that*

$$L = h_T \left(L_1 \cap L_2 \cap \Gamma^+ \right) = h_T \left((L_1 \setminus \{\lambda\}) \cap (L_2 \setminus \{\lambda\}) \right).$$

Proof. Let L be given according to the characterization given above in Proposition 2.1 as $h_T (h_1 (E(h_1, h_2)) \cap R)$, and R be given by the dfa $M = (Q, \Sigma_1, \delta, q_0, F)$. We now define the following two strictly minimal linear grammars G_1 and G_2 generating L_1 and L_2, respectively, in a similar way as already shown for the corresponding linear grammars in Theorem 4.1:

$$G_i = (\{S\}, \Gamma, P_i \cup \{S \to \lambda\}, S), \; i \in \{1, 2\},$$
$$\Gamma = \Sigma_1 \cup \tilde{\Sigma}_1 \cup Q \cup \{C, E, X, Y, Z\},$$
$$P_1 = \{S \to q_{i_{m+1}} \tilde{c}_m q_{i_m} \ldots q_{i_2} \tilde{c}_2 q_{i_1} q_{i_1} \tilde{c}_1 q_{i_0} Sb_1 CZb_2 C \ldots Zb_k CZ \mid a \in \Sigma_2,$$
$$h_1(a) = b_1 \ldots b_k, h_2(a) = c_1 \ldots c_m, \delta\left(q_{i_j}, c_j\right) = q_{i_{j+1}}, 0 \le j \le m\}$$
$$\cup \{S \to Eq_f SE \mid q_f \in F\} \cup \{S \to q_0 XSYYZ\},$$
$$P_2 = \{S \to qqSZ \mid q \in Q\} \cup \left\{S \to \tilde{b} SbC \mid b \in \Sigma_1\right\}$$
$$\cup \{S \to ESE, S \to XYSY\}.$$

The projection $g_T : \Gamma^* \to T^*$ is defined by $h_T(a) = \begin{cases} a, & \text{if } a \in T, \\ \lambda, & \text{if } a \in \Gamma \setminus T. \end{cases}$

The first production to be applied (only once) in G_1 is $S \to Eq_f SE$, the first one to be applied (only once) in G_2 is $S \to ESE$; in this way, the left delimiter E and the right delimiter E are introduced. The last step of a derivation before applying the final rule $S \to \lambda$ in G_1 is to apply $S \to q_0 XSYYZ$ and in G_2 is to apply $S \to XYSY$. These rules, which are the only ones involving the symbols X and Y, can also be applied exactly once in the last but one step of the derivations in G_1 and G_2, respectively, because otherwise the strings derived in G_1 and G_2 cannot fit together anyway (which is the reason for putting S at different positions of the substring XYY in the productions in P_1 and P_2, respectively).

The main work for obtaining the required representation is done by the rest of the productions in P_1 and P_2 in a similar way as described for the adjoining rules of the bidirectional sticker system constructed in the proof of Theorem 3.1. Finally, we have to take care of eliminating λ from L_1 and L_2 (e.g., by taking the intersection with Γ^+), because otherwise the empty word would be in L in any case. $\square$

The result of the derivation in a strictly minimal grammar can easily be represented by the control word of the derivation:

Let $G = (\{S\}, T, P \cup \{S \to \lambda\}, S)$ be a strictly minimal grammar with the productions in $P \cup \{S \to \lambda\}$ being labelled in a unique way such that

- $S \to \lambda$ is labelled by 0 and

- $P = \{k : S \to u_k S v_k \mid 1 \leq k \leq n\}$.

Hence the control word of a terminal derivation in G yielding x is of the form $w0$ with $w \in [1...n]^*$. The resulting terminal word x is obtained by considering the rules of w only, i.e., x can be written as $h^{(l)}(w) h^{(r)}\left(w^R\right)$, where

- $h^{(l)} : [1...n]^* \to T^*$ with $h^{(l)}(k) = u_k$, $1 \leq k \leq n$, and

- $h^{(r)} : [1...n]^* \to T^*$ with $h^{(r)}(k) = v_k$, $1 \leq k \leq n$.

From this representation of the words in $L(G)$ we immediately infer that there is a one-to-one correspondence between the control words $w \in [1...n]^*$ and the derivation of terminal words $x \in L(G)$ with x being represented as $h^{(l)}(w) h^{(r)}\left(w^R\right)$. By taking the barred version of the control symbols from $[1...n]$, we obtain another representation of x as $x = h(w) h\left(\overline{w}^R\right)$ with $h : \left\{k, \overline{k} \mid 1 \leq k \leq n\right\}^* \to T^*$ and $h(k) = h^{(l)}(k)$ and $h\left(\overline{k}\right) = h^{(r)}(k)$.

Using this representation of terminal words derived in a strictly minimal grammar by using the corresponding control word of the derivation, from Theorem 4.2 we infer the following result:

Corollary 4.1 *For any recursively enumerable language L, $L \subseteq T^*$, there exist an alphabet V, two morphisms $h_1 : V^* \to \Gamma^*$ and $h_2 : V^* \to \Gamma^*$ as well as a projection $h_T : \Gamma^* \to T^*$ such that $L = h_T(h_1(L_0) \cap h_2(L_0))$ where $L_0 = \left\{w\overline{w}^R \mid w \in V^+\right\}$.*

Proof. Starting from the representation given in Theorem 4.2, we mainly have to prove that we can represent the two strictly minimal languages L_1 and L_2 by $h_1(V^*)$ and $h_2(V^*)$, respectively, for some alphabet V. As already pointed out above, $L_i = h_i'(rcp([1...n_i]^*))$ for some morphisms $h_i' : [1...n_i]^* \to \Gamma^*$, $1 \le i \le 2$. In order to obtain a single control alphabet V, we define

$$V = \{k \mid 1 \le k \le n_1\} \cup \{n_1 + m \mid 1 \le m \le n_2\} = [1...(n_1 + n_2)]$$

as well as

$$h_1(j) = \begin{cases} h_1'(j) & \text{for } 1 \le j \le n_1, \\ \lambda & \text{for } n_1 + 1 \le j \le n_1 + n_2, \end{cases}$$

$$h_2(j) = \begin{cases} \lambda & \text{for } 1 \le j \le n_1, \\ h_2'(j - n_1) & \text{for } n_1 + 1 \le j \le n_1 + n_2. \end{cases}$$

From these definitions we obviously obtain $L_i = h_i(L_0')$, $1 \le i \le 2$, with $L_0' = rcp(V^*)$. Finally, we observe that for $L_0 = rcp(V^+)$ we obtain

$$(h_1(L_0') \setminus \{\lambda\}) \cap (h_2(L_0') \setminus \{\lambda\}) = h_1(L_0) \cap h_2(L_0),$$

which completes the proof. $\square$

As every natural number k can be represented by $10^k 1$, from the preceding representations of recursively enumerable languages we get the following representations based on a unique generator language by using deterministic generalized sequential machines:

Theorem 4.3 *For any recursively enumerable language L, $L \subseteq T^*$, there exist two dgsm mappings $g_1 : \{0,1\}^* \to \Gamma^*$ and $g_2 : \{0,1\}^* \to \Gamma^*$ as well as a projection $h_T : \Gamma^* \to T^*$ such that $L = h_T(g_1(L_0) \cap g_2(L_0))$, where either*

- $L_0 = \{w\overline{w}^R \mid w \in \{0,1\}^+\}$ *or even*

- $L_0 = \{ww^R \mid w \in \{0,1\}^+\}.$

Proof. For $L_0 = \{w\overline{w}^R \mid w \in \{0,1\}^+\}$, the desired representation of the given language L is immediately obtained from the representation proved in Corollary 4.1 by designing

$$g_i = (Q_i, \{0,1\}, \Gamma, s_0, \{s_f\}, \delta_i), \quad i \in \{1,2\},$$

in such a way that $s10^k1 \Rightarrow^*_{g_i} h_i(k) s_f$ for $s \in \{s_0, s_f\}$, i.e., by taking the control alphabet $V = [1...n]$ and the $h_i(k)$, $1 \le k \le n$, as in Corollary 4.1, we define

$$Q_i = \{s_k \mid 0 \le k \le n+3\}, \quad s_f = s_{n+2}, \quad s_t = s_{n+3},$$

as well as

$$\delta_i(s_0, 1) = \delta_i(s_f, 1) = (s_1, \lambda)$$

and

$$\delta_i(s_k, 0) = (s_{k+1}, \lambda), \quad \delta_i(s_{k+1}, 1) = (s_f, h_i(k)), \quad \text{for } 1 \le k \le n;$$

moreover, we have the following trap rules leading to the trap state s_t:

$$\delta_i(s_0, 0) = \delta_i(s_f, 0) = \delta_i(s_1, 1) = \delta_i(s_{n_i+1}, 0)$$
$$= \delta_i(s_t, 0) = \delta_i(s_t, 1) = (s_t, \lambda).$$

Yet the power of deterministic generalized sequential machines not only allows us to encode the control words over a given arbitrarily large control alphabet, but also can help to avoid the use of barred symbols. As pointed out above, control words representing valid derivations in the underlying strictly minimal linear grammars have the unique specific control label 0 indicating the use of the final production $S \to \lambda$, which information can be used to deterministically find the middle of the control word where we have to change from producing the left parts of the productions to producing the right ones. Hence, for $L_0 = \{ww^R \mid w \in \{0,1\}^+\}$ we have to design the g_i in the following way:

As long as no two symbols 1 occur in a sequence, the output of g_i is the left part of the production labelled by k, i.e., $s10^k1 \Rightarrow^*_{g_i} h_i^{(l)}(k) s'$, for $k \ge 1$ and some states $s, s' \in Q_i$. Whenever g_i detects a sequence 11 representing the label 0, i.e., the final production $S \to \lambda$, then another sequence 11 must occur and then the second part of the interpretation of the control word begins, where the output of g_i has to be the right part of the production, i.e., $s10^k1 \Rightarrow^*_{g_i} h_i^{(r)}(k) s'$ for $k \ge 1$ and some states $s, s' \in Q_i$. Observe that the maximum of the numbers k to be considered only depends on the number of productions not equal to $S \to \lambda$ occurring in the two production sets and therefore is bounded for a given language L. The details of the constructions of the g_i are obvious and therefore omitted. □

We just would like to mention that when using deterministic generalized sequential machines we can avoid the output of the empty word, which means that the preceding theorem also holds true for the following generator languages:

- $L_0 = \left\{ w\overline{w}^R \mid w \in \{0,1\}^* \right\}$

- $L_0 = \left\{ ww^R \mid w \in \{0,1\}^* \right\}$

4.2   Copy Languages

In the second part of this section we are going to show that any recursively enumerable language can also be represented by using copy languages instead of reversed copy languages. For the ground representation of recursively enumerable languages instead of minimal linear grammars we now take regular matrix grammars with axioms, which by using similar ideas as in Theorems 4.1 and 4.2 allows us to obtain the following representations of recursively enumerable languages:

Theorem 4.4 *For any recursively enumerable language L, $L \subseteq T^*$, there exist two languages $L_1, L_2 \subseteq \Gamma^*$ generated by regular matrix grammars with axioms G_1 and G_2, respectively, as well as a projection $h_T : \Gamma^* \to T^*$ such that*
$$L = h_T \left(L_1 \cap L_2 \cap \Gamma^+ \right) = h_T \left((L_1 \setminus \{\lambda\}) \cap (L_2 \setminus \{\lambda\}) \right).$$

Proof. Let L be given according to the characterization given above in Proposition 2.1 as $h_T \left(h_1 \left(E \left(h_1, h_2 \right) \right) \cap R \right)$, and R be given by the dfa $M = (Q, \Sigma_1, \delta, q_0, F)$. We now define the following two regular matrix grammars with axioms G_1 and G_2 :

$$G_i = (\{LR\}, \Gamma, P_i \cup \{[L \to \lambda, R \to \lambda]\}, \{LR\}),\ i \in \{1, 2\},$$
$$\Gamma = \Sigma_1 \cup \tilde{\Sigma}_1 \cup Q \cup \{C, E, Z\},$$
$$P_1 = \{[L \to q_{i_0} \tilde{c}_1 q_{i_1} q_{i_1} \tilde{c}_2 q_{i_2} \ldots q_{i_m} \tilde{c}_m q_{i_{m+1}} L,$$
$$\qquad R \to b_1 C Z b_2 C \ldots Z b_k C Z R \mid a \in \Sigma_2,$$
$$\qquad\quad h_1 \left(a \right) = b_1 \ldots b_k, h_2 \left(a \right) = c_1 \ldots c_m, \delta \left(q_{i_j}, c_j \right) = q_{i_{j+1}}, 0 \le j \le m \}$$
$$\qquad \cup \{[L \to E q_0 L, R \to q_f E Z R] \mid q_f \in F\},$$
$$P_2 = \{[L \to qq, R \to Z] \mid q \in Q\} \cup \left\{ \left[L \to \tilde{b}, R \to bC \right] \mid b \in \Sigma_1 \right\}$$
$$\qquad \cup \{[L \to EL, R \to ER]\}.$$

The projection $g_T : \Gamma^* \to T^*$ is defined by $h_T(a) = \begin{cases} a, & \text{if } a \in T, \\ \lambda, & \text{if } a \in \Gamma \setminus T. \end{cases}$

The left part of the languages L_1 and L_2, which in the proofs of Theorems 4.1 and 4.2 was generated in reversed order, now is generated in the "normal" direction. In order to obtain a non-empty word in the intersection of $L(G_1)$ and $L(G_2)$, the first matrix to be applied in G_1 has to be $[L \to Eq_0L, R \to q_fEZR]$ and the first matrix to be applied in G_2 has to be $[L \to EL, R \to ER]$; in both cases, the last matrix to be applied is $[L \to \lambda, R \to \lambda]$. □

We should like to mention that it is easy to construct the regular matrix grammars with axiom(s) G_1 and G_2 in such a way that even

$$L = h_T(L(G_1) \cap L(G_2)).$$

as shown for minimal linear grammars in Theorem 4.1; the details of the proof can easily be deducted from the preceding proof.

As for minimal linear grammars we observe that the result of the derivation in the regular matrix grammars with axioms G_1 and G_2 constructed in Theorem 4.4 can easily be represented by the corresponding control words:
For $G_i = (\{LR\}, \Gamma, P_i \cup \{[L \to \lambda, R \to \lambda]\}, \{LR\})$, $i \in \{1, 2\}$, let the productions in $P_i \cup \{[L \to \lambda, R \to \lambda]\}$ be labelled in a unique way such that

- $[L \to \lambda, R \to \lambda]$ is labelled by 0 and

- $P_i = \{k : [L \to u_kL, R \to v_kR] \mid 1 \le k \le n_i\}.$

Hence the control word of a terminal derivation in G_i yielding x is of the form $w0$ with $w \in [1...n]^*$. The resulting terminal word x is obtained by considering the rules of w only, i.e., x can be written as $h^{(l)}(w)h^{(r)}(w)$, where

- $h^{(l)} : [1...n]^* \to T^*$ with $h^{(l)}(k) = u_k$, $1 \le k \le n$, and

- $h^{(r)} : [1...n]^* \to T^*$ with $h^{(r)}(k) = v_k$, $1 \le k \le n$.

From this representation of the words in $L(G)$ we immediately infer that there is a one-to-one correspondence between the control words $w \in [1...n]^*$ and the derivations of terminal words $x \in L(G)$ with x being represented as $h^{(l)}(w)h^{(r)}(w)$. By taking the barred version of the control symbols from $[1...n]$, we obtain the representation of x as $x = h(w)h(\overline{w})$ with $h : \{k, \overline{k} \mid 1 \le k \le n\}^* \to T^*$ and $h(k) = h^{(l)}(k)$ and $h(\overline{k}) = h^{(r)}(k)$. Hence, from Theorem 4.4 we immediately infer the following result:

Corollary 4.2 *For any recursively enumerable language L, $L \subseteq T^*$, there exist an alphabet V, two morphisms $h_1 : V^* \to \Gamma^*$ and $h_2 : V^* \to \Gamma^*$ as well as a projection $h_T : \Gamma^* \to T^*$ such that $L = h_T \left(h_1 \left(L_0 \right) \cap h_2 \left(L_0 \right) \right)$ where $L_0 = \{ w\overline{w} \mid w \in V^+ \}$.*

By again taking the coding of a natural number k by $10^k 1$, from the preceding representation of recursively enumerable languages we get the following representations based on unique copy languages by using deterministic generalized sequential machines (compare Theorem 4.3):

Theorem 4.5 *For any recursively enumerable language L, $L \subseteq T^*$, there exist two dgsm mappings $g_1 : \{0,1\}^* \to \Gamma^*$ and $g_2 : \{0,1\}^* \to \Gamma^*$ as well as a projection $h_T : \Gamma^* \to T^*$ such that $L = h_T \left(g_1 \left(L_0 \right) \cap g_2 \left(L_0 \right) \right)$, where either*

- $L_0 = \left\{ w\overline{w} \mid w \in \{0,1\}^+ \right\}$,

- $L_0 = \left\{ ww \mid w \in \{0,1\}^+ \right\}$,

- $L_0 = \{ w\overline{w} \mid w \in \{0,1\}^* \}$, *or*

- $L_0 = \{ ww \mid w \in \{0,1\}^* \}$.

Acknowledgements. During the Workshop on Molecular Computing held as a Summer School Course of the Black Sea University in Mangalia in August 1997, a lot of fruitful discussions with the participants gave rise to new ideas. Among the colleagues having influenced the development of this paper, I should like to mention Somenath Biswas, Hendrik Jan Hoogeboom, Valeria Mihalache, and Gheorghe Păun.

References

[1] L. M. Adleman, Molecular computation of solutions to combinatorial problems, *Science*, 226 (1994), 1021 – 1024.

[2] J. Dassow, Gh. Păun, *Regulated Rewriting in Formal Language Theory*, Springer-Verlag, Berlin, Heidelberg, 1989.

[3] R. Freund, Gh. Păun, G. Rozenberg, A. Salomaa, Bidirectional sticker systems, *Third Annual Pacific Symposium on Biocomputing*, Hawaii, January 1998.

[4] L. Kari, Gh. Păun, G. Rozenberg, A. Salomaa, S. Yu, DNA computing, sticker systems, and universality, *Acta Informatica*, in press.

[5] Gh. Păun, Some open problems in molecular computing, manuscript for the *Workshop on Molecular Computing,* Mangalia, 1997.

[6] G. Rozenberg, A. Salomaa, eds., *Handbook of Formal Languages,* Springer-Verlag, Berlin, 1997.

Universality Results for Finite H Systems and for Watson-Crick Finite Automata[1]

Carlos MARTÍN-VIDE

Research Group on Mathematical Linguistics and Language Engineering
Rovira i Virgili University
Pça. Imperial Tàrraco, 1, 43005 Tarragona, Spain
cmv@astor.urv.es

Gheorghe PĂUN

Institute of Mathematics of the Romanian Academy
PO Box 1 – 764, 70700 Bucureşti, Romania
gpaun@imar.ro

Grzegorz ROZENBERG

Department of Computer Science, Leiden University
PO Box 9512, 2300 RA Leiden, The Netherlands
rozenber@wi.leidenuniv.nl

Arto SALOMAA

Academy of Finland and Department of Mathematics, Turku University
20014 Turku, Finland
asalomaa@utu.fi

Abstract. Because "computers" based on uncontrolled splicing with respect to a finite set of splicing rules can only compute at the level of finite automata, it is of interest to have universality results at this level of computability. We find finite automata which are universal for the class of finite automata with a bounded number of states (and of input symbols). This directly leads to universal extended H systems with finite sets of rules applied without any additional control. Another "implementation" of this result is in terms of so-called Watson-Crick finite automata, a kind of devices similar to finite automata

[1]Research supported by the Direcció General de Recerca, Generalitat de Catalunya (PIV), by the Academy of Finland – Project 137358, and the National Institute for Biological Research, Bucharest, Romania

and working on double stranded tapes. Also Watson-Crick automata which are universal for the class of Watson-Crick finite automata with a bounded number of states and of symbols are found. The constructions are significantly simpler when one of the heads of the Watson-Crick automaton works in a two-way manner (the other works in the usual left-to-right one-way manner) and when one uses a triple-stranded tape (on one of them we work in the two-way manner).

1 Introduction

The starting point of our investigations here are the following two important results in DNA computing based on the splicing operation: (1) when we use a finite set of splicing rules, applied in the free manner, without any control on their use, then we get only regular languages (*all* regular languages if we also use a squeezing mechanism in the form of a terminal alphabet); however, (2) when a regular set of splicing rules is applied in the free manner, or a finite set of rules are applied under certain conditions (permitting or forbidding contexts, target languages, priorities, time-varying, and so on; see, for instance, [11], [12], [15], [17]), then we get a characterization of recursively enumerable languages. The proofs are constructive; in the last case, starting from universal type-0 Chomsky grammars or from universal Turing machines, we obtain universal H systems with finite sets of rules used in a controlled manner. Always in the cases investigated up to now, the control is of a type as known in the formal language theory, for instance in the regulated rewriting area, or in grammar system theory. That is, at least in this moment, there is no practical way to implement such a control in a reliable and efficient way.

Free splicing is a "natural" operation, hence we can hope to use it as a basis for computing devices. Yet, if we have to confine ourselves to the level of finite automata (regular languages), then at least we have to find universal computers of this type. However, one knows no universality result for finite automata. It is fairly evident that a one-head one-tape (one-way or even two-ways) finite automaton cannot be universal for the class of all finite automata in the usual sense: M_u is universal if it accepts an input string $code(M)x$ if and only if M, any specific finite automaton, accepts x, where $code(M)$ is an encoding of M. Two reasons support this. First, one cannot encode the very way of using a finite automaton in terms of a finite automaton (we have to remember symbols in the input string without marking them, and this cannot be done with a finite set of states). Second, we cannot codify the states of any given finite automaton in such a way that the finite set of states of the universal automaton can handle them (an arbitrarily large set of states will lead to arbitrarily long codes, hence again the information carried by them cannot be handled by a finite set of states). More technically: if M_u has n states and it can accept, say, $code(M)a^{n+1}$, then, because of obvious pumping properties, it will accept all strings $code(M)a^{ki+p}$, for some $k \geq 1, p \geq 0$, for all $i \geq 0$, irrespective whether or not M accepts $a^{ki+p}, i \geq 0$.

Then, we reduce our ambitions: let us consider the class of finite automata with at most n states and at most m symbols in the input alphabet. Can we find a universal finite automaton for this class ? The answer is affirmative ! The matter relies, in fact, in the mode of defining the universality. If we work with a rather complex "program" to be run on the universal automaton (as usual, the "program" contains a code of the automaton we want to simulate and an input string, but this time these two are shuffled in a way we will specify below), then the universality is obtained. What is interesting (and pleasant for DNA computing based on splicing) is that this universal finite automaton leads directly to a universal extended finite H system which produces exactly the strings we want to recognize, cleaned of all additional codes, markers, etc.

Another elegant "implementation" of the universal finite automaton is in terms of Watson-Crick finite automata, recently introduced in [7]. Such machineries are finite automata with the tape consisting of a double stranded sequence; each strand is scanned by a separate head. Details will be given in Section 5 below. Writing the code of the particular finite automaton on a strand and the string to be recognized on the other one, we get a Watson-Crick finite automaton which is universal in the restricted sense described above.

What then about Watson-Crick finite automata which are universal for the Watson-Crick finite automata themselves ? When we consider only the restricted class of automata with a bounded number of states and of input symbols, then, as expected, universality results are obtained. An easy way to write such a machinery (easy in the sense that the input string is of a low complexity) is by using triple-stranded tapes, one of them used in the two-way manner. It is interesting to note that there are proteins whose molecules are triple-stranded: collagen is an example. (Unfortunately, the collagen cannot be handled in such a nice way like DNA; in particular, one knows no splicing-like operation for it.)

2 Some Language and Automata Theory Prerequisites

We recall here some notions, notations and results needed in the sequel; we refer to [20] for further details.

For an alphabet V, we denote by V^* the free monoid generated by V under the operation of concatenation; the empty string is denoted by λ. Moreover, $V^+ = V^* - \{\lambda\}$ and $|x|$ is the length of $x \in V^*$. By *FIN, REG, RE* we denote the families of linear, regular, and recursively enumerable languages, respectively.

A *finite automaton* will be presented either in the form $M = (K, V, s_0, F, \delta)$, where K is the set of states, V is the alphabet, s_0 is the initial state, F is the set of final (accepting) states, and $\delta : K \times V \longrightarrow 2^K$ is the transition mapping, or in the form $M = (K, V, s_0, F, P)$, where the components K, V, s_0, F are as above and $P \subseteq KV \times VK$ is the set of transition rules written as rewriting rules: $s' \in \delta(s, a)$ is equivalent with $sa \to as'$. For $s, s' \in K, a \in V, x, x' \in V^*$ we write $xsax' \Longrightarrow xas'x'$ iff $sa \to as' \in P$. The language recognized by M is defined by

$$L(M) = \{x \in V^* \mid s_0 x \Longrightarrow^* xs_f, \text{ for some } s_f \in F\}.$$

3 Universal Finite Automata

Let us consider a sort of "partial universality" for finite automata.

Take an alphabet V and a set K of states.

We construct the following finite automaton

$$M_u = (K_u, V \cup K \cup \{c_1, c_2\}, q_{0,u}, F_u, P_u),$$

where

$$K_u = \{q_{0,u}, q'_{0,u}, q''_{0,u}, q_f\}$$
$$\cup \{[s], (s), (s)', (s)'', \overline{(s)}, \overline{(s)}', \overline{(s)}'', \overline{(s)}''' \mid s \in K\}$$
$$\cup \{[sa], [sas'], [sas']', [sas']'', [sas']''', [sas']^{iv} \mid s, s' \in K, a \in V\},$$
$$F_u = \{q_f\},$$

and P_u contains the following transition rules:

1. $q_{0,u}s \to sq'_{0,u}, \ s \in K,$

      $q'_{0,u}a \to aq''_{0,u}, \ a \in V,$

      $q''_{0,u}s \to sq_{0,u}, \ s \in K,$

2. $q_{0,u}s_0 \to s_0[s_0],$

      $[s_0]a \to a[s_0a], \ a \in V,$

      $[s_0a]s \to s[s_0as], \ s \in K, a \in V,$

3. $[sas']s'' \to s''[sas']', \ s, s', s'' \in K, a \in V,$

      $[sas']'b \to b[sas']'', \ s, s' \in K, a, b \in V,$

      $[sas']''s'' \to s''[sas'], \ s, s', s'' \in K, a \in V,$

      $[sas']c_1 \to c_1[sas']''', \ s, s' \in K, a \in V,$

      $[sas']'''s'' \to s''[sas']''', \ s, s', s'' \in K, a \in V,$

      $[sas']'''c_2 \to c_2[sas']^{iv}, \ s, s' \in K, a \in V,$

4. $[sas']^{iv}a \to a(s'), \ s, s' \in K, a \in V,$

5. $(s)s' \to s'(s)', \ s, s' \in K,$

      $(s)'a \to a(s)'', \ s \in K, a \in V,$

      $(s)''s' \to s'(s), \ s, s' \in K,$

6. $(s)s \to s[s], \ s \in K,$

      $[s]a \to a[sa], \ s \in K, a \in V,$

      $[sa]s' \to s'[sas'], \ s, s' \in K, a \in V,$

7. $(s)s' \to s'\overline{(s)}', \ s, s' \in K,$

      $\overline{(s)}s' \to s'\overline{(s)}', \ s, s' \in K,$

      $\overline{(s)}'a \to a\overline{(s)}'', \ s \in K, a \in V,$

      $\overline{(s)}''s' \to s'\overline{(s)}, \ s, s' \in K,$

      $\overline{(s)}c_1 \to c_1\overline{(s)}''', \ s \in K,$

      $\overline{(s)}'''s' \to s'\overline{(s)}''', \ s, s' \in K, s \neq s',$

      $\overline{(s)}'''s \to sq_f, \ s \in F,$

      $q_f s \to sq_f, \ s \in K,$

      $q_f c_2 \to c_2 q_f.$

This automaton is universal for the class of finite automata of the form $M = (K', V', s_0, F, P)$ with $K' \subseteq K, V' \subseteq V$, in the following sense:

For a finite automaton $M = (K', V', s_0, F, P)$ with $V' \subseteq V, K' \subseteq K$, we consider the string

$$code(M) = s_1 a_1 s_1' s_2 a_2 s_2' \ldots s_n a_n s_n' c_1 s_{f1} s_{f2} \ldots s_{fm} c_2,$$

where $s_i a_i \to a_i s_i' \in P, 1 \le i \le n$, each string $s_i a_i s_i'$ appears only once, $s_{f1}, s_{f2}, \ldots, s_{fm}$ are the elements of F, and c_1, c_2 are new symbols.

For two strings $z, x \in V^*$ with $x = a_1 a_2 \ldots a_p, a_i \in V, 1 \le i \le p$, we define the *block shuffle operation* of z, x by

$$bls(z, x) = z a_1 z a_2 \ldots z a_p z.$$

Then,

$$bls(code(M), x) \in L(M_u) \text{ iff } x \in L(M).$$

Indeed, M_u works as follows: in the initial state $q_{0,u}$ and in each state (s), one parses $code(M)$ in such a way that some blocks $s_i a_i s_i'$ are skipped, then one block of this type is memorized (when starting, we must have $s_i = s_0$) in the state $[s_i a_i s_i']$, then further blocks $s_j a_j s_j'$ are skipped after passing also over $c_1 s_{f_1} \ldots s_{f_m} c_2$ (thus reaching the state $[s_i a_i s_i']^{iv}$); the rule $s_i a_i \to a_i s_i'$ of M is simulated by a rule of type 4, returning to a state of the form (s); the process can be iterated; using rules in group 7, we reach q_f only when we have simulated in M_u a parsing in M ending in a state of F.

Thus, we may state the following result.

Theorem 3.1 *There are universal finite automata, in the sense defined above, for the class of finite automata with a bounded number of states and a bounded number of input symbols.*

It is now clear why we have called this universality "partial": M_u is universal only for finite automata with the alphabet being a subset of V and the set of states being a subset of K; moreover, the input string of M_u is the complex $bls(code(M), x)$, containing $|x| + 1$ times the code of M.

The fact that a fixed maximal set of input symbols is used is not restrictive: we may work with only two symbols, codifying in a suitable way the symbols of an arbitrary alphabet we want to use. This amounts at preprocessing the input, codifying it, and postprocessing the result of the computation of M_u (decodification). On the contrary, the limitation of the number of states is restrictive: one knows that the number of states induces an infinite hierarchy of regular languages, for each integer n there is a language L_n which can be recognized by a finite automaton with n states but not by automata with less than n states.

4   Universal Finite H Systems

The previous (partially) universal finite automaton can be easily implemented in terms of extended H systems. We start by defining this basic notion of DNA computing.

Consider an alphabet V and two special symbols, $\#, \$$, not in V.

A *splicing rule* over V is a string $u_1\#u_2\$u_3\#u_4$, where $u_1, u_2, u_3, u_4 \in V^*$. For a splicing rule $r = u_1\#u_2\$u_3\#u_4$ and four strings $x, y, w, z \in V^*$ we write

$$(x, y) \vdash_r (w, z) \quad \text{iff} \quad x = x_1u_1u_2x_2, \ y = y_1u_3u_4y_2,$$
$$w = x_1u_1u_4y_2, \ z = y_1u_3u_2x_2,$$
$$\text{for some } x_1, x_2, y_1, y_2 \in V^*.$$

We say that we *splice* the strings x, y at the *sites* u_1u_2, u_3u_4, respectively.

A pair $\sigma = (V, R)$, where V is an alphabet and R is a set of splicing rules over V is called an *H scheme*. With respect to a splicing scheme $\sigma = (V, R)$ and a language $L \subseteq V^*$ we define

$$\sigma(L) = \{w \in V^* \mid (x, y) \vdash_r (w, z) \text{ or } (x, y) \vdash_r (z, w), \text{ for } x, y \in L, r \in R\},$$
$$\sigma^0(L) = L,$$
$$\sigma^{i+1}(L) = \sigma^i(L) \cup \sigma(\sigma^i(L)), \ i \geq 0,$$
$$\sigma^*(L) = \bigcup_{i \geq 0} \sigma^i(L).$$

An *extended H system* is a construct $\gamma = (V, T, A, R)$, where V is an alphabet, $T \subseteq V, A \subseteq V^*$, and $R \subseteq V^*\#V^*\$V^*\#V^*$. ($T$ is the *terminal alphabet*, A is the set of *axioms*, and R is the set of *splicing rules*.) When $T = V$, the system is said to be non-extended. The pair $\sigma = (V, R)$ is the *underlying H scheme* of γ.

The language generated by γ is defined by $L(\gamma) = \sigma^*(A) \cap T^*$. (We iterate the splicing operation according to rules in R, starting from strings in A, and we keep only the strings composed of terminal symbols.)

We denote by $EH(F_1, F_2)$ the family of languages generated by extended H systems $\gamma = (V, T, A, R)$, with $A \in F_1, R \in F_2$, where F_1, F_2 are two given families of languages. (Note that R is a language, hence the definition makes sense.)

Two basic results concerning the generative power of extended H systems are the following.

Theorem 4.1 $EH(FIN, FIN) = REG.$

Theorem 4.2 $EH(FIN, REG) = RE$.

The inclusion $EH(FIN, FIN) \subseteq REG$ is proved in [5], [18], the inclusion $REG \subseteq EH(FIN, FIN)$ is proved in [14]. Theorem 4.2 is proved in [11].

By starting the proof of the inclusion $REG \subseteq EH(FIN, FIN)$ in [14] from the universal finite automaton in the proof of Theorem 3.1, we obtain an extended H system with finite components equivalent with the universal one. However, as we have seen in the section above, on the tape of the universal automaton M_u there remain both the string x and the $|x| + 1$ copies of $code(M)$, where M is the particular finite automaton which we want to simulate.

We cannot directly simulate M_u by an extended H system with finite components in such a way to obtain the string x without any other additional symbol. However, we can do this, thus obtaining a "partially universal" extended H system with finite components if we start not from $bls(code(M), x)$ as above, but from a string containing one more copy of x added at the end of the string: using $bls(code(M), x)$ we check whether or not $x \in L(M)$ (as above) and only in the affirmative case we remove all auxiliary symbols, producing the terminal string x. Here is such a universal H system associated with $M_u = (K_u, V \cup K \cup \{c_1, c_2\}, q_{0,u}, F_u, P_u)$:

$$\gamma_u = (W, V, A_u, R_u),$$

with

$$
\begin{aligned}
W &= V \cup K \cup K_u \cup \{B_0, B, E, Z, c_1, c_2\}, \\
A_u &= \{BqZ \mid q \in K_u\} \cup \{ZZ\}, \\
R_u &= \{Bq_{0,u}\#Z\$B_0\#\} \\
&\quad \cup \{Bq'\#Z\$Bq\alpha\# \mid q\alpha \to \alpha q' \in P_u, \text{ for } q, q' \in K_u, \\
&\qquad \alpha \in V \cup K \cup \{c_1, c_2\}\} \\
&\quad \cup \{\#ZZ\$BqE\# \mid q \in K\}.
\end{aligned}
$$

If we add to A_u the axiom

$$w_0 = B_0 \, bls(code(M), x) E x,$$

then we get an extended H system $\gamma'_u = (W, V, A'_u, R_u)$ such that $L(\gamma'_u) = L(M)$. Indeed, the only way to obtain a string in V^*, that is without symbols in $K \cup K_u \cup \{B_0, B, E, Z, c_1, c_2\}$, is to simulate the rules in P_u on the prefix B_0 $bls(code(M), x)E$ of the axiom w_0, step by step, from the left (this is ensured by the fact that all splicing rules require the presence of B_0 or of B), until reducing this string to BqE, for some $q \in K$ (in fact, by the mode of work of M_u we have $q \in F_u$); then also the block BqE can be removed by the splicing rule $\#ZZ\$BqE\#$ and the string x remains.

We also formulate the important (for DNA computing) conclusion of this discussion in the form of a theorem.

Theorem 4.3 *There are extended H systems with finite sets of rules which are universal for the class of finite automata with a bounded number of states and a bounded number of input symbols.*

Note that M is "run" on the "computer" γ_u via the "program" w_0, which is a unique string (associated with both M and x, hence it contains both the "algorithm" and the "input data"). This "program" is not very simple/short, it even has a non-context-free character, because of the presence of several copies of the code of M. However, the string $w_0 = B_0$ $bls(code(M), x)Ex$ can be generated from simpler strings by splicing: construct the string $z = code(M)$, produce copies of it (by amplification), then produce arbitrarily many strings of the form $X_i z X_i', i \geq 1$ (in fact, we need exactly $n + 2$ such copies, where $n = |x|$). If $x = a_1 a_2 \ldots a_n, a_i \in V, 1 \leq i \leq n$, consider also the strings $Y_i a_i Y_i', 1 \leq i \leq n$. Finally, consider the strings $B_0 Z_1, Z_2 Ex$. It is now a simple task to devise splicing rules which can build the string $w_0 = B_0$ $bls(code(M), x)Ex$ starting from the blocks mentioned above; the symbols $X_i, X_i', Y_i, Y_i', Z_1, Z_2$ can control the operations in such a way that when none of them is present in a string, then that string is equal to w_0. We leave this task to the reader and we pass now to Watson-Crick automata.

5 Watson-Crick Finite Automata

First, let us fix some new notation and terminology.

Consider an alphabet V and a symmetric relation ρ on V, $\rho \subseteq V \times V$ (of *complementarity*, as a model of the Watson-Crick complementarity).

Besides the monoid V^*, of strings over V, we associate to V also the product monoid, $V^* \times V^*$. In accordance with the way of representing DNA molecules, where one considers the two strands as placed one over the other, we write the elements (x_1, x_2) of $V^* \times V^*$ in the form $\begin{pmatrix} x_1 \\ x_2 \end{pmatrix}$. Therefore, the concatenation of two pairs $\begin{pmatrix} x_1 \\ x_2 \end{pmatrix}, \begin{pmatrix} y_1 \\ y_2 \end{pmatrix}$ is $\begin{pmatrix} x_1 y_1 \\ x_2 y_2 \end{pmatrix}$. We write $\begin{pmatrix} V^* \\ V^* \end{pmatrix}$ instead of $V^* \times V^*$.

We also denote:

$$\begin{bmatrix} V \\ V \end{bmatrix}_\rho = \{ \begin{bmatrix} a \\ b \end{bmatrix} \mid a, b \in V, (a, b) \in \rho \}, \text{ and}$$

$$WK_\rho(V) = \begin{bmatrix} V \\ V \end{bmatrix}_\rho^+.$$

The set $WK_\rho(V)$ is called the *Watson-Crick domain* associated to V and ρ. The elements $\begin{bmatrix} w_1 \\ w_2 \end{bmatrix} \in WK_\rho(V)$ are called well-formed double stranded sequences, or simply *double stranded sequences*, or *molecules*, in order to remind the reality they are modeling. When used as input or output sequences for automata, such objects are also called *Watson-Crick tapes*. The two component strings, w_1, w_2, are called strands, the upper strand and the lower one, respectively.

Note that $\begin{bmatrix} \lambda \\ \lambda \end{bmatrix}$ is not a molecule (it has no biochemical counterpart). However, $\begin{pmatrix} \lambda \\ \lambda \end{pmatrix}$ is the identity of $\begin{pmatrix} V^* \\ V^* \end{pmatrix}$, it is identified with λ and, when not significant in a given context, it is not explicitly written.

We shall also use below "incomplete molecules" from the set

$$W_\rho(V) = \begin{bmatrix} V \\ V \end{bmatrix}_\rho^* (\begin{pmatrix} \lambda \\ V^* \end{pmatrix} \cup \begin{pmatrix} V^* \\ \lambda \end{pmatrix}).$$

The elements $\begin{bmatrix} w_1 \\ w_2 \end{bmatrix} \begin{pmatrix} x_1 \\ x_2 \end{pmatrix}$ of $W_\rho(V)$ with $\begin{bmatrix} w_1 \\ w_2 \end{bmatrix} \in WK_\rho(V)$ are called *well-started double stranded* sequences. Clearly, $WK_\rho(V) \subset W_\rho(V)$.

Based on the idea of Watson-Crick complementarity and the operations defined above, a language generating mechanism has been defined in [9], [13], called a *sticker system*. This is a completely different mechanism from those in [21], [22], but essentially based on the annealing operation also used in [1], [10], [21], [22]: one starts from a finite set of well-started sequences and one builds molecules by iteratively sticking upper blocks, lower blocks, or "dominoes" of arbitrary forms, according to certain restrictions. In general, either characterizations of regular languages or characterizations of recursively enumerable languages are obtained, depending on the way of controlling the operation and on the squeezing mechanisms used (see [9], [13]). The Watson-Crick automata, introduced in [7], are automata counterparts of these grammar-like devices.

A *Watson-Crick finite automaton* is a construct

$$M = (K, V, \rho, s_0, F, \delta),$$

where K and V are disjoint finite alphabets, $\rho \subseteq V \times V$ is a symmetric relation, $s_0 \in K$, $F \subseteq K$, and $\delta : K \times V^* \times V^* \longrightarrow 2^K$ is a mapping such that $\delta(s, x, y) \neq \emptyset$ only for finitely many triples $(s, x, y) \in K \times V^* \times V^*$.

The elements of K are called *states*, V is the (input) alphabet, ρ is a complementarity relation on V, s_0 is the initial state, F is the set of final states, and δ is the transition mapping; the interpretation of $s' \in \delta(s, x_1, x_2)$ is: in state s, the automaton passes over x_1 in the upper level strand and over x_2 in the lower level strand of a double stranded sequence, and enters the state s'. A transition determined by $s' \in \delta(s, x, y)$ can also be written in the form $s \begin{pmatrix} x \\ y \end{pmatrix} \to \begin{pmatrix} x \\ y \end{pmatrix} s'$, with the same meaning.

For $\begin{pmatrix} u_1 \\ u_2 \end{pmatrix}, \begin{pmatrix} v_1 \\ v_2 \end{pmatrix}, \begin{pmatrix} w_1 \\ w_2 \end{pmatrix} \in \begin{pmatrix} V^* \\ V^* \end{pmatrix}$ such that $\begin{bmatrix} u_1 v_1 w_1 \\ u_2 v_2 w_2 \end{bmatrix} \in WK_\rho(V)$, and $s, s' \in K$, we write

$$\begin{pmatrix} v_1 \\ v_2 \end{pmatrix} s \begin{pmatrix} u_1 \\ u_2 \end{pmatrix} \begin{pmatrix} w_1 \\ w_2 \end{pmatrix} \Longrightarrow \begin{pmatrix} v_1 \\ v_2 \end{pmatrix} \begin{pmatrix} u_1 \\ u_2 \end{pmatrix} s' \begin{pmatrix} w_1 \\ w_2 \end{pmatrix} \quad \text{if } s' \in \delta(s, u_1, u_2).$$

The language recognized by a Watson-Crick finite automaton is defined by

$$L(M) = \{ w_1 \in V^* \mid s_0 \begin{bmatrix} w_1 \\ w_2 \end{bmatrix} \Longrightarrow^* \begin{bmatrix} w_1 \\ w_2 \end{bmatrix} s_f, \text{ for } s_f \in F,$$
$$\text{and } w_2 \in V^*, \begin{bmatrix} w_1 \\ w_2 \end{bmatrix} \in WK_\rho(V) \}.$$

(Other languages are also defined in [7] – of molecules, of strings in the lower level strand, of control words associated to computations – but we do not consider them here.)

We emphasize the fact that the work of Watson-Crick automata is defined for elements of $WK_\rho(V)$ only, that is for double stranded sequences of elements in V paired according to the complementarity relation ρ. We can represent such a machine as consisting of a double tape on which an element of $WK_\rho(V)$ is written, a finite memory able to store a state from a finite set of states, and two read only heads, one of them scanning the upper level and the other one scanning the lower level of the tape. One starts with the two heads placed in front of the first symbol of each level, in state s_0. The two heads move to the right, according to the current state of the machine, as indicated by the transition mapping (the transition rules); a transition means to move the two heads across blocks defined by the transition rules. One stops and accepts the initial sequence when both heads reach the right hand end of the sequence written on the tape, entering a final state. Figure 1 illustrates this representation.

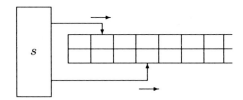

Figure 1: A Watson-Crick finite automaton

Several variants of Watson-Crick finite automata are introduced in [7], but we do not consider them here (*stateless*, if $K = F = \{s_0\}$; *all-final*, if $F = K$; *simple*, if for all (s, x_1, x_2) such that $\delta(s, x_1, x_2) \neq \emptyset$ we have either $x_1 = \lambda$ or $x_2 = \lambda$).

We denote by UWK the family of languages $L(M)$ recognized by unrestricted (U) Watson-Crick finite automata.

The Watson-Crick finite automata – as well as the sticker systems – make an essential use of the intrinsic computational completeness of DNA as pointed out in [19]: due to the Watson-Crick complementarity, there is a direct isomorphism between the set of DNA molecules and the so-called twin-shuffle language (over the alphabet $\{0, 1\}$), which characterizes the recursively enumerable languages modulo a sequential transducer (see [6]). Thus, the following result is as expected; a proof can be found in [7].

Theorem 5.1 *Each recursively enumerable language is the weak coding of a language in the family UWK.*

6 A Watson-Crick Finite Automaton which is Universal for Finite Automata

In view of the result in Theorem 5.1, it is of a clear interest to find universal Watson-Crick finite automata: modulo a weak coding, they will be universal for the whole class of Turing machines.

We postpone the discussion about this subject for the next section and we remark here that a Watson-Crick finite automaton can be an elegant implementation of a universal finite automaton as that in the proof of Theorem 3.1. Let $M = (K, V, s_0, F, P)$ be a finite automaton and x be a string in V^*. Consider the string $w = code(M)^n$, where $n = |x|$, as well as the string xc^m, for $m = |w| - |x|$ (that is, $|w| = |xc^m|$). Consider the relation $\rho = (K \cup V \cup \{c, c_1, c_2\}) \times (K \cup V \cup \{c, c_1, c_2\})$ (the total relation). Write the string xc^m on the upper strand and the string w on the lower strand of a Watson-Crick tape. The way of working of the universal finite automaton in the proof of Theorem 3.1 suggests a way of defining the transitions of a Watson-Crick automaton which works on the Watson-Crick tape mentioned above and recognizes the strings xc^m if and only if $x \in L(M)$: the lower head scans a copy of the code of M in the lower strand of the tape, nondeterministically chooses a move $sa \rightarrow as'$ of M, according to it the occurrence of the symbol a read by the upper head is scanned, and the state of the Watson-Crick automaton checks the correct linking of states of M, memorizing them. When reaching the first occurrence of the symbol c in the upper strand (there is at least one such an occurrence), a final state of M must be reached, in order to correctly finish the parsing. We leave the straightforward details of the construction to the reader (see also the construction below).

This time, the "program" of M is rather simple, a sequence of $|x|$ copies of a single string, the code of M, separated from the "input data" (the string x). However, there are here two drawbacks: the length of this "program" is still rather large, while the string in the first strand, that recognized by our universal machine, ends with a tail of symbols c which is also rather long. Both these drawbacks can be eliminated if we allow the lower head of the Watson-Crick automaton to move in both directions. Then one copy of $code(M)$ is enough, hence the tape can be of the form $\begin{bmatrix} xc^m \\ code(M)c^p \end{bmatrix}$, where m and p are integers such that $m \geq 1$ and $|xc^m| = |code(M)c^p|$ (at least one occurrence of c is present in the upper strand, to mark the end of the tape, while the shorter strand, whichever it is, is completed with additional occurrences of c). Again the construction is similar in essence to that in the proof of Theorem 3.1, but we present it in full details because it can be of interest to see a concrete Watson-Crick finite automaton which is universal in the sense considered above.

First, let us specify a notation: one move in a Watson-Crick finite automaton with a two-way lower head is given in the form of two rewriting rules, one specifying the move of the upper head and one specifying the move of the lower head; to be more suggestive, we write these rules one above the other; always, the states involved in the two moves should be identical (the state belongs to the automaton, not to each of the heads).

Let K and V be the set of states and the alphabet which are maximal for the finite automata we want to simulate. We construct the Watson-Crick finite automaton

$$M_u = (K_u, V \cup K \cup \{c, c_1, c_2\}, \rho_u, q_{0,u}, \{q_f\}, P_u),$$

where

$$\rho_u = (V \cup K \cup \{c, c_1, c_2\}) \times (V \cup K \cup \{c, c_1, c_2\}),$$
$$K_u = \{q_{0,u}, q'_{0,u}, q''_{0,u}, q_f\}$$
$$\cup \{[s], (s), (s)', (s)'', (s)''', (s)^{iv}, \overline{[s]} \mid s \in K\}$$
$$\cup \{[sa] \mid s \in K, a \in V\},$$

and the set P_u consists of the following transition rules (rules of the form $s\lambda \to s'\lambda$ in the upper positions mean that the upper strand head does not move):

1. $\begin{pmatrix} q_{0,u}\lambda \to q'_{0,u}\lambda \\ q_{0,u}s \to sq'_{0,u} \end{pmatrix}$, $s \in K$,

2. $\begin{pmatrix} q'_{0,u}\lambda \to q''_{0,u}\lambda \\ q'_{0,u}a \to aq''_{0,u} \end{pmatrix}$, $a \in V$,

3. $\begin{pmatrix} q''_{0,u}\lambda \to q_{0,u}\lambda \\ q''_{0,u}s \to sq_{0,u} \end{pmatrix}$, $s \in K$,

4. $\begin{pmatrix} q_{0,u}\lambda \to [s_0]\lambda \\ q_{0,u}s_0 \to s_0[s_0] \end{pmatrix}$,

5. $\begin{pmatrix} [s]a \to a[sa] \\ [s]a \to a[sa] \end{pmatrix}$, $s \in K, a \in V$,

6. $\begin{pmatrix} [sa]\lambda \to (s')\lambda \\ [sa]s' \to s'(s') \end{pmatrix}$, $s, s' \in K, a \in V$,

7. $\begin{pmatrix} (s)\lambda \to (s)'\lambda \\ (s)s' \to s'(s)' \end{pmatrix}$, $s, s' \in K$,

8. $\begin{pmatrix} (s)'\lambda \to (s)''\lambda \\ (s)'a \to a(s)'' \end{pmatrix}$, $s \in K, a \in V$,

9. $\begin{pmatrix} (s)''\lambda \to (s)\lambda \\ (s)''s' \to s'(s) \end{pmatrix}$, $s, s' \in K$,

10. $\begin{pmatrix} (s)\lambda \to [s]\lambda \\ (s)s \to s[s] \end{pmatrix}$, $s \in K$,

11. $\begin{pmatrix} (s)\lambda \to (s)'''\lambda \\ s'(s) \to (s)'''s' \end{pmatrix}$, $s, s' \in K$,

12. $\begin{pmatrix} (s)'''\lambda \to (s)^{iv}\lambda \\ a(s)''' \to (s)^{iv}a \end{pmatrix}$, $s \in K, a \in V$,

13. $\begin{pmatrix} (s)^{iv}\lambda \to (s)\lambda \\ s'(s)^{iv} \to (s)s' \end{pmatrix}$, $s, s' \in K$,

14. $\begin{pmatrix} (s)c \to c\overline{[s]} \\ (s)c_1 \to c_1\overline{[s]} \end{pmatrix}$, $s \in K$,

15. $\begin{pmatrix} \overline{[s]}\lambda \to \overline{[s]}\lambda \\ \overline{[s]}s' \to s'\overline{[s]} \end{pmatrix}$, $s, s' \in K, s \neq s'$,

16. $\begin{pmatrix} \overline{[s]}\lambda \to q_f\lambda \\ \overline{[s]}s \to sq_f \end{pmatrix}$, $s \in F$,

17. $\begin{pmatrix} q_f\lambda \to q_f\lambda \\ q_f s \to sq_f \end{pmatrix}$, $s \in K$,

18. $\begin{pmatrix} q_f c \to cq_f \\ q_f\lambda \to q_f\lambda \end{pmatrix}$,

19. $\begin{pmatrix} q_f\lambda \to q_f\lambda \\ q_f c_2 \to c_2 q_f \end{pmatrix}$,

20. $\begin{pmatrix} q_f\lambda \to q_f\lambda \\ q_f c \to cq_f \end{pmatrix}$.

Assume now that we start with a finite automaton $M = (K', V', s_0, F, P)$, with $K' \subseteq K, V' \subseteq V$, and we write a string xc^m in the upper strand and $code(M)c^p$ in the lower strand of the input tape of M_u, with m, p as specified above.

While the upper head of M_u remains in the same place of its strand, the lower one looks for a transition of M which can parse the currently read symbol of x. At the beginning, this must be a transition of the form $s_0 a \to as$. By rules 5, 6 one then simulates a step of the work of M. The state (s') can go to the right (rules 7, 8, 9) or to the left (rules 11, 12, 13), looking for a valid continuation (rule 10). When the string x is finished and one reaches c in the upper strand (we may assume that at the same moment we reach c_1 in the lower strand, because we can move freely the lower head), we pass to checking whether or not the current state is final with respect to M. The work of M_u stops correctly only in the affirmative case (rule 16). Therefore, $xc^m \in L(M_u)$ if and only if $x \in L(M)$. (Rules

17 – 20 are used for scanning the suffixes c^m, c_2c^p of the two strands.)

7 Universality for Watson-Crick Automata

The proof of Theorem 5.1 above proceeds along the following phases: (1) starting from a type-0 Chomsky grammar G, one constructs three morphisms h_1, h_2, h_3 such that $L(G) = h_3(h_1(EQ(h_1, h_2)) \cap R)$, where $EQ(h_1, h_2)$ is the equality set of h_1, h_2 (that is, the set of words w such that $h_1(w) = h_2(w)$), and R is a regular language (see [9]); (2) for h_1, h_2 as above, one constructs a sticker system (of a certain type: *coherent*) which generates a set of double stranded sequences having the language $h_1(EQ(h_1, h_2))$ in their upper strands (see [9]); finally, (3) a Watson-Crick automaton can be constructed starting from this sticker system, recognizing the language $h_1(EQ(h_1, h_2)) \cap R$ ([7]); thus, a further morphism (h_3 above, which is in fact a weak coding) suffices to characterize RE. All the three steps are constructive. If we start from a universal type-0 grammar G_u instead of G, then we obtain a unique Watson-Crick automaton which should be universal in a natural sense. However, the above path from a universal type-0 grammar to a universal Watson-Crick automaton is too long and indirect, the result will be too complex (we even do not see an easy way to write the "program" to be run on such a universal machine). The task of finding *simple* Watson-Crick finite automata which are universal for the whole class of such automata, hence for the class of Turing machines, remains as a *research topic*. As in the case of finite automata, we consider here only the easier task of finding Watson-Crick finite automata which are universal for the class of automata with a bounded number of states and of input symbols.

In order to solve this problem, we need a normal form result for Watson-Crick finite automata. In [7], the transition mapping is defined on $K \times V^* \times V^*$. This can be significantly simplified.

We say that a Watson-Crick finite automaton is 1-*limited* if its transition mapping is defined on $(K \times V \times \{\lambda\}) \cup (K \times \{\lambda\} \times V)$ (in each step, exactly one symbol is scanned, either in the upper strand or in the lower strand of the tape).

As expected, the following normal form result holds.

Lemma 7.1 *For each Watson-Crick finite automaton M there is a 1-limited Watson-Crick automaton M' such that $L(M) = L(M')$.*

Proof. Starting from $M = (K, V, \rho, s_0, F, P)$, we construct $M' = (K', V, \rho,$

s_0, F, P') as follows. For each transition rule

$$t : s \begin{pmatrix} a_1 a_2 \dots a_n \\ b_1 b_2 \dots b_m \end{pmatrix} \rightarrow \begin{pmatrix} a_1 a_2 \dots a_n \\ b_1 b_2 \dots b_m \end{pmatrix} s'$$

in P, $n \geq 0, m \geq 0, n + m \geq 2$, we introduce in P' the transitions

$$s \begin{pmatrix} a_1 \\ \lambda \end{pmatrix} \rightarrow \begin{pmatrix} a_1 \\ \lambda \end{pmatrix} s_{t,1},$$

$$s_{t,i} \begin{pmatrix} a_{i+1} \\ \lambda \end{pmatrix} \rightarrow \begin{pmatrix} a_{i+1} \\ \lambda \end{pmatrix} s_{t,i+1}, \ 1 \leq i \leq n - 1,$$

$$s_{t,n} \begin{pmatrix} \lambda \\ b_1 \end{pmatrix} \rightarrow \begin{pmatrix} \lambda \\ b_1 \end{pmatrix} s'_{t,1},$$

$$s'_{t,i} \begin{pmatrix} \lambda \\ b_{i+1} \end{pmatrix} \rightarrow \begin{pmatrix} \lambda \\ b_{i+1} \end{pmatrix} s'_{t,i+1}, \ 1 \leq i \leq m - 2,$$

$$s'_{t,m-1} \begin{pmatrix} \lambda \\ b_m \end{pmatrix} \rightarrow \begin{pmatrix} \lambda \\ b_m \end{pmatrix} s';$$

all states $s_{t,i}, s'_{t,i}$ are introduced in K', together with all states in K.

One can easily see that the obtained automaton is equivalent with M (the new states control the work of M' in a deterministic way) and 1-limited. □

For a Watson-Crick finite automaton in the 1-limited normal form, $M = (K, V, \rho, s_0, F, P)$, we can consider a codification of it of the form

$$code(M) = [s_1 \begin{pmatrix} \alpha_1 \\ \beta_1 \end{pmatrix} s'_1] \dots [s_n \begin{pmatrix} \alpha_n \\ \beta_n \end{pmatrix} s'_n][s_{f1}] \dots [s_{fm}],$$

where each $[s_i \begin{pmatrix} \alpha_i \\ \beta_i \end{pmatrix} s'_i]$ is a symbol corresponding to a move $s_i \begin{pmatrix} \alpha_i \\ \beta_i \end{pmatrix} \rightarrow \begin{pmatrix} \alpha_i \\ \beta_i \end{pmatrix} s'_i$ in P (hence one of α_i, β_i is a symbol, the other one is empty), and each $[s_{fi}]$ is a symbol associated to a final state in F.

When parsing a molecule $x = \begin{bmatrix} a_1 a_2 \dots a_r \\ b_1 b_2 \dots b_r \end{bmatrix}$, $a_i, b_i \in V, 1 \leq i \leq r, r \geq 1$, we scan one symbol at a time, in either of the two strands, hence we perform $2r$ steps. The "speed" of the two heads is different, the distance between them can be arbitrarily large. Thus, we have to merge the code of M with the symbols in each strand of x, considering the molecule

$$w_0 = \begin{bmatrix} code(M)a_1 \ code(M)a_2 \dots code(M)a_r \ code(M)c \ code(M) \\ code(M)b_1 \ code(M)b_2 \dots code(M)b_r \ code(M)c \ code(M) \end{bmatrix}$$

$$= \begin{bmatrix} bls(code(M), a_1 a_2 \dots a_r c) \\ bls(code(M), b_1 b_2 \dots b_r c) \end{bmatrix}.$$

(The complementarity relation contains all pairs in ρ plus all pairs (α, α), with α appearing in $code(M) \cup \{c\}$.)

Now, a universal Watson-Crick finite automaton can be constructed following the same idea as in the case of usual finite automata (Section 3): scan an occurrence of $code(M)$ in either of the two strands, choose a move of the form $s\begin{pmatrix}\alpha \\ \beta\end{pmatrix} \rightarrow \begin{pmatrix}\alpha \\ \beta\end{pmatrix} s'$ encoded by some symbol $[s\begin{pmatrix}\alpha \\ \beta\end{pmatrix} s']$, simulate this move (clearly, when working in the upper strand, we must have $\beta = \lambda$, and when working in the lower strand we must have $\alpha = \lambda$); the states of the universal automaton will ensure the correct linking of states of M, thus controlling the parsing in the two strands exactly as the states of M do; we advance with different speeds along the two strands; when reaching the column $\begin{pmatrix}c \\ c\end{pmatrix}$ we continue by checking whether or not the current state is final with respect to M (that is a state identified by a symbol $[s]$ in $code(M)$; we end in a final state of the universal automaton only in the affirmative case.

Denoting by M_u the Watson-Crick automaton whose construction is sketched above, we obtain

$$w_0 \in L(M_u) \quad \text{iff} \quad x \in L(M),$$

hence the universality property.

The "program" w_0 above (it also contains the data to be processed, the molecule x, intercalated with copies of $code(M)$) is rather complex (of a non-context-free type, because of the repeated copies of $code(M)$ in each of the two strands). A way to reduce the complexity of the starting molecule of this universal Watson-Crick automaton is the same as in the case of usual finite automata (where we have passed to Watson-Crick automata): consider one more strand of the tape, that is, work with Watson-Crick automata with triple-stranded tapes and three heads scanning them, controlled by a common state.

First, in such a case we can simplify the shape of the "program" w_0: write $a_1 a_2 \ldots a_r$ in the first strand, $b_1 b_2 \ldots b_r$ in the second strand, maybe followed by a number of occurrences of the symbol c such that these two strands are of the same length as the third one, where we write $2r$ copies of $code(M)$. The head scanning the third strand can go from left to right in the usual way, identifying in each occurrence of $code(M)$ a move which is simulated in one of the other two strands. Note that this time the "algorithm" $(code(M))$ is separated from data (molecule x).

Second, if we allow the head in the third strand to work in a two-way manner (the other heads remain usual one-way heads), then only one copy of *code*(*M*) suffices (see the construction in Section 5 of a two-way two-strands Watson-Crick finite automaton which is universal for the class of finite automata with a bounded number of states and of symbols).

We do not enter here into technical details concerning universal Watson-Crick automata (with two or three strands in their tapes). On the one hand, such details can be easily filled-in by the interested reader, on the other hand, working with triple-stranded molecules (as in the case of collagen) is not at all possible in this moment to allow us to have some hope of implementing our machines; on the third hand (in the genetic engineering area any number of hands is possible...), more important than the universality results of the kinds above are universality results with respect to the whole class of Watson-Crick finite automata, because of their equivalence (modulo a weak coding: Theorem 5.1) with Turing machines. But, as we have already mentioned above, this remains as a *research topic*.

8 Final Remarks

It is worth noting that universal finite automata were looked for (and found) also in [3], [4]. However, there one works with finite automata with outputs, the notion of universality is different from ours (the automata *simulate* one another in a well defined way), and one looks for universal automata in a certain extension of a given finite set of finite automata. In these circumstances, one proves that for *any* finite set of finite automata there is an universal automaton in a *completion* of the set. The connection between the results in [3], [4] and DNA computing remains to be investigated.

References

[1] L. M. Adleman, Molecular computation of solutions to combinatorial problems, *Science*, 226 (Nov. 1994), 1021 – 1024.

[2] E. Baum, D. Boneh, P. Kaplan, R. Lipton, J. Reif, N. Seeman, eds, *Second Annual DIMACS Meeting on DNA Based Computing*, Princeton, 1996.

[3] C. Calude, E. Calude, B. Khoussainov, Deterministic automata: Simulation, universality and minimality, *Annals of Pure and Applied Logic*, 1105 (1997).

[4] E. Calude, M. Lipponen, Deterministic incomplete automata: Simulation, universality and complementarity, in *Unconventional Models of Computation* (C. S. Calude, J. Casti, M. J. Dinneen, eds.), Springer-Verlag, Singapore, 1998, 131 – 149.

[5] K. Culik II, T. Harju, Splicing semigroups of dominoes and DNA, *Discrete Appl. Math.*, 31 (1991), 261 – 277.

[6] J. Engelfriet, G. Rozenberg, Fixed point languages, equality languages, and representations of recursively enumerable languages, *Journal of the ACM*, 27 (1980), 499 – 518.

[7] R. Freund, Gh. Păun, G. Rozenberg, A. Salomaa, Watson-Crick finite automata, *Proc. of the Third Annual DIMACS Meeting on DNA Based Computers*, Philadelphia, 1997, 305 – 317.

[8] T. Head, Gh. Păun, D. Pixton, Language theory and molecular genetics. Generative mechanisms suggested by DNA recombination, chapter 7 in vol. 2 of [20], 295 – 360.

[9] L. Kari, Gh. Păun, G. Rozenberg, A. Salomaa, S. Yu. DNA computing, sticker systems, and universality, *Acta Informatica*, to appear.

[10] R. J. Lipton, Using DNA to solve NP-complete problems, *Science*, 268 (Apr. 1995), 542 – 545.

[11] Gh. Păun, Regular extended H systems are computationally universal, *J. Automata, Languages, Combinatorics*, 1, 1 (1996), 27 – 36.

[12] Gh. Păun, Five (plus two) universal DNA computing models based on the splicing operation, in [2], 67 – 86.

[13] Gh. Păun, G. Rozenberg, Sticker systems, *Theoretical Computer Sci.*, 1998.

[14] Gh. Păun, G. Rozenberg, A. Salomaa, Computing by splicing, *Theoretical Computer Science*, 168, 2 (1996), 321 – 336.

[15] Gh. Păun, G. Rozenberg, A. Salomaa, Computing by splicing. Programmed and evolving splicing systems, *IEEE Intern. Conf. on Evolutionary Computing*, Indianapolis, 1997, 273 – 277.

[16] Gh. Păun, G. Rozenberg, A. Salomaa, *DNA Computing. New Computing Paradigms*, Springer-Verlag, Heidelberg, 1998.

[17] Gh. Păun, A. Salomaa, DNA computing based on the splicing operation, *Mathematica Japonica*, 43, 3 (1996), 607 – 632.

[18] D. Pixton, Regularity of splicing languages, *Discrete Appl. Math.*, 69 (1996), 101 – 124.

[19] G. Rozenberg, A. Salomaa, Watson-Crick complementarity, universal computations and genetic engineering, *Techn. Report* 96-28, Dept. of Computer Science, Leiden Univ., Oct. 1996.

[20] G. Rozenberg, A. Salomaa, eds., *Handbook of Formal Languages*, Springer-Verlag, Heidelberg, 1997, 3 volumes.

[21] S. Roweis, E. Winfree, R. Burgoyne, N. Chelyapov, M. Goodman, P. Rothe-mund, L. Adleman, A sticker based architecture for DNA computation, in [2], 1 – 27.

[22] E. Winfree, X. Yang, N. Seeman, Universal computation via self-assembly of DNA; some theory and experiments, in [2], 172 – 190.

Simulating Turing Machines
by Extended mH Systems

Pierluigi FRISCO, Giancarlo Mauri, Claudio FERRETTI

Dipartimento di Scienze dell'Informazione
Università di Milano
via Comenico 39, 20135 Milano, Italy
ferretti@dsi.unimi.it

Abstract. We describe a method of translation from a deterministic Turing machine to an extended mH splicing system. The accuracy of this translation is proved by a theorem which shows that the only string present in a single copy is a coding of a configuration of the simulated Turing machine. A simpler translation from a universal Turing machine with three tapes to an extended mH system is given, too.

1 Introduction

Recently, many different formal models of computation have been proposed under the inspiration of biological processes. DNA based computation, generally speaking, considers using transformations of biological molecules as computational steps. It is interesting to compare any of these models to the classical well known computational models, like grammars or Turing machines.

Splicing systems are generative mechanisms based on the splicing operation introduced by Tom Head as a model of DNA recombination. This is a rather new research area which is based on the observation that the incredibly complex structure of a living being results from applying a few simple operations (local mutation, copying and splicing, recombination, crossing over) to an initial DNA sequence. The complexity of the output implies that these operations are powerful and of a very fundamental nature. The

221

basic universal theorem, described in [6], proves that these systems may generate the class of recursively enumerable languages.

On the basis of that proof, suitably modified, using finitely many rules and with the application regulated by certain additional mechanism, it is possible to obtain mechanisms of the same power, but starting from any given Turing machine. We here approach this path, leading to a concrete and formally sound description of a splicing system universal with respect to Turing machines.

2   Turing Machines

We denote by $\mathbf{Z}$ the set of integers, and by $\mathbf{N}$ the set of non-negative numbers.

A Turing machine (TM) T is composed as usual by:

- an infinite tape divided into cells, represented by numbers $x \in \mathbf{Z}$;

- a writing-reading head that may indicate only one cell in the tape at each time;

- a set of symbols $M = \{a_1, \ldots, a_k\}$, where a_k is called blank (also indicated with $\flat$);

- a set of states $S = \{A_0, \ldots, A_p\}$, where A_0 is called initial state; in the subset F of S, there are the final states;

- a set of rules (called also instructions) $R = \{I_1, \ldots, I_u\}$, where each rule is a quintuple $\langle A_i, a_j, a_n, D, A_m \rangle$, where A_i, A_m are states, a_j, a_n are symbols, and D is the direction, that is an element of $\{+1, -1\}$ (or of the set $\{\text{right}, \text{left}\}$).

We consider that T stops when no rule can be applied anymore to the tape; we say that T *accepts* the starting input if it stops in a final state.

In addition to considering Turing machines which would perform various other mathematical operations, Turing gave a description of a *universal Turing machine* (UTM) which could perform any operation which any other ordinary Turing machine could perform.

This ability to make a relatively simple machine act like a more complicated machine is achieved by giving the simple machine complicated instructions. In particular, the UTM has in one part of its tape a complete symbolic description of the machine it is expected to imitate. Then the universal machine stores on another part of its tape a copy of the tape that would be on the imitated machine, and makes the changes on this part of the tape which the imitated machine would make. The remaining part of the tape must be used for intermediate scratch work, for instance to record what state the imitated machine is in, what part of the tape it is scanning, etc. The internal structure of the UTM has to include instructions to use these various kinds of data, and to move back and forth between the different parts of the tape. Since the method of storing all of this information on the tape is rather complicated, the internal structure of the UTM which Turing described is also rather complicated, requiring a large number of states.

In a next section we will give the description of a UTM with three tapes and using fifteen states. It is described in [5], and it will be used to obtain a universal extended mH system simulating this UTM.

3   Extended H Systems

We use the following notations obtained by [3]: V^* is the free monoid generated by the alphabet V, λ is the empty string, $V^+ = V^* - \{\lambda\}$, $|x|$ is the length of $x \in V^*$, FIN, REG, RE are the families of finite, regular, and recursively enumerable languages, respectively. For general formal language theory prerequisites we refer to [4], [7], and for regular rewriting to [1].

An *extended H system* is a quadruple

$$\gamma = (V, L, X, U),$$

where: V is an alphabet, $L \subseteq V$, $X \subseteq V^*$, $U \subseteq V^*\#V^*\$V^*\#V^*$; $\#$ and $\$$ are special symbols not in V.

V is the *alphabet* of γ, L is the *terminal alphabet*, X is the *set of axioms*, and U is the *set of splicing rules*; the symbols in L are called *terminals* and those in $V - L$ are called *non terminals*.

For $x, y, z, w \in V^*$ and $r = u_1\#u_2\$u_3\#u_4$ in U we define $(x, y) \vdash_r (z, w)$ if and only if $x = x_1 u_1 u_2 x_2, y = y_1 u_3 u_4 y_2, z = x_1 u_1 u_4 y_2, w = y_1 u_3 u_2 x_2$, for some $x_1, x_2, y_1, y_2 \in V^*$.

The strings x, y are called the *terms* of the splicing.

For an H system $\gamma = (V, L, X, U)$ and for any language $G \subseteq V^*$, we write

$$\sigma(G) = \{z \in V^* \mid (x, y) \vdash_r (z, w) \text{ or } (x, y) \vdash_r (w, z) \text{ for } x, y \in G, r \in U\}$$

and we define

$$\sigma^*(G) = \bigcup_{i \geq 0} \sigma^i(G),$$

where

$$\sigma^0(G) = G, \sigma^{i+1}(G) = \sigma^i(G) \cup \sigma(\sigma^i(G)), \text{ for } i \geq 0.$$

The *language generated* by the H system γ is defined by

$$L(\gamma) = \sigma^*(X) \cap L^*.$$

Then, for two families of languages, F_1, F_2, we denote

$$EH(F_1, F_2) = \{L(\gamma) \mid \gamma = (V, L, X, U), X \in F_1, U \in F_2\}.$$

In the definition above, the rule to be used and the position where the terms of the splicing shall be cut are not prescribed, they are chosen in a nondeterministic way. Moreover, after splicing two strings x, y and obtaining two strings z and w, we may use again x or y as a term of a splicing, possibly the second one being z or w, but also the new strings are supposed to appear in infinitely many copies, they are not "consumed" by splicing. Probably more realistic is the assumption that at least part of the strings are available in a limited number of copies. This leads to consider *multisets*, i.e., sets with multiplicities to their elements.

In the style of [2], a multiset over V^* is a function $Q : V^* \to \mathbf{N} \cup \infty$; $Q(x)$ is the number of copies of $x \in V^*$ in the multiset Q. The set $\{w \in V^* \mid Q(w) > 0\}$ is called the *support* of Q and it is denoted by $\mathrm{supp}(Q)$. A usual set $B \subseteq V^*$ is interpreted as the multiset defined by $B(x) = 1$ for $x \in B$, and $B(x) = 0$ for $x \notin B$. For two multisets Q_1, Q_2 we define their union by $(Q_1 \cup Q_2)(x) = Q_1(x) + Q_2(x)$, and their difference by $(Q_1 - Q_2)(x) = Q_1(x) - Q_2(x), \forall x \in V^*$, provided $Q_1(x) \geq Q_2(x), \forall x \in V^*$.

Usually a multiset with finite support, Q, is presented as a set of pairs $(x, Q(x))$, for $x \in \mathrm{supp}(Q)$.

An *extended mH system* is a quadruple $\gamma = (V, L, X, U)$, where V, L, U are as in an extended H system and X is a multiset over V^*.

For such an mH system and two multisets Q_1, Q_2 over V^* we define

$$
\begin{aligned}
Q_1 \Rightarrow Q_2 \quad &\text{iff} \quad \exists x, y, z, w \in V^* \text{ such that} \\
&\text{i)} \quad Q_1(x) \geq 1, Q_1(y) \geq 1, \\
&\qquad \text{and if } x = y, \text{ then } Q_1(x) \geq 2, \\
&\text{ii)} \quad x = x_1 u_1 u_2 x_2, y = y_1 u_3 u_4 y_2, \\
&\qquad z = x_1 u_1 u_4 y_2, w = y_1 u_3 u_2 x_2,
\end{aligned}
$$

$$\text{for } x_1, x_2, y_1, y_2 \in V^*, u_1\#u_2\$u_3\#u_4 \in U$$

iii) $Q_2 = (Q_1 - \{(x,1),(y,1)\}) \cup \{(z,1),(w,1)\}.$

The *language generated* by an extended mH system γ is

$$L(\gamma) = \{w \in L^* \mid w \in \text{supp}(Q) \text{ for some } Q \text{ such that } X \Rightarrow^* Q\},$$

where $\Rightarrow^*$ is the reflexive and transitive closure of $\Rightarrow$.

For two families of languages F_1, F_2, we denote

$$EH(mF_1, F_2) = \{L(\gamma) \mid \gamma = (V, L, X, U)$$
$$\text{is an mH system with supp}(X) \in F_1, U \in F_2\}.$$

An H system can be interpreted like an mH system working with multiset of the form $Q(x) = \infty$ for all x such that $Q(x) \neq 0$. Such multiset are called ω-multiset and the corresponding H systems are called ωH systems. The corresponding families will be also denoted by $EH(\omega F_1, F_2)$.

An H system $\gamma = (V, L, X, U)$ with $V = L$ is called *non-extended*; the families of languages generated by such systems, corresponding to $EH(mF_1, F_2)$ and $EH(\omega F_1, F_2)$ are denoted by $H(mF_1, F_2)$ and $H(\omega F_1, F_2)$, respectively. The family of languages generated by mH systems $\gamma = (V, L, X, U)$ with $\text{card}(\text{supp}(X)) \leq k$, for a given k, and $U \in F_2$, is denoted by $EH(m[k], F_2)$. In a similar way, by $EH(\omega[k], F_2)$ we denote the family of languages generated by ωH systems with at most k axioms.

4   About Extended mH Systems

In the previous paragraph we have described non-deterministic systems based on splicing operation. For *deterministic H systems* we mean the ones that for each splicing step may use only one splicing rule with only a couple of strings. The ones that may use the same rule with more than one couple of strings or more than one rule with one or more couples of strings, will be defined *non-deterministic*.

The concept of determinism and non-determinism may be used for all systems based on splicing operation.

The basic universality theorem ([6]) gives a translation from a type-0 grammar to a non-deterministic extended H system; non-determinism is inborn in the simulated grammar and that cannot be excluded with the translation. For instance, let us consider a grammar generating $a^n b^n$, with $n \geq 1$, when it is arrived at the creation of a certain number, $m \geq 1$, of consecutive a and b, it may stop or continue writing the $m+1$ a and b, and this is non-deterministic. A deterministic TM has an univocal computation and this may be held with a translation in a system based on splicing operation.

The choice of the kind of control system to use in the H system (with multisets, with permitting context conditions, etc.) comes out from TM's characteristics. This theoretical mechanism uses the tape, the work support, and rules that may be considered like the "memory" of the operations that has to be done on the tape. Translating these concepts in a H system, and examining its characteristics, we may consider the tape as a string that has to be present in single copy, because considering more than a work support obliges us to choose the one we want to use; the elements with memory functions may be present in how many copies we want (at least one) and so we may consider them – from a theoretical point of view – infinite.

The outcome of the elaboration of a TM is written on the tape, so we have to find a way to distinguish in the H system the string containing this informations from others. This is made using TM's final states and it will be described in the next section.

From these considerations we have chosen to use extended mH systems having the string with the tape information present in a unique copy, and the ones describing TM's rules in infinite copies.

5   A Simulator for Deterministic TM

A configuration of a TM, constituted by the working area of the tape, the state of the machine and the symbol pointed by the head, has been coded in the extended mH system by a string (called *work string*) containing all these informations. Now we make an example of this code.

Let a be the symbol pointed by the head and changed with s if the TM writes; A is the state of the TM, changed with N if the machine changes state and is always at the right of the symbol pointed by the head; l represents the character to the left of the symbol pointed by the head, and d the one at the right; x and y indicate the strings of symbols to the left of l and to the right of d (without the ends of infinite blanks), and y' (when present) is the first symbol to the left of y.

The work string is delimited with a couple of characters ($<<$ at the left and $>>$ at the right) not present in the alphabet of the TM and that are used to simulate the ends of infinite blanks.

So the work string representing a configuration of a TM has the form

$<< xlaAdy >>$.

TM's rules are divided in two sets: the first contains the rules where the machine writes a symbol, changes state and moves the writing-reading head to the right (for instance $\langle A, a, s, \text{right}, N \rangle$), the second contains the rules where the machine writes a symbol, changes state and moves the writing-reading head to left (for instance $\langle A, a, s, \text{left}, N \rangle$). If the machine does not change state and/or does not write, we consider the state changing and the writing with the particularity that the new state and the new symbol are similar to the previous ones.

The translations for the sets of instructions of a TM in splicing rules are given below; there are also the ones regarding blanks additions to the left and to the right of the work string. These last two are not properly translations of an instruction of a TM, but they are necessary because it is impossible to know first the part of the tape used by the TM. We simulate the non-finite tape adding one blank (indicated with $\flat$) when the writing-reading head moves to the ends of infinite blanks.

The axioms we need to simulate the instructions with movement to the right are delimited with square brackets, the ones used to simulate instructions with movement to the left with curly brackets, the ones used to add one blank are delimited with round brackets.

The symbols $<, >, [,], \lfloor, \rfloor, (,)$ are not in the alphabet of the TM. Now we describe the working of the basic steps of our simulation.

Change state, write and move to the right: from $<< xlaAdy'y >>$ to $<< xlsdNy'y >>$ using $r_1 = l\#aAd\$[\#sdN]$ and $r_2 = lsdN\#]\$[aAd\#y'$

$$
\begin{array}{ccc}
<< xlaAdy'y >> & << xlsdN] & << xlsdNy'y >> \\
[sdN] & \vdash_{r_1} \quad [aAdy'y >> \quad \vdash_{r_2} & [aAd]
\end{array}
$$

Add a blank to the right: from $<< xl > A >$ to $<< xl \ \flat A >>$ using $r_1 = l\# > A > \$(\# \ \flat)A$ and $r_2 = l \ \flat\#)A\$\#A >>$ for each $A \in S$

$$
\begin{array}{ccc}
<< xl > A > & << xl \ \flat)A & << xl \ \flat A >> \\
(\flat)A & \vdash_{r_1} \quad (> A > \quad \vdash_{r_2} &)A \\
A >> & A >> & (> A >
\end{array}
$$

Change state, write and move to the left: from $<< xlaAdy >>$ to $<< xlNsdy >>$ using $r_1 = aA\#d\$\lfloor Ns\#\rfloor$ and $r_2 = l\#aA\rfloor\$\lfloor\#Nsd$

$$
\begin{array}{ccc}
<< xlaAdy >> & << xlaA\rfloor & << xlNsdY >> \\
\lfloor Ns\rfloor & \vdash_{r_1} \quad \lfloor Nsdy >> \quad \vdash_{r_2} & \lfloor aA\rfloor
\end{array}
$$

Add a blank to the left: from $<< Ady >>$ to $<< \flat bAdy >>$ using $r_1 =< A\#d\$ < \flat A\#$ for each $A \in S$

$$<< Ady >> \qquad << \flat Ady >>$$
$$<< \flat A \qquad \vdash_{r_1} \qquad << A$$

S represents the set of the states of the TM. Creating as many splicing rules to add blank as many are the states in S is needed because it is impossible to know first which states will go over the work area going in the ends of infinite blanks.

It has been noted that the generated strings like $[aAd],)A, (< A >, \lfloor aA \rfloor, << A$ may not be used in any splicing.

The only axiom present in a single copy is that one coding the initial configuration of the TM, all the others are in an infinite number of copies.

The extended mH system will stop when the TM will enter a final state for which it will not exist a quintuple beginning with that state and the symbol pointed by the head.

6   The Similarity Theorem

We have defined a TM using splicing rules, now we will give a formal description of this translation and the proof that it is consistent.

Let T be a deterministic TM with only a tape infinite in both directions, defined by $\langle M, S, R \rangle$ where:

- $M = \{a_1, \ldots, a_k\}$, set of symbols, where $a_k = \flat$ is the blank;

- $S = \{A_0, \ldots, A_p\}$, set of states, where A_0 is the initial state and $F = \{A_i, \ldots, A_{i+m}\}$ where $0 \leq i \leq p$ and $0 \leq m \leq p-i$, $F \subseteq S$, set of final states for which do not exist instructions of the type $\langle A_j, -, -, -, - \rangle$ with $A_j \in \{A_i, \ldots, A_{i+m}\}$;

- $R = \{I_1, \ldots, I_u\}$ set of instructions where each instruction is a quintuple $\langle A_y, a_l, a_n, D, A_0 \rangle$ with $A_y, A_0 \in S$, $a_l, a_n \in M$, and $D \in \{\text{right}, \text{left}\}$.

Let $L(T)$ be the language accepted by T.

The extended mH system simulating this TM is defined by $\gamma = \langle V, L, X, U \rangle$ where:

- $V = M \cup S \cup \{[,], \lfloor, \rfloor, (,), <, >\}$, alphabet;

- $L = M \cup F \cup \{<,>\}$ (note that $L \subseteq V$), set of terminals;
- $X \subseteq V^*$, set of axioms, support of the multiset function $Q : X \to \mathbb{N} \cup \{\infty\}$, and composed with:

  - axiom for state changing, writing and moving to the right:

    $$[a_n d A_0], \quad \forall d \in V,$$
    $$\forall a_n \in M, A_0 \in S \text{ s.t. } \langle A_i, a_l, a_n, \text{right}, A_0 \rangle \in R,$$
    $$Q([a_n d A_0]) = \infty;$$

  - axioms to add a blank to the right:

    $$(\flat)A, \quad \forall A \in S, Q((\flat)A) = \infty;$$
    $$A >>, \quad \forall A \in S, Q(A >>) = \infty;$$

  - axiom for state changing, writing and moving to the left:

    $$\lfloor A_0 a_n \rfloor, \quad \forall a_n \in M, A_0 \in S \text{ s.t. } \langle A_i, a_l, a_n, \text{left}, A_0 \rangle \in R;$$
    $$Q(A_0 a_n) = \infty;$$

  - axiom to add a blank to the left:

    $$<< \flat A, \quad \forall A \in S, Q(<< \flat A) = \infty;$$

- $U \subseteq V^* \# V^* \$ V^* \# V^*$ set of splicing rules, formed with:

  - rules for state changing, writing and moving to the right:

    $$v_1 \# a_l A_i w_1 \$ [\# a_n w_1 A_0], \ v_2 a_n w_2 A_0 \#] \$ [a_l A_i w_2 \# v_3,$$
    $$\forall w_1, w_2, v_1, v_2, v_3 \in M \cup \{<,>\},$$
    $$\forall a_n, a_l \in M; A_0, A_i \in S \text{ s.t. } \langle A_i, a_l, a_n, \text{right}, A_0 \rangle \in R.$$

  - rules to add a blank to the right:

    $$v_1 \# > A > \$ (\# \flat) A, \ v_2 \ \flat \#) A \$ \# A >>,$$
    $$\forall v_1, v_2 \in M \cup \{<,>\}, \forall A \in S,$$

  - rules for state changing, writing and moving to the left:

    $$a_l A_i \# v_1 \$ A_0 a_n \#, \ v_2 \# a_l A_i \$ \# A_0 a_n v_3,$$
    $$\forall v_1, v_2, v_3 \in M \cup \{<,>\},$$
    $$\forall a_n, a_l \in M; A_0, A_i \in S, \text{ s.t. } \langle A_i, a_l, a_n, \text{left}, A_0 \rangle \in R.$$

– rules to add a blank to the left:

$$< A\#v\$ < \flat A\#, \quad \forall v \in M \cup \{<, >\}, \forall A \in S.$$

If the TM receives as input the string $a_{q_1} \ldots a_{q_y}$, with $a_{q_1} \ldots a_{q_y} \in M$, and the head reads the symbol a_{q_1}, then the set of axioms X completes with the work axiom

$$<< a_{q_1} A_0 a_{q_2} \ldots a_{q_y} >>, \quad \text{where } a_{q_1}, \ldots, a_{q_y} \in M,$$
$$A_0 \in S, Q(<< a_{q_1} A_0 a_{q_2} \ldots a_{q_y} >>) = 1.$$

If we define the language accepted by the TM as:

$$\begin{aligned} L(T) \;=\; & \{w \mid w \in M^*, w = a_{q_1} \ldots a_{q_y}, \\ & (\ldots \flat a_{q_1} \ldots a_{q_y} \flat \ldots, a_{q_1}, A_0) \Rightarrow^* (v, a, A_z) \\ & \text{where } v \in M^*, a \in M, A_z \in F\}, \end{aligned}$$

and the one accepted by γ as:

$$L(\gamma) = \{v \mid v \in L^*, v \in \text{supp}(B) \text{ s.t. } X \vdash^* B\}.$$

we may demonstrate that γ simulates T, that is if we include the work axiom obtained by w, with $w \in L(T)$, we will obtain $v \in L(\gamma)$.

Theorem 6.1 (*Similarity theorem*) *The extended mH system described before simulates the TM T from which it is created, that is: if and only if a code of $w \in L(T)$ is included in γ, it will produce $v \in L(\gamma)$ describing the work area of the tape of the TM when it accepts w. Moreover all strings obtained beginning with $<<$ and ending with $>>$ represent the work area of T in its configurations.*

Proof. Since the TM is deterministic, each configuration has only one next configuration, so for each computational step there is only a quintuple that may be used. This univocal elaboration is kept also in the extended mH system where, for each quintuple, we have created some splicing rules.

Let us see how for each quintuple only a type of splicing rules is created.

- Generic instructions: $\langle A_i, a_l, a_n, \text{right}, A_0 \rangle$: The created splicing rules are

$$v_1 \# a_l A_i w_1 \$[\# a_n w_1 A_0], v_2 a_n w_2 A_0 \#]\$[a_l A_i w_2 \# v_3,$$
$$\forall w_1, w_2, v_1, v_2, v_3 \in M \cup \{<, >\}; \forall a_n, a_l \in M, A_0, A_i \in S.$$

Note that the state-symbol pair, which makes unique that quintuple, is present in all splicing rules created by it. Since no quintuple has the same state-symbol pair, no one of the other instructions may create splicing rules with that pair.

For instructions like $\langle A_i, a_l, a_n, \text{left}, A_0 \rangle$ it holds the analogous result.

Splicing rules necessary to add a blank to the right or to the left are not translation of any instruction, and then they may not be obtained by any quintuple.

We prove now that the extended mH system created is deterministic and that it simulates the TM, we prove also that all strings obtained during the process beginning with $<<$ and ending with $>>$, represent the work area of the TM for some configuration it had during elaboration.

- Apply the rule regarding quintuple of the type: $\langle A_i, a_l, a_n, \text{right}, A_0 \rangle$: without loss of generality we may suppose to have $<< a_1 \ldots a_{l-1} a_l A_i a_{l+1} a_{l+2} \ldots a_z >>$ as a work string present in a unique copy.

Considering the translation given above it is possible to see that in the splicing rules created from the quintuple we also have:

$$a_{l-1} \# a_l A_i a_{l+1} \$ [\# a_n a_{l+1} A_0], \, a_{l-1} a_n a_{l+1} A_0 \#] \$ [a_l A_i a_{l+1} \# a_{l+2}$$

and the axiom $[a_n a_{l+1} A_0]$ has been created in an infinite number of copies.

For what we have said before only that quintuple may create these splicing rules.

It is possible to see that only the first one of these two rules may be used with the work string and the axiom; so, there will be no ambiguity, because there is only the work string having $a_{l-1} a_l A_i a_{l+1}$ and only a type of axiom of the form $[a_n a_{l+1} A_0]$. Using this rule we obtain two strings both present in a unique copy:

$$<< a_1 \ldots a_{l-1} a_n a_{l+1} A_0], [a_l A_i a_{l+1} a_{l+2} \ldots a_z >>$$

We have obtained two strings, both with the state-symbol pair but now it is possible to use only the second splicing rule, because it is the only rule considering the two pairs being delimited by a character not present in $M \cup \{<, >\}$ (in this case [and]).

Using the second rule we obtain $<< a_1 \ldots a_{l-1} a_n a_{l+1} A_0 a_{l+2} \ldots a_z >>$ and $[a_l A_i a_{l+1}]$, both in a unique copy. The second one, even if it contains the state-symbol pair, may not be used because it has a form unusable by any splicing rule. Words like this one are called *useless* strings.

The details regarding splicing rules for state changing, writing and movement to the left are similar and will not be given; in this case useless strings obtained are of the form $\lfloor a_l A_i \rfloor$ for each $A_i \in S$, for each $a_l \in M \cup \{<, >\}$, such that $\langle A_i, a_l, a_n, \text{left}, A_0 \rangle \in R$.

Rules to add blanks to both sides of the work string maintain the determinism in the extended mH system; in particular, we consider here the ones to add a blank to the right side. Axioms $(\flat) A_i$ and $A_i >>$ are created in infinitely many copies for each $A_i \in S$; for each $A_i \in M \cup \{<, >\}$ there are the splicing rules $a_i \# > A_i > \$(\# \flat) A_i$ and $a_i \flat \#) A_i \$ \# A_i >>$.

Let us consider to have the work string $<< a_1 \ldots a_{l-1} a_l > A_i >$ in unique copy. No splicing rules obtained as a translation of instructions of the TM may be used, because $> A_i$ is not a state-symbol pair ($>$ is not a symbol of the TM), so the only type of instructions that may be used is that one regarding the blank addition to the right. Among all splicing rules, those considering a_i at the left of the first $>$ and A_i as a state will be used. So, using $a_i \# > A_i > \$(\# \flat)$, the two strings $<< a_1 \ldots a_{l-1} a_l \flat) A_i$ and $(> A_i >$ are obtained.

The second one is not in a usable form because to the left of the first $>$ there is ($\notin M$; with the first string and with $A_i >>$ it may be used a rule in the second group, in particular that one considering a_i as symbol and A_i as state. With this rule we obtain $<< a_1 \ldots a_{l-1} a_l \flat A_i >>$ and $) A_i$, where the first one is the work string and the second an useless one.

The demonstration regarding the blank adding to the left side of the work string is similar and will not be given. In this case the useless strings obtained have the form: $<< A_i$ for each $A_i \in S$.

If we consider the string

$$<< a_1 \ldots a_{l-1} a_l A_i a_{l+1} a_{l+2} \ldots a_z >>$$

from where we left, and we have obtained

$$<< a_1 \ldots a_{l-1} a_n a_{l+1} A_0 a_{l+2} \ldots a_z >>$$

by using the two splicing rules in the first part of the proof, it is easy to see that, considering the coding of the work area of the tape of the TM we gave, the step made is the same as made by the simulated instruction. No others axiom or useless string begin with $<<$ and end with $>>$, so only the strings with this characteristics are a representation of the work area of the tape of the simulated TM.

Let us assume by absurd that the coding of a string $w \notin L(T)$ leads to $v \in L(\gamma)$, i.e., by using splicing rules obtained from the instructions of T, with the coding of w, we obtain a string including a final state. For what we have said in the previous part of the proof, splicing rules come in a unique way from the instructions of the TM, so, the same instructions used in the TM have to bring the machine to a final state, that is $w \in L(T)$. But this is contradictory and then the hypothesis is wrong, so for each $w \notin L(T)$, γ will not produce any string including a final state. □

7 Complexity

Let us consider the space and time complexity of an extended mH system created as a function of the simulated TM. If the TM has:

- $\text{card}(M) = k, \text{card}(S) = p + 1, \text{card}(F) = m,$

- i move to the right instructions,

- j move to the left instructions,

the extended mH system will have:

- $\text{card}(V) = k + p + 9 = v, \text{card}(L) = k + m + 2$, from the respective definitions,

-
    - vi axioms for the instructions with move to the right, given v possible elements d,
    - two axioms for each $p + 1$ state, to add blanks to the right,
    - j axioms for the instructions with move to the left, one for each instruction,
    - $p + 1$ axioms to add blanks to the left, one for each state,
    - and one work axioms, for a total of

    $vi + 2(p + 1) + j + p + 1 + 1 = vi + 3p + j + 4$ axioms,

-
    - $i(k + 2)^2 + i(k + 2)^3$ rules for the instructions with move to the right,
    - $(k + 2)(p + 1) + (k + 2)(p + 1)$ rules to add blanks to the right,
    - $j(k + 2) + j(k + 2)^2$ rules for the instructions with movement to the left,
    - $(k + 2)(p + 1)$ rules to add blanks to the left, for a total of

    $i(k + 2)^3 + (i + j)(k + 2)^2 + (3p + 3 + j)(k + 2)$ rules.

For any TM the extended mH system will create a number of splicing rules proportional to the cube of the number of the elements of the alphabet of the machine. This complexity represents a limit for an implementation in a computer.

If the TM receives as input a string long l cells, stopping its elaboration after p steps with $t \geq 1$ maximal number of cells used in the work area, using q cells to the left of the input string, and r to the right $(q+l+r = t)$, the extended mH system will stop after $2s + q + 2r$ steps (2 splicings for each step of the TM, one splicing for each blank added to the left, and 2 splicing operations for each blank added to the right) obtaining a work string made with $t + 4$ elements of the alphabet V (including $<<$ to the left and $>>$ to the right).

8   A Simplified Universal Turing Machine with Three Tapes

From the similarity theorem it is possible to see that if we translate a UTM with the characteristics we have defined (for instance the one defined by Turing in [8]), we obtain a universal deterministic extended mH system.

Furthermore, we have obtained a translation from a UTM with three tapes (described in [5]) demonstrating in this way the extreme versatility of the splicing operation. A short summary of the UTM we have translated is given and some parts of the obtained universal deterministic extended mH system are described.

Let us consider a TM having three tapes, which will be called T_1, T_2 and T_3, on each of which only the symbols 0 or 1 can occur, and fifteen states (from q_1 to q_{15}). The description of this machine is given by a list of sextuples, with the group of the first four items in each of them called the *determinant* (the state-symbols group), which in this model must consist of the internal state and the symbol scanned on each of the three tapes. For instance $q_8 101$ is the determinant which indicates that the sextuple in which it occurs has to have effect only if the machine is in internal state q_8, scans 1 on T_1, 0 on T_2 and 1 on T_3. As an added convention $*$ is used in the determinant to indicate that the sextuple is to have effect regardless of whether the corresponding tape shows a 0 or a 1.

The fifth symbol in each sextuple is the operation which is to be performed next. It may be: move to the right (indicated by the letter R), move to the left (indicated by the letter L), write 0 (indicated by 0), write 1 (indicated by 1). These operation are given as in the case of an ordinary

TM, except that a subscript indicates which tape it is applied to. Thus R_1 indicates a step to the right along T_1, and 0_3 indicates the printing of a 0 on T_3.

Finally, the sixth symbol in each sextuple gives the next state of the TM.

We do not give the complete list of the sextuples (that may be found in [5]) but only some of these and their translation into splicing rules and axioms. In order to use this machine, the tape T_1 should be a loop of tape, containing the description of the machine being imitated, T_2 should be an infinite tape which is blank except for containing a coding of the determinant of the machine being imitated, and T_3 should be a copy of the infinite tape which would be on the machine being imitated. The three tapes are used for the purposed indicated above at each step, except that T_2 will contain the sequence of all the past determinants of the machine being imitated, only the most recent of which is used at any step.

In states q_1 through q_6 the UTM is searching along T_1 to find the quadruple that will be pertinent to the next operation of the machine it is imitating. It is done by searching along T_1 to find a block of 1's having the same length as the block on T_2.

- $q_1 11 * R_2 q_2$, $q_2 1 * * R_1 q_1$: the machine steps along to the right ot T_1 and T_2 alternately, starting in state q_1 at the left end of the block of 1's on each tape.

- $q_1 10 * L_2 q_3$: if the machine reaches 0 on T_2 while within the block of 1's on T_1, the block on T_1 was longer, and the machine begins preparation for the comparing the next block on T_1 with the same block on T_2 by moving back one space to return to the block on T_2.

- $q_3 11 * R_1 q_3$: the machine move along toward the right end of the block of 1's on T_1.

- $q_3 01 * L_2 q_4$: when the machine has just passed the right end of the block of 1's on T_1, it moves back toward the beginning of the block on T_2.

- $\ldots$

Now we (partially) describe the universal deterministic extended mH system $\gamma = (V, L, X, U)$ simulating this machine.

-

$$V = \{A, B, C, D, E, F, G, H, I, L, M, N, O, P, Q, R, S, T, U, V,$$
$$W, X, Z, a, b, c, d, e, f, g, h, i, l, m, n, q, r, s, t, u, v, w, x,$$
$$y, z, 0, 1, (,), [,], \lfloor, \rfloor, <, >\},$$

- $L = \{0, 1, <, >, X\}$, set of terminal symbols;

- $Y \subseteq V^*$, set of axioms, support of the multiset function $Q : Y \to$ $\mathbf{N} \cup \{\infty\}$, defined as $Y = O \cup S$, where:

$$O = \{[1B], [1A], [A1], [B1], [E], \lfloor E1 \rfloor, \lfloor 1E \rfloor, EB],$$
$$[B11], [A], [1], O >>, O1], [EA\rfloor, [EA], [B], [C], [1A1\rfloor,$$
$$[1C1], [1C0], [10D], [00D], [01A], [11D], [0D1\rfloor, [0D0],$$
$$[1D0], [11A\rfloor, [11), (B1A], [11 >>, OEA0, F], F00G,$$
$$[00H], [00I], [00L], [0G0], [G1], [1P, [0H0], [1N,$$
$$[0I0], [01M], [01R], [N, [], SRT, SG1], [1MT, S1P,$$
$$[0MT, SH0N, SH1P, abc, eZf, ef, ed, e0d, aZ], abf, eIc,$$
$$< f, << 0q, aZ <], [0b1f, [0b1q, ghi, ml, mZn, gn, Li,$$
$$mZ], [0h1l, mLi, l >>, l0 >>, rQs, t1Bu, t11Bu, rs, r1s,$$
$$vw, x1Qy, tMBw, tRBw, xQy, zB1A], zB\rfloor, vQu, [00U],$$
$$[0L0], VAXY, VUZY\},$$

such that for each $m \in O, Q(m) = \infty$; and $S = \lfloor [11A] \rfloor$;

- $U \subseteq V^* \# V^* \$ V^* \# V^*$ composed by:

  - splicing rules to step along the blocks of 1's in T_1 and T_2:

$$< \#1A1\$[\#11A], < 11A\#]\$[1A1\#1, < 11A\#]\$[1A1\#0,$$
$$1\#1A1\$[\#11A], 111A\#]\$[1A1\#1, 111A\#]\$[1A1\#0,$$
$$0\#1A1\$[\#11A], 011A\#]\$[1A1\#1, [1\#A1]\$[\#1B],$$
$$[\#11B\$ < \#1B1, < 11B\#]\$[1B1\#1, [\#11B\$1\#1B1,$$
$$111B\#]\$[1B1\#1, 111B\#]\$1B1\# >, 0\#1B1\$[\#11B],$$
$$011B\#]\$[1B1\#1, [1\#B1]\$[\#1A];$$

  - splicing rules to use if the block of 1's in T_1 is longer than the one in T_2: the system changes state in T_2 (from B to E) to permit the head to go back:

$$[1\#1B]\$1B\# >>, 11\#B1B]\$[\#E], 11E\#]\$[1\# >>;$$

- splicing rule to move to left T_2's heads:

$$1E\# > \$\lfloor E1\# \rfloor, 1\#1E\rfloor\$\lfloor \#E1 >, 1E\#1\$\lfloor E1\# \rfloor,$$
$$1\#1E\rfloor\$\lfloor \#E11, < \#1E\rfloor\$\lfloor \#E11, 0\#1E\rfloor\$\lfloor \#E11;$$

- splicing rule to put again B in T_2:

$$< \#E1\$[\#1B], , 0\#E1\$[\#1B], < 1B\#]\$[E1\#, 01B\#]\$[E1\#;$$

- splicing rules to change A with C in T_1 and to go to the next block of 1's:

$$[E\#1]\$[B1\#B], [E\#B]\$[\#C], 11\#A0\$[E\#C],$$
$$11\#A1\$[E\#C], 11C\#]\$[EA\#0, 11C\#]\$[EA\#1;$$

- splicing rules to change C with D in T_1 and to go to the next block of 1's:

$$1\#1C1\$[\#11D], 11\#1C0\$[\#10D], 111D\#]\$[1C1\#1,$$
$$111D\#]\$[1C0\#0, 111D\#]\$[1C1\#0, 111D\#]\$[1C0\#1,$$
$$110D\#]\$[1C1\#1, 110D\#]\$[1C0\#0, 110D\#]\$[1C1\#0,$$
$$110D\#]\$[1C0\#1, 1\#1D1\$[\#11D], 111D\#]\$[1A1\#1,$$
$$111D\#]\$[1A1\#0, 1\#0D0\$[\#00D], 100D\#]\$[0D0\#1,$$
$$100D\#]\$[0D0\#0, 1\#1D0\$[\#10D], 110D\#]\$[1D0\#1,$$
$$110D\#]\$[1D0\#0, 0\#0D00\$[\#00D], 000D\#]\$[0D0\#1,$$
$$000D\#]\$[0D0\#0;$$

- splicing rules to change D with A in T_1 when arrived to the end of the block of 1's:

$$0\#0D1\$[\#01A], 1\#0D1\$[\#01A], 001A\#]\$[0D1\#1,$$
$$101A\#]\$[0D1\#1;$$

- splicing rule to permit to begin scanning of the blocks of 1's in T_1 and T_2:
$$[0D1\#]\$[11A\#\rfloor.$$

- ...

The input of the TM, composed by what is written on three tapes, is part of the extended mH system like three axioms present in single copy. As each of these axioms represents a tape, they are represented in the same way used before, that is: bounded with $<<$ to the left and $>>$ to the right, and with a character to the right of the symbol pointed by the head. The state of the UTM is represented with another axiom present in a unique copy. There is no biunivocal correspondence between the state of the UTM and the axioms representing them: a state is represented with more than one axiom. The first axiom, referred to as the circular tape T_1, is circular too.

If we consider that these three strings are in S we may say that for each $s \in S$ we have $Q(s) = 1$.

9   Final Remark

A software implementation of this translation technique has been developed, accepting as input a symbolic description both of the specific Turing machine to be reproduced, and of the fixed translation rules. The output is a step by step execution of the resulting deterministic mH system.

References

[1] J. Dassow, Gh. Păun, *Regulated Rewriting in Formal Language Theory*, Springer-Verlag, Berlin, Heidelberg, 1989.

[2] S. Eilenberg, *Automata, Languages and Machines*, A, Academic Press, New York, 1974.

[3] R. Freund, L. Kari, Gh. Păun, DNA computing based on splicing: the existence of universal computers, Technical Report 185-2/FR-2/95, TU Wien, 1995.

[4] M. A. Harrison, *Introduction to Formal Language Theory*, Addison-Wesley, Reading, Mass., 1978.

[5] E. F. Moore, A simplified universal Turing machine, *Proceedings of the ACM*, 1952.

[6] Gh. Păun, Gh. Păun, Regular extended H systems are computationally universal, *J. Automata, Languages and Combinatorics*, 1, 1 (1996), 27 – 36.

[7] A. Salomaa, *Formal Languages*, Academic Press, New York, 1973.

[8] A. M. Turing, On computable numbers, with an application to the Entscheidungs problem, *Proc. London. Math. Soc.*, Series 2, Vol. 24, 230–265, 1936.

Controlled and Distributed H Systems of a Small Diameter

Andrei PĂUN, Mihaela PĂUN

Faculty of Mathematics, University of Bucharest
Str. Academiei 14, 70109 Bucharest, Romania
paun@sandwich.math.unibuc.ro

Abstract. This paper is a direct continuation of [9]. Characterizations of recursively enumerable languages are given, by means of H systems with permitting contexts or with target languages and by means of communicating distributed H systems having splicing rules of small size (that is, involving short context strings). Representations of context-free languages are also obtained in certain particular cases.

1 Extended H Systems with Permitting Contexts

The extended H systems were introduced in [14], as generative mechanisms based on the *splicing operation* introduced in [7]. A splicing rule corresponds to two restriction enzymes producing sticky ends which match, hence can be pasted by a ligation reaction. Because the enzymes have recognizing patterns of a small length, it is of interest to consider H systems using splicing rules of a small dimension. This parameter, called the *radius* of a splicing rule, was already considered in several papers; see, e.g., [8], [11]. A refinement is considered in [9] for H systems with permitting contexts.

A *splicing rule* (over an alphabet V) is a string $r = u_1 \# u_2 \$ u_3 \# u_4$, where $u_1, u_2, u_3, u_4 \in V^*$ and $\#, \$$ are two special symbols not in V. (V^* is the free monoid generated by the alphabet V under the operation of concatenation; the empty string is denoted by λ; the length of $x \in V^*$ is denoted by $|x|$.)

239

For $x, y, w, z \in V^*$ and r as above we write

$$(x, y) \vdash_r (w, z) \quad \text{iff} \quad x = x_1 u_1 u_2 x_2, \ y = y_1 u_3 u_4 y_2,$$
$$w = x_1 u_1 u_4 y_2, \ z = y_1 u_3 u_2 x_2,$$
$$\text{for some } x_1, x_2, y_1, y_2 \in V^*.$$

We say that we splice x, y at the *sites* $u_1 u_2$, $u_3 u_4$. These sites encode the patterns recognized by restrictions enzymes able to cut the DNA sequences between u_1, u_2, respectively between u_3, u_4. The strings x, y are called the *terms* of the splicing.

The *radius* of a splicing rule $u_1 \# u_2 \$ u_3 \# u_4$ is the length of the longest string u_1, u_2, u_3, u_4.

An *extended H system* ([14]) is a quadruple $\gamma = (V, T, A, R)$, where V is an alphabet, $T \subseteq V$, A is a finite subset of V^*, and $R \subseteq V^* \# V^* \$ V^* \# V^*$, for $\#, \$$ being special symbols not in V. (V is the total alphabet, T is the terminal alphabet, A is the set of axioms, R is the set of splicing rules.) The pair $\sigma = (V, R)$ is sometimes called the underlying *H scheme* associated to γ.

For any $L \subseteq V^*$ and $\gamma = (V, T, A, R)$ we define

$$\sigma(L) = \{w \mid (x, y) \vdash_r (w, z) \text{ or } (x, y) \vdash_r (z, w), \text{ for } x, y \in L, r \in R\},$$
$$\sigma^*(L) = \bigcup_{i \geq 0} \sigma^i(L), \text{ for}$$
$$\sigma^0(L) = L,$$
$$\sigma^{i+1}(L) = \sigma^i(L) \cup \sigma(\sigma^i(L)), \ i \geq 0.$$

Then, the language generated by γ is

$$L(\gamma) = \sigma^*(A) \cap T^*.$$

(We iteratively splice strings, starting from axioms, and we keep in the generated language only strings consisting of terminal symbols.)

Convention. Two generative mechanisms are considered equivalent if the languages generated by them differ by at most the empty string.

We denote by *CF, RE* the families of context-free and of recursively enumerable languages, respectively. We refer to [16] for basic elements of formal language theory.

The extended H systems with finitely many splicing rules generate only regular languages ([5], [15]); when a regular set of splicing rules is used we can characterize the family of recursively enumerable languages ([10]). However, a computing device with infinitely many rules, even constituting a regular language, is not of practical interest. In view of the result in [5], [15], in order to increase the power of finite extended H systems beyond the power of finite automata we have to impose a control on the splicing operation. Several such control mechanisms were considered in [3], [6], [12], [13], etc. We discuss here only one of them.

An *extended H system with permitting contexts* is a quadruple $\gamma = (V, T, A, R)$, where V, T, A are as above, and R is a finite set of triples (we call them rules with permitting contexts) $p : (r = u_1 \# u_2 \$ u_3 \# u_4; C_1, C_2)$, where r is a usual splicing rule and $C_1, C_2 \subseteq V$.

With respect to such a triple p, for $x, y, w, z \in V^*$ we write $(x, y) \vdash_p (w, z)$ iff $(x, y) \vdash_r (w, z)$, all symbols of C_1 appear in x, and all symbols of C_2 appear in y. Then, the language generated by γ is defined as usual.

We denote by $EH(p[m])$ the family of languages generated by extended H systems with permitting contexts, having rules with the maximal radius equal to $m, m \geq 1$. In [3], [6] it is proved that $EH(p[m]) = RE, m \geq 3$. The result is improved in [11]:

Theorem 1.1 $EH(p[m]) = RE, m \geq 2$.

That is to say, systems of radius two suffice. Stronger versions of this result were proved in [9].

2 The Diameter of an H System

Let $\gamma = (V, T, A, R)$ be an extended H system (with permitting contexts). We define the *diameter* of γ (in [9] it is called the *width* of γ) by

$$dia(\gamma) = (n_1, n_2, n_3, n_4),$$

where

$$n_i = \max\{|u_i| \mid u_1 \# u_2 \$ u_3 \# u_4 \in R\}, \ 1 \leq i \leq 4.$$

The family of languages generated by extended H system with permitting contexts having the diameter less than or equal to $(n_1, n_2, n_3, n_4), n_i \geq 0$, $1 \leq i \leq 4$, is denoted by $EH(p[n_1, n_2, n_3, n_4])$ (the vectors are ordered is the componentwise manner).

The construction in [11] in the proof of Theorem 1.1 provides the equality $RE = EH(p[2, 2, 2, 2])$. However, there are $3^4 = 81$ families

$EH(p[n_1, n_2, n_3, n_4])$ such that $(n_1, n_2, n_3, n_4) \leq (2, 2, 2, 2)$. It is proved in [9] that many of them are equal to RE.

The following lemma is useful to this aim and counterparts of it will be valid also for the classes of H systems used in the subsequent sections; for the sake of completeness, we recall here its simple proof.

Lemma 2.1 $EH(p[n_1, n_2, n_3, n_4]) = EH(p[n_3, n_4, n_1, n_2])$, for all $n_i \geq 0$, $1 \leq i \leq 4$.

Proof. Consider an extended H system $\gamma = (V, T, A, R)$ and construct the system $\gamma' = (V, T, A, R')$ with

$$R' = \{(u_3\#u_4\$u_1\#u_2; C_2, C_1) \mid (u_1\#u_2\$u_3\#u_4; C_1, C_2) \in R\}.$$

Because $(x, y) \vdash_p (w, z)$ by $p = (u_1\#u_2\$u_3\#u_4; C_1, C_2)$ if and only if $(y, x) \vdash_{p'} (z, w)$ by $p' = (u_3\#u_4\$u_1\#u_2; C_2, C_1)$, we obtain $L(\gamma) = L(\gamma')$. Clearly, if $dia(\gamma) = (n_1, n_2, n_3, n_4)$, then $dia(\gamma') = (n_3, n_4, n_1, n_2)$. □

Proofs of the following lemmas can be found in [9]. All of them are based on the rotate-and-simulate construction from [10], [3], [6], modified in such a way to keep the diameter of the obtained H system as small as possible.

Lemma 2.2 $RE \subseteq EH(p[0, 2, 1, 0])$.

Lemma 2.3 $RE \subseteq EH(p[2, 0, 0, 1])$.

Lemma 2.4 $RE \subseteq EH(p[1, 2, 0, 1])$.

Lemma 2.5 $RE \subseteq EH(p[2, 1, 1, 0])$.

Synthesizing these results, we obtain the following characterizations of RE.

Theorem 2.1 $RE = EH(p[n_1, n_2, n_3, n_4])$, for all (n_1, n_2, n_3, n_4) (*componentwise*) *greater than or equal to any of the following vectors*

$$(0, 2, 1, 0), \ (1, 0, 0, 2), \ (2, 0, 0, 1), \ (0, 1, 2, 0),$$
$$(1, 2, 0, 1), \ (0, 1, 1, 2), \ (2, 1, 1, 0), \ (1, 0, 2, 1).$$

Theorem 2.1 gives a precise characterization of 49 families of the form $EH(p[n_1, n_2, n_3, n_4])$; thus, out of the 81 families of this type, 32 remain to be further investigated. It is *conjectured* in [9] that all these remaining families are strictly included in RE.

This conjecture is related to the *open problem* whether or not systems of radius 1 can characterize the family RE. In [11] one formulates the conjecture that this is not the case. More specifically, in [11] it is conjectured that $EH(p[1]) = CF$. The inclusion $CF \subseteq EH(p[1])$ is confirmed in [1], the converse inclusion is still open.

The construction in [1] proves the relation $CF \subseteq EH(p[1, 1, 1, 1])$. A stroger result is proved in [9]:

Theorem 2.2 $CF \subseteq EH(p[0, 1, 1, 0]) = EH(p[1, 0, 0, 1])$.

3 Extended H Systems with Targets

In the systems considered above, the control is formulated on the terms of the splicing. Another possibility is to impose restrictions on the result of a splicing operation.

An extended H system *with local targets* is a construct $\gamma = (V, T, A, R)$, where V is an alphabet, $T \subseteq V$ (the terminal alphabet), A is a finite language over V (axioms), and R is a finite set of pairs $p = (r, Q_p)$, where $r = u_1 \# u_2 \$ u_3 \# u_4$ is a splicing rule over V and Q_p is a regular language over V. For $x, y, z, w \in V^*$ and $p = (r, Q_p)$ in R we write $(x, y) \vdash_p (z, w)$ if and only if $(x, y) \vdash_r (z, w)$ and $z, w \in Q_p$ (the results of the splicing with respect to r belong to Q_p).

We denote by $EH(lt[m]), m \geq 1$, the family of languages generated by extended H systems with local targets having splicing rules of radius at most m and by $EH(lt[n_1, n_2, n_3, n_4])$ the family of languages generated by H systems with local targets and of diameter at most $(n_1, n_2, n_3, n_4), n_i \geq 0, 1 \leq i \leq 4$.

In [12] it is proved that $EH(lt[2]) = RE$; in our terminology, in [12] one proves that $EH(lt[2, 2, 2, 2]) = RE$. We improve here this result.

Lemma 3.1 $RE = EH(lt[0, 0, 1, 0]) = EH(lt[1, 0, 0, 0]$.

Proof. Consider a type-0 grammar $G = (N, T, S, P)$ in Kuroda normal form. Denote by P_1 the set of context-free rules in P and by P_2 the set of non-context-free rules in P. We assume the rules in P labelled in a one-to-one manner; denote $U = N \cup T \cup \{B\}$, where B is a new symbol. We construct the extended H system with local targets

$$\gamma = (V, T, A, R),$$

with

$$\begin{aligned}
V = \ & N \cup T \cup \{B, X, X', Y, Z, Z', Z_\lambda, Z'_\lambda\} \\
& \cup \{Z_r \mid r : u \to v \in P\} \cup \{Y_\alpha, Z_\alpha, Z'_\alpha \mid \alpha \in U\}, \\
A = \ & \{XBSY, ZY, XZ', Z_\lambda, Z'_\lambda\} \cup \{Z_r vY \mid r : u \to v \in P\} \\
& \cup \{Z_\alpha Y_\alpha, X'\alpha Z'_\alpha \mid \alpha \in U\},
\end{aligned}$$

and the following splicing rules (we give at the same time the target languages associated to them):

Simulate : 1. $\#\$Z_r\#$, $Q_r = XU^*Y \cup \{Z_r CY\}$, $r : C \to x \in P_1$,
            2. $\#\$Z_r\#$, $Q_r = XU^*Y \cup \{Z_r CDY\}$, $r : CD \to EF \in P_2$,
Rotate : 3. $\#\$Z_\alpha\#$, $Q_\alpha = XU^*Y_\alpha \cup \{Z_\alpha \alpha Y\}$,
            4. $\#\$X\#$, $Q'_\alpha = X'U^*Y_\alpha \cup \{XZ'_\alpha\}$,
            5. $\#\$Z\#$, $Q''_\alpha = X'U^*Y \cup \{ZY_\alpha\}$, $\alpha \in U$,
            6. $\#\$X'\#$, $Q_6 = XU^*Y \cup \{X'Z'\}$,
Terminate : 7. $\#\$B\#$, $Q_7 = T^*Y \cup \{XBZ_\lambda\}$,
            8. $\#\$Z'_\lambda\#$, $Q_8 = T^* \cup \{Z'_\lambda Y\}$.

The splicing rules of type 1 simulate the context-free rules of P, the splicing rules of type 2 simulate the non-context-free rules of P (in the right end of strings of the form XwY, with $w \in U^*$). The rules in groups 3, 4, 5, 6 permute circularly the strings bounded by X and Y:

$$\begin{aligned}
(Xw \mid \alpha Y, Z_\alpha \mid Y_\alpha) &\vdash_3 (XwY_\alpha, Z_\alpha \alpha Y), \\
(X' \alpha \mid Z'_\alpha, X \mid wY_\alpha) &\vdash_4 (X'\alpha wY_\alpha, XZ'_\alpha), \\
(X'\alpha w \mid Y_\alpha, Z \mid Y) &\vdash_5 (X'\alpha wY, ZY_\alpha), \\
(X \mid Z', X' \mid \alpha wY) &\vdash_6 (X\alpha wY, X'Z'),
\end{aligned}$$

for $\alpha \in U, w \in U^*$.

In this way, any sentential form of G can be reproduced in γ, in a circularly permuted form: if $S \Longrightarrow^* w_1 w_2$ in G, then the string $Xw_2 Bw_1 Y$ can be produced in γ. Note that always we have only one occurrence of B in the strings generated in γ.

Finally, the splicing rules of types 7, 8 can remove the auxiliary symbols X, Y, B. This is possible only when B is adjacent to X (hence the string is in the same permutation as in G) and no nonterminal symbol is present (otherwise, the target language restriction is not fulfilled).

The inclusion $L(G) \subseteq L(\gamma)$ follows.

Also the converse inclusion is true: no terminal string which is not in $L(G)$ can be produced by γ. This is due to the control ensured by the target languages and the precise association of auxiliary symbols $Z, Z', Z_r, Z_\alpha, Z'_\alpha, Z_\lambda, Z'_\lambda$ with the splicing rules. No splicing operation which does not correspond to the simulation of rules in P, the rotation of strings, and the final steps described above are possible, excepting splicings which involve "by-product" strings obtained by splicings as mentioned above, that is, strings containing symbols Z with subscripts and primes or not. For instance, using rule 3 we can perform

$$(Z_\alpha | \alpha Y, Z_\alpha | \alpha Y) \vdash_3 (Z_\alpha \alpha Y, Z_\alpha \alpha Y),$$

hence the input strings are reproduced. The same effect is obtained in all cases of splicings of this type.

Thus, $L(G) = L(\gamma)$, which proves the inclusion $RE \subseteq EH(lt[0, 0, 1, 0])$. An equality as that in Lemma 2.1 is true also for extended H systems with target languages, hence we also get $RE = EH(lt[1, 0, 0, 0])$. $\qquad \square$

Lemma 3.2 $RE = EH(lt[0, 1, 0, 0]) = EH(lt[0, 0, 0, 1])$.

Proof. Consider a type-0 grammar $G = (N, T, S, P)$ in Kuroda normal form. Denote by P_1 the set of context-free rules in P and by P_2 the set of non-context-free rules in P. We assume the rules in P labelled in a one-to-one manner; denote $U = N \cup T \cup \{B\}$, where B is a new symbol. We construct the extended H system with local targets

$$\gamma = (V, T, A, R),$$

with

$$V = N \cup T \cup \{B, X, Y, Y', Z, Z', Z_\lambda, Z'_\lambda\}$$
$$\cup \{Z_r \mid r : u \to v \in P\} \cup \{X_\alpha, Z_\alpha, Z'_\alpha \mid \alpha \in U\},$$
$$A = \{XSBY, Z'Y, XZ, Z_\lambda, Z'_\lambda\} \cup \{XvZ_r \mid r : u \to v \in P\}$$
$$\cup \{Z'_\alpha \alpha Y', X_\alpha Z_\alpha \mid \alpha \in U\},$$

and the following splicing rules (we give at the same time the target languages associated to them):

$Simulate$:   1. $\#Z_r\$\#$, $Q_r = XU^*Y \cup \{XCZ_r\}$, $r : C \rightarrow x \in P_1$,
                2. $\#Z_r\$\#$, $Q_r = XU^*Y \cup \{XCDZ_r\}$, $r : CD \rightarrow EF \in P_2$,
$Rotate$:   3. $\#Z_\alpha\$\#$, $Q_\alpha = X_\alpha U^*Y \cup \{X\alpha Z_\alpha\}$,
                4. $\#Y\$\#$, $Q'_\alpha = X_\alpha U^*Y' \cup \{Z'_\alpha Y\}$,
                5. $\#Z\$\#$, $Q''_\alpha = XU^*Y' \cup \{X_\alpha Z\}$, $\alpha \in U$,
                6. $\#Y'\$\#$, $Q_6 = XU^*Y \cup \{Z'Y'\}$,
$Terminate$: 7. $\#B\$\#$, $Q_7 = XT^* \cup \{Z_\lambda BY\}$,
                8. $\#Z'_\lambda\$\#$, $Q_8 = T^* \cup \{XZ'_\lambda\}$.

One can see that this is the same construction as in the proof of Lemma 3.1; the only change is that the string is rotated in the opposite direction. The proof that $L(G) = L(\gamma)$ is analogous. As we said in the proof of Lemma 3.1, we have an equality as that in Lemma 2.1 for extended H systems with target languages, hence we also get $RE = EH(lt[0,0,0,1])$.   □

Using Lemma 3.1 and Lemma 3.2 we get the next theorem:

Theorem 3.1 $EH(lt[1,0,0,0]) = EH(lt[0,1,0,0]) = EH(lt[0,0,1,0]) = EH(lt[0,0,0,1]) = RE$.

It is interesting to note that the only possible improvement of this result is to prove the equality $RE = EH(lt[0,0,0,0])$. Note also the similarity between Theorems 2.1 and 3.1, while Theorem 3.1 refers to splicing systems of a smaller diameter than those in Theorem 2.1. The explanation lies in the power of the target language control, which can check contexts as substrings of the strings resulting from splicing, not only scattered contexts, as the control by permitting contexts does.

4   Communicating Distributed H Systems

Because several enzymes cannot work together (each enzyme needs specific reaction conditions) it is important to diminish the number of splicing rules which have to work together. This leads to distributed architectures of the kinds used in grammar system area, [2]. We consider here only the communicating distributed H systems, introduced in [4].

A *communicating distributed H system* (of degree n, $n \geq 1$) is a construct

$$\Gamma = (V, (A_1, R_1, V_1), \ldots, (A_n, R_n, V_n)),$$

where V is an alphabet, A_i is a finite subset of V^*, R_i is a finite subset of $V^*\#V^*\$V^*\#V^*$, and $V_i \subseteq V$, $1 \leq i \leq n$.

Each triple (A_i, R_i, V_i) is called a *component* of the system, or a *test tube*; A_i is the set of axioms of the tube i, R_i is the set of splicing rules of the tube i, V_i is the *selector* of the tube i.

We denote

$$B = V^* - \bigcup_{i=1}^{n} V_i^*.$$

The pair $\sigma_i = (V, R_i)$ is the underlying H scheme associated to the component i of the system.

An n-tuple $(L_1, \ldots, L_n)$, $L_i \subseteq V^*$, $1 \leq i \leq n$, is called a *configuration* of the system; L_i is also called the *contents* of the ith tube.

For two configurations $(L_1, \ldots, L_n)$, $(L_1', \ldots, L_n')$, we define

$$(L_1, \ldots, L_n) \Longrightarrow (L_1', \ldots, L_n') \text{ iff, for each } i, 1 \leq i \leq n,$$

$$L_i' = \bigcup_{j=1}^{n} (\sigma_j^*(L_j) \cap V_i^*) \cup (\sigma_i^*(L_i) \cap B).$$

In words, the contents of each tube is spliced according to the associated set of rules (we pass from L_i to $\sigma_i^*(L_i)$, $1 \leq i \leq n$), and the result is redistributed among the n components according to the selectors $V_1, \ldots, V_n$; the part which cannot be redistributed (does not belong to some V_k^*, $1 \leq k \leq n$) remains in the tube. Because we have imposed no restriction on the alphabets V_i, for example, we did not suppose that they are pairwise disjoint, when a string in $\sigma_j^*(L_j)$ belongs to several languages V_i^*, then copies of this string will be distributed to all tubes i with this property.

The *language generated* by Γ is

$$L(\Gamma) = \{w \in V^* \mid w \in L_1 \text{ for } (A_1, \ldots, A_n) \Longrightarrow^* (L_1, \ldots, L_n), t \geq 0\},$$

where $\Longrightarrow^*$ is the reflexive and transitive closure of the relation $\Longrightarrow$.

The diameter of a communicating distributed H system is defined in the natural way. We denote by $CDH([n_1, n_2, n_3, n_4])$ the family of languages generated by communicating distributed H systems with the diameter less than or equal to (n_1, n_2, n_3, n_4), $n_i \geq 0$, $1 \leq i \leq 4$.

Results as those in Theorems 2.1 and 3.1 hold true also for these variants of H systems:

Theorem 4.1 $RE = CDH([n_1, n_2, n_3, n_4])$, for all (n_1, n_2, n_3, n_4) (componentwise) greater than or equal to any of the following vectors

$$(0, 2, 1, 0), \ (1, 0, 0, 2), \ (2, 0, 0, 1), \ (0, 1, 2, 0),$$
$$(1, 2, 0, 1), \ (0, 1, 1, 2), \ (2, 1, 1, 0), \ (1, 0, 2, 1).$$

This theorem can be obtained by a series of lemmas corresponding to Lemmas 2.2 – 2.4. We present here only two of them, with a complete construction but without proving the correctness of this construction; this task is left to the reader.

Lemma 4.1 $RE \subseteq CDH([0, 2, 1, 0])$.

Proof. Consider a type-0 grammar $G = (N, T, S, P)$ in Kuroda normal form. Denote by P_1 the set of context-free rules in P and by P_2 the set of non-context-free rules in P. We assume the rules in P labelled in a one-to-one manner; denote $U = N \cup T \cup \{B\}$, where B is a new symbol. We construct the system γ with the alphabet

$$
\begin{aligned}
V = \ & N \cup T \cup \{B, X, X', Y, Z_X, Z_Y, Z_\lambda, Z'_\lambda\} \\
& \cup \{Z_r \mid r : C \to x \in P_1\} \\
& \cup \{Y_r, Z_r, Z'_r \mid r : CD \to EF \in P_2\} \\
& \cup \{Y_\alpha, Z_\alpha, Z'_\alpha \mid \alpha \in U\},
\end{aligned}
$$

and the following components:

$$
\begin{aligned}
A_1 = \ & \emptyset, \\
R_1 = \ & \emptyset, \\
V_1 = \ & T, \\
A_2 = \ & \{XBSY\} \cup \{Z_r xY \mid r : C \to x \in P_1\} \\
& \cup \{Z_r Y_r, Z'_r EFY \mid r : CD \to EF \in P_2\} \\
& \cup \{Z_\alpha Y_\alpha \mid \alpha \in U\}, \\
R_2 = \ & \{\#CY\$Z_r\# \mid r : C \to x \in P_1\} \\
& \cup \{\#DY\$Z_r\#, \#CY_r\$Z'_r\# \mid r : CD \to EF \in P_2\} \\
& \cup \{\#\alpha Y\$Z_\alpha\# \mid \alpha \in U\}, \\
V_2 = \ & N \cup T \cup \{X, Y, B\}, \\
A_{3,\alpha} = \ & \{X'\alpha Z'_\alpha\}, \\
R_{3,\alpha} = \ & \{\#Z'_\alpha\$X\#\},
\end{aligned}
$$

$$V_{3,\alpha} = N \cup T \cup \{X, Y_\alpha, B\}, \text{ for } \alpha \in U,$$
$$A_4 = \{Z_Y Y\},$$
$$R_4 = \{\#Y_\alpha \$ Z_Y \# \mid \alpha \in U\},$$
$$V_4 = N \cup T \cup \{X, B\} \cup \{Y_\alpha \mid \alpha \in U\},$$
$$A_5 = \{X Z_X\},$$
$$R_5 = \{\# Z_X \$ X' \#\},$$
$$V_5 = N \cup T \cup \{X', Y, B\},$$
$$A_6 = \{Z_\lambda, Z'_\lambda\},$$
$$R_6 = \{\# BY \$ Z_\lambda \#, \# Z'_\lambda \$ X \#\},$$
$$V_6 = T \cup \{X, Y, B\}.$$

The work of the system γ is very much similar to that of an extended H system with permitting contexts: the use of a terminal alphabet is ensured by the selection imposed by V_1 (the first tube has only the duty of colleting the terminal strings produced by the other components), while the filters control the process in the same way as permitting contexts do. The reader can check that the equality $L(G) = L(\gamma)$ holds. □

In a quite similar way, but simulating the rules of the initial grammar in the left end of the strings produced by the communicating distributed H system and, moreover, rotating the string in the opposite direction (cut a symbol from the left end and add it in the right hand end), we can also prove:

Lemma 4.2 $RE \subseteq CDH([2, 0, 0, 1])$.

Moreover, we have:

Lemma 4.3 $RE \subseteq CDH([1, 2, 0, 1])$.

Proof. We start by a type-0 grammar $G = (N, T, S, P)$ in Kuroda normal form as in the proof of Lemma 4.1 and we construct the system γ with the alphabet

$$\begin{aligned}
V = {}& N \cup T \cup \{B, X, X', Y, Z_X, Z_Y, Z_\lambda, Z'_\lambda\} \\
& \cup \{Z_r \mid r : C \to x \in P_1\} \\
& \cup \{Y_r, Z_r, Z'_r \mid r : CD \to EF \in P_2\} \\
& \cup \{Y_v, Z_v \mid u \to v \in P\}, \\
& \cup \{Y_\alpha, Z_\alpha, Z'_\alpha \mid \alpha \in U\},
\end{aligned}$$

and the following components:

$$A_1 = \emptyset,$$
$$R_1 = \emptyset,$$
$$V_1 = T,$$
$$A_2 = \{XBSY\} \cup \{Z_r Y_x \mid r : C \to x \in P_1\}$$
$$\cup \{Z_r Y_r, Z'_r Y_{EF} \mid r : CD \to EF \in P_2\}$$
$$\cup \{Z_\alpha Y_\alpha \mid \alpha \in U\},$$
$$R_2 = \{\#CY\$\#Y_x \mid r : C \to x \in P_1\}$$
$$\cup \{\#DY\$\#Y_r, \#CY_r\$\#Y_{EF} \mid r : CD \to EF \in P_2\}$$
$$\cup \{\#\alpha Y\$\#Y_\alpha \mid \alpha \in U\},$$
$$V_2 = N \cup T \cup \{X, Y, B\},$$
$$A_{3,v} = \{Z_v vY\},$$
$$R_{3,v} = \{Z_v \#\$\#Y_v\},$$
$$V_{3,v} = N \cup T \cup \{X, Y_v, B\}, \text{ for } u \to v \in P,$$
$$A_{4,\alpha} = \{X'\alpha Z'_\alpha\},$$
$$R_{4,\alpha} = \{X\#\$\#Z'_\alpha\},$$
$$V_{4,\alpha} = N \cup T \cup \{X, Y_\alpha, B\}, \text{ for } \alpha \in U,$$
$$A_5 = \{Z_Y Y\},$$
$$R_5 = \{\#Y_\alpha\$\#Y \mid \alpha \in U\},$$
$$V_5 = N \cup T \cup \{X', B\} \cup \{Y_\alpha \mid \alpha \in U\},$$
$$A_6 = \{XZ_X\},$$
$$R_6 = \{X'\#\$\#Z_X\},$$
$$V_6 = N \cup T \cup \{X', Y, B\},$$
$$A_7 = \{ZY'_B, Z_\lambda, Z'_\lambda\},$$
$$R_7 = \{\#BY\$\#Y'_B, Z_\lambda\#\$\#Y'_B, X\#\$\#Z'_\lambda\},$$
$$V_7 = T \cup \{X, Y, B\}.$$

We obtain $L(G) = L(\gamma)$. $\qquad\qquad\qquad\qquad\qquad\qquad\qquad\qquad$ □

Systems of radius one are sufficient in order to generate all context-free languages:

Theorem 4.2 $CF \subseteq CDH([0, 1, 1, 0]) = CDH([1, 0, 0, 1]).$

Proof. We use the same idea as in [1], with the modifications in [9], adapted for the case of communicating distributed H systems.

Consider a context-free grammar $G = (N, T, S, P)$ in the strong Chomsky normal form (see, for instance, [17]), that is with the rules in P of the forms $X \to a$, $X \to YZ$, for $X, Y, Z \in N, a \in T$, and with the additional restrictions:

1. if $X \to YZ$ is in P, then $Y \neq Z$,

2. if $X \to YZ$ is in P, then for each rule $X \to Y'Z'$ in P we have $Z' \neq Y$ and $Y' \neq Z$.

Moreover, we assume the rules in P labelled in a one-to-one manner.

We construct the system with the alphabet

$$V = T \cup \{B_1, B_2, B_2', Z_{B_1}, Z_{B_2}, Z_{B_2'} \mid B \in N\}$$
$$\cup \{Z_\lambda, Z_\lambda'\},$$

the components

$$A_1 = \emptyset,$$
$$R_1 = \emptyset,$$
$$V_1 = T,$$
$$A_2 = \{B_1 a B_2 \mid B \to a \in P, a \in T\},$$
$$R_2 = \emptyset,$$
$$V_2 = \emptyset,$$
$$A_7 = \{Z_\lambda, Z_\lambda'\},$$
$$R_7 = \{\#S_2\$Z_\lambda\#, \#Z_\lambda'\$S_1\#\},$$
$$V_7 = T \cup \{S_1, S_2\},$$

as well as the following components for each rule $r : B \to CD \in P$

$$A_{3,r} = \emptyset,$$
$$R_{3,r} = \{\#C_2\$D_1\#\},$$
$$V_{3,r} = T \cup \{C_1, C_2, D_1, D_2\},$$
$$A_{4,r} = \{Z_{B_2'} B_2'\},$$
$$R_{4,r} = \{\#D_2\$Z_{B_2'}\#\},$$
$$V_{4,r} = T \cup \{C_1, D_2\},$$
$$A_{5,r} = \{B_1 Z_{B_1}\},$$
$$R_{5,r} = \{\#Z_{B_1}\$C_1\#\},$$

$$V_{5,r} = T \cup \{C_1, B_2'\},$$
$$A_{6,r} = \{Z_{B_2} B_2\},$$
$$R_{6,r} = \{\# B_2' \$ Z_{B_2} \#\},$$
$$V_{6,r} = T \cup \{B_1, B_2'\}.$$

A string $B_1 w B_2$, with $w \in T^+$, $B \in N$, is generated in γ if and only if $B \Longrightarrow^* w$ in the grammar G. When $B = S$, then also w can be generated in γ.

The first component only selects the terminal strings, the second component introduces the axioms corresponding to the terminal rules in P. The components associated with a rule $r : B \to CD$ in P simulate the application of this rule in the folowing way.

Let us consider two strings $C_1 w_1 C_2$, $D_1 w_2 D_2$ produced by γ (note that the axioms in A_2 are of this form, all other axioms are of a different form). We can perform, in components $(3, r), (4, r), (5, r), (6, r)$, respectively:

$$(C_1 w_1 | C_2, D_1 | w_2 D_2) \vdash (C_1 w_1 w_2 D_2, D_1 C_2),$$
$$(C_1 w_1 w_2 | D_2, Z_{B_2'} | B_2') \vdash (C_1 w_1 w_2 B_2', Z_{B_2'} D_2),$$
$$(B_1 | Z_{B_1}, C_1 | w_1 w_2 B_2') \vdash (B_1 w_1 w_2 B_2', C_1 Z_{B_1}),$$
$$(B_1 w_1 w_2 | B_2', Z_{B_2} | B_2) \vdash (B_1 w_1 w_2 B_2, Z_{B_2} B_2').$$

The use of the rule $B \to CD$ in P has been simulated by the passing from $C_1 w_1 C_2$, $D_1 w_2 D_2$ to $B_1 w_1 w_2 B_2$.

The filters ensure the communication of strings in such a way that no "illegal" splicing is possible.

When we obtain a string of the form $S_1 w S_2$, it can be communicated to the seventh component, where the symbols S_1 and S_2 can be removed:

$$(S_1 w | S_2, Z_\lambda |) \vdash (S_1 w, Z_\lambda S_2),$$
$$(| Z_\lambda', S_1 | w) \vdash (w, S_1 Z_\lambda').$$

Consequently, $L(G) = L(\gamma)$; as $dia(\gamma) = (0, 1, 1, 0)$, we have the inclusion $CF \subseteq CDH([0, 1, 1, 0])$. Because $CDH([0, 1, 1, 0]) = CDH([1, 0, 0, 1])$, the proof is complete.   □

5   Final Remarks

Results similar to those in [9] and in the previous sections are expected for all types of H systems which characterize the family of recursively enumerable languages and for which this characterization is based on the rotate-and-simulate idea. It remains *open* the case of classes of H systems where the characterization of RE is based on other ideas, for instance, on using multisets. In [3], [6] one uses H systems of this type of radius 5, but this bound can probably be improved.

References

[1] V. T. Chakaravarthy, K. Krithivasan, A note on extended H systems with permitting/forbidding contexts of radius one, *Bulletin of the EATCS*, 62 (June 1997), 208 – 213.

[2] E. Csuhaj-Varju, J. Dassow, J. Kelemen, Gh. Păun, *Grammar Systems. A Grammatical Approach to Distribution and Cooperation*, Gordon and Breach, London, 1994.

[3] E. Csuhaj-Varju, L. Freund, L. Kari, Gh. Păun, DNA computing based on splicing: universality results, *First Annual Pacific Symp. on Biocomputing*, Hawaii, Jan. 1996 (L. Hunter, T. E. Klein, eds.), World Sci. Publ., Singapore, 1996, 179 – 190.

[4] E. Csuhaj-Varju, L. Kari, Gh. Păun, Test tube distributed systems based on splicing, *Computers and AI*, 15, 2-3 (1996), 211 – 231.

[5] K. Culik II, T. Harju, Splicing semigroups of dominoes and DNA, *Discrete Appl. Math.*, 31 (1991), 261 – 277.

[6] R. Freund, L. Kari, Gh. Păun, DNA computing based on splicing: The existence of universal computers, Technical Report 185-2/FR-2/95, Technical Univ. Wien, 1995.

[7] T. Head, Formal language theory and DNA: an analysis of the generative capacity of specific recombinant behaviors, *Bull. Math. Biology*, 49 (1987), 737 – 759.

[8] T. Head, Gh. Păun, D. Pixton, Language theory and molecular genetics. Generative mechanisms suggested by DNA recombination, chapter 7 in volume 2 of *Handbook of Formal Languages* (G. Rozenberg, A. Salomaa, eds.), Springer-Verlag, Berlin, Heidelberg, 1997, 295 – 360.

[9] A. Păun, Controlled H systems of small radius, *Fundamenta Informaticae*, 31, 2 (1997), 185 – 193.

[10] Gh. Păun, Regular extended H systems are computationally universal, *J. Automata, Languages and Combinatorics*, 1, 1 (1996), 27 – 36.

[11] Gh. Păun, Computing by splicing: How simple rules ?, *Bulletin of the EATCS*, 60 (1996), 145 – 150.

[12] Gh. Păun, Splicing systems with targets are computationally complete, *Inform. Processing Letters*, 59 (1996), 129 – 133.

[13] Gh. Păun, G. Rozenberg, A. Salomaa, Restricted use of the splicing operation, *Intern. J. Computer Math.*, 60 (1996), 17 – 32.

[14] Gh. Păun, G. Rozenberg, A. Salomaa, Computing by splicing, *Theoretical Computer Science*, 168, 2 (1996), 321 – 336.

[15] D. Pixton, Regularity of splicing languages, *Discrete Appl. Math.*, 69 (1996), 101 – 124.

[16] G. Rozenberg, A. Salomaa, eds., *Handbook of Formal Languages*, 3 volumes, Springer-Verlag, Berlin, Heidelberg, 1997.

[17] D. Wood, *Theory of Computation*, Harper and Row Publishers, New York, 1987.

The Power of H Systems:
Does Representation Matter ?

Hendrik Jan HOOGEBOOM, Nikè van VUGT

Department of Computer Science, Leiden University
P. O. Box 9512, 2300 RA Leiden, The Netherlands
hoogeboom/nvvugt@wi.leidenuniv.nl

Abstract. Splicing rules of the form (u, v, w, x) are usually represented as strings of the form $u\#v\$w\#x$. The effect of splicing with rules from several families of languages has been determined in the literature.

We investigate whether the results about splicing systems obtained in this way are indeed properties of the *splicing system* and not of the specific *string representation* of the rules. We study in detail single and iterated splicing systems, by considering the alternative string representation $u\#w\$v\#x$, and indeed obtain the same classifications as for the standard representation. We briefly discuss some related representations.

1   Introduction

Analogous to the splicing of two molecules with the help of restriction enzymes to produce one or two other molecules, two strings $x = x_1 u_1 v_1 y_1$ and $y = x_2 u_2 v_2 y_2$ can be spliced according to a splicing rule $r = (u_1, v_1, u_2, v_2)$ to give another string $z = x_1 u_1 v_2 y_2$:

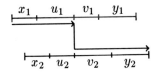

In the literature on splicing sytems, r is usually represented by the string $u_1 \# v_1 \$ u_2 \# v_2$, and so a set of splicing rules is a language. A classification of the generating power of splicing systems with rules defined by the six families of the Chomsky hierarchy is given in [5].

Although this string representation of splicing rules is very natural, other representations are possible. The question arises whether these results are properties of the *splicing systems* or of the specific *representation* of splicing rules that is chosen. For example, would we get different results if we first write the left contexts of the rule and then the right contexts, choosing $u_1 \# u_2 \$ v_1 \# v_2$ instead of $u_1 \# v_1 \$ u_2 \# v_2$ as the string representation of (u_1, v_1, u_2, v_2) ? We will answer this question in detail for single and iterated splicing systems, and show that the families of splicing languages we consider are not influenced by this change in representation. We briefly discuss some other, related, string representations.

A preliminary version of these results was presented at the workshop on Molecular Computing, August 1997 in Mangalia, Romania. We have since included new results, most notably those from Section 5.2.

2   Preliminaries

We use classical notions from formal language theory, see, e.g, [6]. For notions related to splicing theory, the reader may consult the survey [5]. We give some definitions here, mainly to fix our notation.

2.1   Formal Language Theory

For two sets A and B, $A \subset B$ denotes the proper inclusion of A in B, and $A \subseteq B$ denotes the inclusion of A in B (where A and B may be equal).

We denote the empty word by λ.

The quotient of two languages L_1 and L_2, denoted L_1/L_2, is defined as the set $\{x \mid \text{there exists } y \text{ in } L_2 \text{ such that } xy \text{ is in } L_1\}$.

A generalized sequential machine (gsm) is a 6-tuple $\mathcal{A} = (Q, V_1, V_2, \delta, q_0, F)$, where Q is the finite set of states, V_1 is the input alphabet, V_2 is the output alphabet, $q_0 \in Q$ is the initial state, $F \subseteq Q$ is the set of final states, and δ is a mapping from $Q \times V_1$ into finite subsets of $Q \times V_2^*$, the transition relation. By omitting all reference to the output alphabet V_2, we obtain a finite automaton.

We use FIN, REG, LIN, CF, CS and RE to denote the families of finite, regular, linear (context-free), context-free, context-sensitive and recursively enumerable languages, respectively. These families form the Chomsky hierarchy.

2.2 Splicing Theory

We give, in an adapted form, some definitions concerning splicing rules and H systems from [5].

Definition 2.1 *A splicing rule (over an alphabet V) is an element of $(V^*)^4$. For such a rule $r = (u_1, v_1, u_2, v_2)$ and strings $x, y, z \in V^*$ we write*

$$(x, y) \vdash_r z \quad \text{iff} \quad x = x_1 u_1 v_1 y_1, \ y = x_2 u_2 v_2 y_2, \ \text{and}$$
$$z = x_1 u_1 v_2 y_2, \ \text{for some } x_1, y_1, x_2, y_2 \in V^*.$$

The string z is said to be obtained by splicing *the strings x and y using the rule r.*

Definition 2.2 *An H system (or splicing system) is a triple $h = (V, L, R)$ where V is an alphabet, $L \subseteq V^*$ is the initial language and $R \subseteq (V^*)^4$ is a set of splicing rules, the splicing relation. For a given H system $h = (V, L, R)$ we define*

$$\sigma(h) = \{z \in V^* \mid (x, y) \vdash_r z \text{ for some } x, y \in L \text{ and } r \in R\}$$

to be the (single splicing) *language generated by h.*

In [5], splicing rules are represented as strings rather than 4-tuples : a splicing rule $r = (u_1, v_1, u_2, v_2)$ is given as the string $Z(r) = u_1 \# v_1 \$ u_2 \# v_2$ ($\#$ and $\$$ are special symbols not in V), i.e., Z is a mapping from $(V^*)^4$ to $V^* \# V^* \$ V^* \# V^*$, that gives a *string representation* of each splicing rule. We extend Z in the natural way to a mapping from splicing relations to languages. The name of this mapping is suggested by a more graphical notation for the splicing rule r, that used in [9] :

$$\frac{u_1 \mid v_1}{u_2 \mid v_2}$$

and by the way the diagram is then read to get the $u_1 \# v_1 \$ u_2 \# v_2$ notation.

Since a splicing relation R is now represented by the language $Z(R)$, we can consider the effect of splicing with rules from a certain family of languages: for instance, what is the result of splicing linear languages with linear splicing rules?

	FIN	REG	LIN	CF	CS	RE
FIN	FIN	FIN	FIN	FIN	FIN	FIN
REG	REG	REG	REG, LIN	REG, CF	REG, RE	REG, RE
LIN	LIN, CF	LIN, CF				
CF	CF	CF				
CS				RE		
RE						

Table 1: The position of $S(\mathcal{F}_1, \mathcal{F}_2)$ in the Chomsky hierarchy

Example 2.1 Let $L = (L_1 \cdot d) \cup (d \cdot L_2)$, where $L_1 = \{a^n b^n \mid n \geq 1\} \in LIN$ and $L_2 = \{b^n c^n \mid n \geq 1\} \in LIN$. Let $h = (\{a, b, c, d\}, L, R)$ be a splicing system with splicing relation $R = \{(a, b^i d, d, b^i c) \mid i \geq 1\}$. Then $\mathsf{Z}(R) = \{a\#b^i d\$d\#b^i c \mid i \geq 1\} \in LIN$. The language generated by h is

$$\sigma(h) = \{a^n b^n c^n \mid n \geq 1\} \notin LIN.$$

Given two families of languages, $\mathcal{F}_1$ and $\mathcal{F}_2$, a family $S(\mathcal{F}_1, \mathcal{F}_2)$ of single splicing languages (obtained by splicing $\mathcal{F}_1$ languages with $\mathcal{F}_2$ rules) is defined in the obvious way :

$$S(\mathcal{F}_1, \mathcal{F}_2) = \{\sigma(h) \mid h = (V, L, R) \text{ with } L \in \mathcal{F}_1 \text{ and } \mathsf{Z}(R) \in \mathcal{F}_2\}.$$

The families $S(\mathcal{F}_1, \mathcal{F}_2)$ are investigated in [7] and [10], for $\mathcal{F}_1$ and $\mathcal{F}_2$ in the Chomsky hierarchy : $FIN, REG, LIN, CF, CS, RE$. An overview of these results is presented in [5], upon which we base our discussions. When $S(\mathcal{F}_1, \mathcal{F}_2)$ was not found to be equal to one of these six families, the greatest lower bound $\mathcal{F}_3$ and the smallest upper bound $\mathcal{F}_4$ among them are given: $\mathcal{F}_3 \subset S(\mathcal{F}_1, \mathcal{F}_2) \subset \mathcal{F}_4$.

These results are collected in Table 1 from [5], which we repeat here. $\mathcal{F}_1$ is listed from top to bottom, $\mathcal{F}_2$ from left to right. As an example, the optimal classification of splicing LIN languages with REG rules is $LIN \subset S(LIN, REG) \subset CF$.

3   The Problem

Splicing with 'matching' splicing sites may be modelled by the set of rules $R = \{(a^n, b^m, a^n, b^m) \mid m, n \geq 1\}$. This set is not 'context-free', as a consequence of the specific representation Z that we use.

Let us now consider an alternative representation: first writing the left contexts, and then the right contexts of the splicing sites. Formally we use the mapping $\mathsf{\Pi} : (V^*)^4 \to V^*\#V^*\$V^*\#V^*$, with

$\mathcal{M}(u_1, v_1, u_2, v_2) = u_1\#u_2\$v_1\#v_2$. Then, for the 'matching' rules above, $\mathcal{M}(R) = \{a^n\#a^n\$b^m\#b^m \mid m, n \geq 1\}$ is a context-free language!

Consequently the question arises whether the results stated in Table 1 are properties of the splicing systems or of the specific representation of splicing rules that was chosen. In particular, would Table 1 look different if we chose $\mathcal{M}$ instead of Z as our string representation?

To find the answer to this question, we slightly extend the notation for the families of single splicing languages. Let $\rho : (V^*)^4 \to W^*$ be a given string representation of splicing rules over the alphabet V, for some alphabet W. Then define

$$S_\rho(\mathcal{F}_1, \mathcal{F}_2) = \{\sigma(h) \mid h = (V, L, R), \text{ with } L \in \mathcal{F}_1 \text{ and } \rho(R) \in \mathcal{F}_2\}.$$

Hence, by definition, $S(\mathcal{F}_1, \mathcal{F}_2) = S_Z(\mathcal{F}_1, \mathcal{F}_2)$ for the standard string representation Z.

We will also directly consider the family of *splicing relations* defined by the family of languages $\mathcal{F}$ under the representation ρ,

$$\mathcal{R}_\rho(\mathcal{F}) = \{R \subseteq (V^*)^4 \mid \rho(R) \in \mathcal{F}\}.$$

In the next section we investigate whether the splicing relations defined by the language families from the Chomsky hierarchy are changed when we move from the Z-representation to the $\mathcal{M}$-representation. In other words, we determine whether or not $\mathcal{R}_Z(\mathcal{F}_2) = \mathcal{R}_\mathcal{M}(\mathcal{F}_2)$. It is clear from the example above that for $\mathcal{F}_2 = CF$ this is *not* the case. Obviously $\mathcal{R}_Z(\mathcal{F}_2) = \mathcal{R}_\mathcal{M}(\mathcal{F}_2)$ implies $S_Z(\mathcal{F}_1, \mathcal{F}_2) = S_\mathcal{M}(\mathcal{F}_1, \mathcal{F}_2)$ for all $\mathcal{F}_1$.

In Section 5 we consider the remaining cases (for which $\mathcal{R}_Z(\mathcal{F}_2) \neq \mathcal{R}_\mathcal{M}(\mathcal{F}_2)$) and we prove that even for those families we have $S_Z(\mathcal{F}_1, \mathcal{F}_2) = S_\mathcal{M}(\mathcal{F}_1, \mathcal{F}_2)$.

In Section 6 we investigate the effect of using the $\mathcal{M}$-representation instead of the Z-representation for iterated splicing systems, while in Section 7 we discuss related string representations.

4 Families of Splicing Relations

In this section we compare the families of splicing relations defined by the two string representations of the splicing rules.

Observe that Z and $\mathcal{M}$ define a one-to-one correspondence between a splicing rule (u, v, w, x) and the string representations $u\#v\$w\#x$ and $u\#w\$v\#x$ of this splicing rule, respectively. Hence, when considering the Z-representation, one has $Z(R) = L$ iff $R = Z^{-1}(L)$ and $L \subseteq V^*\#V^*\$V^*\#V^*$. Consequently we may write $\mathcal{R}_Z(\mathcal{F}) = \{Z^{-1}(L) \mid L \subseteq V^*\#V^*\$V^*\#V^*$ and $L \in \mathcal{F}\}$.

Proving that $\mathcal{R}_Z(\mathcal{F}) \subseteq \mathcal{R}_M(\mathcal{F})$ amounts to verifying that $M Z^{-1}(L) \in \mathcal{F}$ for every $L \in \mathcal{F}$ with $L \subseteq V^*\#V^*\$V^*\#V^*$. Note that this is a *closure property* of the family $\mathcal{F}$; closure under the operation $M Z^{-1}$ that maps a string $u\#v\$w\#x$ to the string $u\#w\$v\#x$.

The converse inclusion $\mathcal{R}_Z(\mathcal{F}) \supseteq \mathcal{R}_M(\mathcal{F})$ then follows immediately, as $M Z^{-1} = Z M^{-1}$; the operation is its own inverse.

4.1 FIN, REG, CS and RE Splicing Rules

In this subsection, we prove that $\mathcal{R}_Z(\mathcal{F}) = \mathcal{R}_M(\mathcal{F})$, for $\mathcal{F} \in \{FIN, REG, CS, RE\}$.

For the finite languages, it should be clear that the two representations are equivalent : if R is a finite splicing relation, then both $Z(R)$ and $M(R)$ are finite languages.

For the regular, context-sensitive and recursively enumerable languages, we use the following lemma.

Lemma 4.1 REG, CS and RE are closed under $M Z^{-1}$.

Proof. The mapping $M Z^{-1}$ can be realized by a 2-way deterministic generalized sequential machine (a finite state device with a 2-way input tape and a 1-way output tape, see [1]): on input $u\#v\$w\#x$ it outputs $u\#$, skips v, outputs $w\$$, returns on the input to the first $\#$, outputs $v\#$, skips w and finally outputs x. Since $M Z^{-1}$ is its own inverse, it can also be realized by an inverse 2dgsm mapping.

The result now follows from the closure of REG, CS and RE under inverse 2dgsm mappings, see [1, Theorem 2].

Note that the closure of REG under inverse 2dgsm mappings follows quite easily from the fact that 2-way finite state automata accept only regular languages, cf. [6, Theorem 2.5].   □

Consequently we have equality for the two classes of splicing relations for the families under consideration, and thus for H systems with FIN, REG, CS or RE splicing rules, representation does *not* matter.

Corollary 4.1 $\mathcal{R}_Z(\mathcal{F}) = \mathcal{R}_M(\mathcal{F})$ for $\mathcal{F} = REG, CS, RE$.

Theorem 4.1 $S_Z(\mathcal{F}_1, \mathcal{F}_2) = S_M(\mathcal{F}_1, \mathcal{F}_2)$ for $\mathcal{F}_2 = FIN, REG, CS, RE$.

4.2 *LIN* Splicing Rules

In Section 3 we have already seen that $\mathcal{R}_Z(CF) \neq \mathcal{R}_\mathcal{H}(CF)$. In this subsection we additionally show the same inequality for *LIN*.

Consider the splicing relation $R = \{(a^p, c^q, b^r, d^s) \mid p, q, r, s \geq 1$ and $p + q = r + s\}$. Then the Z-representation of R is a linear language, but the $\mathcal{H}$-representation is not. To prove the latter, we use Lemma 2 from [3], which we repeat here.

Lemma 4.2 *Let $L \subseteq a^+b^+c^+d^+$ be a language such that*

1. *$a^n b^n c^k d^k \in L$ for all $n, k \geq 1$,*

2. *if $a^n b^n c^k d^\ell$ is in L, then $k \leq \ell$, and*

3. *there are integers $t_1, t_2 \geq 1$ such that, if $a^n b^m c^k d^\ell$ is in L and $n > m$, then $(n - m)t_1 \leq (k + \ell)t_2$.*

Then L is not linear context-free.

Lemma 4.3 $L = \{a^p b^r c^q d^s \mid p, q, r, s \geq 1$ and $p + q = r + s\} \notin LIN$.

Proof. Lemma 4.2 is applicable, because obviously $L \subseteq a^+b^+c^+d^+$, conditions (1) and (2) hold, and moreover, taking $t_1 = t_2 = 1$, we can see that (3) also holds. Consequently, L is not linear context-free. $\square$

Since *LIN* is closed under homomorphisms (see [6]), from Lemma 4.3 it follows that $\{a^p \# b^r \$ c^q \# d^s \mid p, q, r, s \geq 1$ and $p + q = r + s\} \notin LIN$ and consequently $\mathcal{R}_Z(LIN) \neq \mathcal{R}_\mathcal{H}(LIN)$.

Theorem 4.2 $\mathcal{R}_Z(\mathcal{F}) \neq \mathcal{R}_\mathcal{H}(\mathcal{F})$ *for $\mathcal{F} = LIN, CF$.*

In a way, for these splicing relations, representation *does* matter!

5 Families of Single Splicing Languages

From Section 4.1 we know that $\mathcal{R}_Z(\mathcal{F}_2) = \mathcal{R}_\mathcal{H}(\mathcal{F}_2)$ for $\mathcal{F}_2 \in \{FIN, REG, CS, RE\}$, and thus that $S_Z(\mathcal{F}_1, \mathcal{F}_2) = S_\mathcal{H}(\mathcal{F}_1, \mathcal{F}_2)$ for these families $\mathcal{F}_2$ and each of the six families $\mathcal{F}_1$ considered here.

For $\mathcal{F}_2 \in \{LIN, CF\}$, however, we have demonstrated that $\mathcal{R}_Z(\mathcal{F}_2) \neq \mathcal{R}_\mathcal{H}(\mathcal{F}_2)$, and consequently, we still have to investigate the situation for these two possibilities for $\mathcal{F}_2$.

Because of the nature of the classifications given in the LIN and CF columns of Table 1, we consider the splicing of non-REG languages apart from the splicing of REG-languages.

For the remainder of this section, let $\mathcal{F}_2 \in \{LIN, CF\}$.

5.1 Splicing Non-REG Languages with LIN or CF Splicing Rules

We show that $S_Z(\mathcal{F}_1, \mathcal{F}_2) = S_И(\mathcal{F}_1, \mathcal{F}_2)$, for $\mathcal{F}_1 \neq REG$, by demonstrating that the results used in [5] to fill the corresponding part of Table 1 also hold when the $И$-representation is used. These results are the following :

1. $S_Z(FIN, \mathcal{F}_2) \subseteq FIN$ (obvious),

2. $\mathcal{F}_1 \subseteq S_Z(\mathcal{F}_1, \mathcal{F}_2)$ [5, Lemma 3.2],

3. $L_1/L_2 \in S_Z(\mathcal{F}_2, \mathcal{F}_2)$ for each $L_1, L_2 \in \mathcal{F}_2$ [5, Lemma 3.7].

The proof of (1) is independent of the splicing rules, therefore $S_И(FIN, \mathcal{F}_2) \subseteq FIN$ holds. In the proof of (2), only one splicing rule is used, (λ, c, c, λ), for which the Z-representation is equal to the $И$-representation. For the splicing relation used to prove (3), which is $R = \{(\lambda, wc, c, \lambda) \mid w \in L_2\}$, where $L_2 \in \mathcal{F}_2, L_2 \subseteq V^*$ and $c \notin V$, it should be clear that both $Z(R) = \#L_2c\$c\#$ and $И(R) = \#c\$L_2c\#$ belong to $\mathcal{F}_2$.

By (1) and (2) we have $FIN \subseteq S_И(FIN, \mathcal{F}_2) \subseteq FIN$. As each RE-language is the quotient of two linear languages, from (3) the inclusion $RE \subseteq S_И(LIN, LIN) \subseteq S_И(LIN, CF)$ follows.

Hence this part of the table does not change, and since exact classifications are found, we have the following result.

Theorem 5.1 $S_Z(\mathcal{F}_1, \mathcal{F}_2) = S_И(\mathcal{F}_1, \mathcal{F}_2)$ *for* $\mathcal{F}_1 \neq REG$, $\mathcal{F}_2 = LIN, CF$.

5.2 Splicing REG Languages with LIN or CF Splicing Rules

We now show that $S_Z(REG, \mathcal{F}_2) = S_И(REG, \mathcal{F}_2)$ also holds, by giving a direct proof.

We start by providing a normal form for splicing systems with regular initial language (and rules from a family that is closed under gsm mappings). According to this normal form, every splicing rule is of the form (u, p, q, x), where u and x are strings, while p and q are *symbols*.

This normal form is suggested by the fact that, when a splicing rule (u_1, v_1, u_2, v_2) is applied, the strings v_1 and u_2 do not appear in the result. We only need the fact that the initial strings have these substrings next to the cutting points, which appears to be a finite state property. However, the interchange of these two strings causes the fact that $\mathcal{R}_Z(LIN) \neq \mathcal{R}_{\mathcal{M}}(LIN)$ and $\mathcal{R}_Z(CF) \neq \mathcal{R}_{\mathcal{M}}(CF)$, as explained in Section 4.2. If we are able to restrict v_1 and u_2 to symbols rather than strings, we do not have this problem.

Let L be a regular language, that is accepted by a finite automaton $\mathcal{A} = (Q, V_1, \delta, q_0, F)$ with $Q \cap V_1 = \emptyset$, which is 'reduced', i.e., every state in Q occurs on a path from the initial state to a final state.

Let $p \in Q$ and $u \in V_1^*$. We use $p \overset{u}{\rightarrow}$ to denote the fact that p has an outgoing path with label u in the state transition diagram of $\mathcal{A}$, i.e., $\delta(p, u) \neq \emptyset$. Similarly, we write $\overset{u}{\rightarrow} p$ if p has an incoming path with label u, i.e., $p \in \delta(q, u)$ for some $q \in Q$. We introduce two copies, Q' and Q'', of the set Q.

Consider the splicing rule (u_1, v_1, u_2, v_2). We replace v_1 and u_2 with symbols that convey essentially the same information, by changing (u_1, v_1, u_2, v_2) into (u_1, p_1', p_2'', v_2) where $p_1' \in Q'$ and $p_2'' \in Q''$ such that $p_1 \overset{v_1}{\rightarrow}$ and $\overset{u_2}{\rightarrow} p_2$. That is, v_1 is replaced by a state in $\mathcal{A}$ from which we can read v_1, and u_2 is replaced by a state in $\mathcal{A}$ where we can arrive after reading u_2.

Having adapted the splicing rules, we change the initial language to fit the new rules. Let $L_{\rightarrow p}$ be the language accepted by $\mathcal{A}_{\rightarrow p} = (Q, V_1, \delta, q_0, \{p\})$, and similarly let $L_{p \rightarrow}$ be the language accepted by $\mathcal{A}_{p \rightarrow} = (Q, V_1, \delta, p, F)$. Now consider the language

$$L' = \bigcup_{p \in Q} ((L_{\rightarrow p} \cdot p') \cup (p'' \cdot L_{p \rightarrow}))$$

over the extended alphabet $Q' \cup Q'' \cup V_1$. Since both $L_{\rightarrow p}$ and $L_{p \rightarrow}$ are regular, L' is a regular language.

Lemma 5.1 *Splicing $L \in REG$ with splicing relation R yields the same language as splicing $L' \in REG$ with the adapted set of splicing rules $R' = \{(u_1, p_1', p_2'', v_2) \mid (u_1, v_1, u_2, v_2) \in R$ and $p_1' \in Q', p_2'' \in Q''$ such that $p_1 \overset{v_1}{\rightarrow}$ and $\overset{u_2}{\rightarrow} p_2\}$.*

Proof. From the construction above, it is clear that $x = x_1 u_1 v_1 y_1$ and $y = x_2 u_2 v_2 y_2$ are in L if and only if $x' = x_1 u_1 \cdot p_1' \in L_{\rightarrow p_1} \cdot p_1'$ and $y' =$

$p_2'' \cdot v_2 y_2 \in p_2'' \cdot L_{p_2 \to}$, for some p_1 such that $p_1 \overset{v_1}{\to}$ and p_2 such that $\overset{u_2}{\to} p_2$. Moreover, $r = (u_1, v_1, u_2, v_2) \in R$ if and only if $r' = (u_1, p_1', p_2'', v_2) \in R'$.

Consequently, $(x, y) \vdash_r x_1 u_1 v_2 y_2$ if and only if $(x', y') \vdash_{r'} x_1 u_1 v_2 y_2$. $\quad\square$

Let $\mathcal{F}$ be a family of languages that is closed under gsm mappings. Consider the Z-representation of a splicing rule, $u_1 \# v_1 \$ u_2 \# v_2$. The translation of $u_1 \# v_1 \$ u_2 \# v_2$ into $u_1 \# p_1' \$ p_2'' \# v_2$ with $p_1' \in Q'$ and $p_2'' \in Q''$ such that $p_1 \overset{v_1}{\to}$ and $\overset{u_2}{\to} p_2$ can be realized by a gsm mapping, that simulates the transition diagram of $\mathcal{A}$. Hence there exists an effective construction that transforms an H system of $(REG, \mathcal{F})$-type into an equivalent H system of $(REG, \mathcal{F})$-type that is in normal form.

Furthermore, the translation of the Z-representation of a rule *in normal form* into the M-representation (i.e., $u_1 \# p_1' \$ p_2'' \# v_2 \mapsto u_1 \# p_2'' \$ p_1' \# v_2$) can also be realized by a gsm mapping. Consequently, for a splicing relation R in normal form, $\mathsf{Z}(R) \in \mathcal{F}$ if and only if $\mathsf{M}(R) \in \mathcal{F}$.

Hence, we have the following result, which is applicable for $\mathcal{F} = LIN, CF$.

Theorem 5.2 $S_\mathsf{Z}(REG, \mathcal{F}) = S_\mathsf{M}(REG, \mathcal{F})$ *whenever $\mathcal{F}$ is closed under gsm mappings.*

Summarizing the results from Sections 4 and 5, we have the equality $S_\mathsf{Z}(\mathcal{F}_1, \mathcal{F}_2) = S_\mathsf{M}(\mathcal{F}_1, \mathcal{F}_2)$ for all $\mathcal{F}_1, \mathcal{F}_2$ in the Chomsky hierarchy.

6   Iterated Splicing Systems

We discuss one other type of splicing system that is treated in [5] : iterated splicing. We start by repeating some definitions, again in a slightly adapted form.

Definition 6.1 *The* iterated splicing language $\sigma^*(h)$ *generated by an H system $h = (V, L, R)$ is defined by*

$$\sigma^0(h) = L,$$
$$\sigma^{i+1}(h) = \sigma^i(h) \cup \sigma(\sigma^i(h)), i \geq 0, \text{ and}$$
$$\sigma^*(h) = \bigcup_{i \geq 0} \sigma^i(h).$$

Let ρ be a given string representation of splicing rules over the alphabet V. Similar to the uniterated case, families of iterated splicing languages

	FIN	*REG*	*LIN*	*CF*	*CS*	*RE*
FIN	*FIN, REG*	*FIN, RE*	*FIN, RE*	*FIN, RE*	*FIN, RE*	*FIN, RE*
REG	*REG*	*REG, RE*	*REG, RE*	*REG, RE*	*REG, RE*	*REG, RE*
LIN	*LIN, CF*	*LIN, RE*	*LIN, RE*	*LIN, RE*	*LIN, RE*	*LIN, RE*
CF	*CF*	*CF, RE*	*CF, RE*	*CF, RE*	*CF, RE*	*CF, RE*
CS	*CS, RE*	*CS, RE*	*CS, RE*	*CS, RE*	*CS, RE*	*CS, RE*
RE	*RE*	*RE*	*RE*	*RE*	*RE*	*RE*

Table 2: The position of $H_Z(\mathcal{F}_1, \mathcal{F}_2)$ in the Chomsky hierarchy

are defined :

$$H_\rho(\mathcal{F}_1, \mathcal{F}_2) = \{\sigma^*(h) \mid h = (V, L, R) \text{ with } L \in \mathcal{F}_1 \text{ and } \rho(R) \in \mathcal{F}_2\},$$

for $\mathcal{F}_1, \mathcal{F}_2 \in \{FIN, REG, LIN, CF, CS, RE\}$. A first notable result was obtained in [4], namely that iterated splicing of REG languages by FIN rules does not lead outside REG : $H_Z(REG, FIN) = REG$. These families were further investigated for $\rho = Z$ in [2], [12], [7], [11] and [8], and the results are listed in Table 2 from [5], which we repeat here. Again, $\mathcal{F}_1$ is listed from top to bottom and $\mathcal{F}_2$ from left to right.

The question is whether this table will change when the $\mathcal{M}$-representation rather than the Z-representation is used. Since we know from Section 4 that $\mathcal{R}_Z(\mathcal{F}) = \mathcal{R}_\mathcal{M}(\mathcal{F})$ for $\mathcal{F} \in \{FIN, REG, CS, RE\}$, while $\mathcal{R}_Z(LIN) \neq \mathcal{R}_\mathcal{M}(LIN)$ and $\mathcal{R}_Z(CF) \neq \mathcal{R}_\mathcal{M}(CF)$, we only have to check whether the results used in [5] to fill the LIN and CF columns of Table 2 also hold when the $\mathcal{M}$-representation is used. Those results are the following :

1. $\mathcal{F}_1 \subseteq H_Z(\mathcal{F}_1, \mathcal{F}_2)$: [5, Lemma 3.12],

2. $\mathcal{F}_1 \subset H_Z(\mathcal{F}_1, \mathcal{F}_2)$ for $\mathcal{F}_1 \in \{REG, LIN, CF, CS\}$: [5, Lemma 3.13],

3. $H_Z(FIN, FIN)$ contains infinite languages: [5, discussion in proof of Theorem 3.3],

4. For all $L \subseteq V^*, L \in \mathcal{F}_1$ and $c, d \notin V$ we have $L' = (dc)^*L(dc)^* \cup c(dc)^*L(dc)^*d \notin H_Z(\mathcal{F}_1, \mathcal{F}_2)$: [5, Lemma 3.16],

5. $H_Z(\mathcal{F}_1, \mathcal{F}_2) \not\subseteq CS$ for $\mathcal{F}_2 \neq FIN$: [5, Lemma 3.15].

In the proofs of (1) and (3) the Z-representation of the splicing relation is in FIN, for (4) it is arbitrary, and for (5) it is in REG. These families are closed under $\mathcal{M}Z^{-1}$, hence the same results are obtained when we use the $\mathcal{M}$-representation.

In the proof of (2) from a splicing relation R a new splicing relation $R' = \{(u_1, cv_1, u_2c, v_2) \mid (u_1, v_1, u_2, v_2) \in R\}$ is constructed. Then $Z(R')$ can be obtained from $Z(R)$ by changing the two symbols $\#$ into $\#c$ and $c\#$, respectively; $\Lambda(R')$ can be obtained from $\Lambda(R)$ by changing the $\$$ into $c\$c$. The families in the Chomsky hierarchy are closed under these operations.

For $\mathcal{F}_1 \neq RE$, by (1), (2), (3) we have $\mathcal{F}_1 \subset H_{\Lambda}(\mathcal{F}_1, \mathcal{F}_2)$, while from (4) and (5) it follows that the smallest upper bound for $H_{\Lambda}(\mathcal{F}_1, \mathcal{F}_2)$ is RE. By (1) immediately $RE \subseteq H_{\Lambda}(RE, \mathcal{F}_2)$.

Consequently, Table 2 does not change when we use the Λ-representation instead of the Z-representation. Note, however, that we have *not* proved that $H_Z(\mathcal{F}_1, \mathcal{F}_2) = H_{\Lambda}(\mathcal{F}_1, \mathcal{F}_2)$, for $\mathcal{F}_2 = LIN, CF$, except for the obvious case in which $\mathcal{F}_1 = RE$ where upper and lower bound coincide with RE.

7   Other Representations

We have considered one alternative string representation for splicing relations. Our representation separates left and right contexts rather than the two initial strings. However, there are 24 possible representations in the '$\#\$\#$'-style, corresponding to the permutations of the four components of the splicing rules. We do not claim that all these permutations have a natural interpretation. In the sequel we discuss the remaining possibilities in a rather informal way.

7.1   Splicing with FIN, REG, CS or RE Rules

For (single or iterated) splicing with FIN, REG, CS, or RE splicing rules, it does not matter which of the '$\#\$\#$'-representations is used: results as those of Section 4.1 hold for each of these representations. In other words, if ρ is one of these representations, then $\mathcal{R}_Z(\mathcal{F}_2) = \mathcal{R}_\rho(\mathcal{F}_2)$ and consequently $S_Z(\mathcal{F}_1, \mathcal{F}_2) = S_\rho(\mathcal{F}_1, \mathcal{F}_2)$ and $H_Z(\mathcal{F}_1, \mathcal{F}_2) = H_\rho(\mathcal{F}_1, \mathcal{F}_2)$, for $\mathcal{F}_2 = LIN, REG, CS, RE$.

7.2   Splicing with LIN or CF Rules

Single splicing non-REG languages with LIN or CF splicing rules yields the same classification in each one of the '$\#\$\#$'-representations, because the results used in [5] to fill the corresponding part of Table 1 are easily seen to hold for all string representations ρ in this style, cf. Section 5.1. Since in these classifications upper and lower bound coincide, this implies that $S_z(\mathcal{F}_1, \mathcal{F}_2) = S_\rho(\mathcal{F}_1, \mathcal{F}_2)$, for $\mathcal{F}_1 \neq REG$ and $\mathcal{F}_2 = LIN, CF$.

In the case of single splicing REG languages with CF splicing rules, we use the normal form of Section 5.2 for the rules, that makes it possible to change the Z-representation $u\#p\$q\#x$ of a splicing rule into the $\mathcal{M}$-representation $u\#q\$p\#x$ by applying a gsm mapping (recall that this is possible only because p and q are symbols). Obviously, there exist gsm mappings that map $u\#p\$q\#x$ into each of the 12 representations in which u precedes x. To see that the other 12 possibilities can also be obtained by operations preserving context-freeness, note that CF is closed under the operation CYCLE, that can move x in front of u [6, Exercise 6.4 c]. Again, this shows that $S_z(REG, CF) = S_\rho(REG, CF)$.

For single splicing REG languages with LIN rules, however, we give an example that shows that the '$\#\$\#$'-representations of a rule (u, v, w, x) in which x precedes u are *not* equivalent to the Z-representation.

Example 7.1 *Consider a splicing sytem* $h = (\{a, b, c, d\}, L, R)$ *with* $L = c \cdot \{a, b\}^* \cdot d$ *and* $R = \{(cb^j, d, c, a^n b^i d) \mid i, j, n \geq 1 \text{ and } i + j = n\}$. *Then the 'reverse' representation of* R *(i.e., the representation that maps* (u, v, w, x) *into* $x\#w\$v\#u$*) is* $R_r = \{a^n b^i d\#c\$d\#cb^j \mid i, j, n \geq 1 \text{ and } i + j = n\}$, *which is a linear language. The single splicing language generated by* h, $\sigma(h) = \{cb^j a^n b^i d \mid i, j, n \geq 1 \text{ and } i + j = n\}$, *is not in* LIN *(by the pumping lemma for linear languages [6], Exercise 6.11). Since* $S_z(REG, LIN) \subset LIN$, *clearly this 'reverse' representation is not equivalent to the* Z*-representation.*

Of course, all representations ρ in which u precedes x *are* equivalent to the Z-representation, by using the same arguments as in the CF case. So for those representations we have $S_z(REG, LIN) = S_\rho(REG, LIN)$.

Again, for iterated splicing, it is easy to see that the results used in [5] to fill the LIN and CF columns of Table 2 hold for every '$\#\$\#$'-representation, by observations as in Section 6. Hence this part of the table does not change either.

Summarizing the results of this final section, we see that for iterated splicing the classifications in Table 2 do not change when we use one of the representations in the '$\#\$\#$'-style, but we do not yet know whether or not all *families* of iterated splicing languages stay the same.

For single splicing systems, however, we have seen that all '$\#\$\#$'-representations are equivalent, except for the twelve LIN cases mentioned above.

Acknowledgements. We wish to thank G. Rozenberg for his support and Jurriaan Hage for his comments on a previous version of this paper.

References

[1] A. V. Aho, J. D. Ullman, A characterization of two-way deterministic classes of languages, *Journal of Computer and System Sciences*, 4, 6 (1970), 523 – 538.

[2] K. Culik II, T. Harju, Splicing semigroups of dominoes and DNA, *Discrete Applied Mathematics*, 31 (1991), 261 – 277.

[3] S. A. Greibach, Linearity is polynomially decidable for realtime pushdown store automata, *Information and Control*, 42, 1 (1979), 27 – 37.

[4] T. Head, Formal language theory and DNA: An analysis of the generative capacity of specific recombinant behaviors, *Bulletin of Mathematical Biology*, 49, 6 (1987), 737 – 759

[5] T. Head, Gh. Păun, D. Pixton, Language theory and molecular genetics: Generative mechanisms suggested by DNA recombination, chapter 7 in vol. 2 of *Handbook of Formal Languages* (G. Rozenberg, A. Salomaa, eds.), Springer-Verlag, Heidelberg, 1997, 295 – 360.

[6] J. E. Hopcroft, J. D. Ullman, *Introduction to Automata Theory, Languages, and Computation*, Addison-Wesley, Reading, Mass., 1979.

[7] Gh. Păun, On the splicing operation, *Discrete Applied Mathematics*, 70, (1996), 57 – 79.

[8] Gh. Păun, Regular extended H systems are computationally universal, *Journal of Automata, Languages and Combinatorics*, 1, 1 (1996), 27 – 36.

[9] Gh. Păun, G. Rozenberg, A. Salomaa, Computing by splicing, *Theoretical Computer Science*, 168 (1996), 321 – 336.

[10] Gh. Păun, G. Rozenberg, A. Salomaa, Restricted use of the splicing operation, *International Journal of Computer Mathematics*, 60 (1996), 17 – 32.

[11] D. Pixton, Linear and circular splicing systems, *Proceedings of the 1st International Symposium on Intelligence in Neural and Biological Systems*, Herndon, 1995, 38 – 45.

[12] D. Pixton, Regularity of splicing languages, *Discrete Applied Mathematics*, 69 (1996), 101 – 124.

Splicing Languages Generated with One Sided Context

Tom HEAD

Mathematical Sciences Department
Binghamton University
Binghamton, NY 13902-6000, USA
tom@math.binghamton.edu

Abstract. The splicing system concept was created in 1987 to allow the convenient representation in formal language theoretic terms of recombinant actions of certain sets of enzymes on double stranded DNA molecules. Characterizations are given here for those regular languages that are generated by splicing systems having splicing rules that test context on only one side. An algorithm is given for deciding whether any arbitrary regular language can be generated by a splicing system in which all splicing rules test context on the same side. Schutzenberger's concept of a constant relative to a language provides the tool for constructing the required splicing rules. To provide a potential biochemical example, the formal generative capacity of the restriction enzyme *Bpm*I in the company of a ligase is discussed. Experimental investigation is invited.

1 Introduction

The splicing system concept was introduced in [6] as a device for representing the string restructuring that takes place during the interactions of linear biopolymers in the presence of precisely specified enzymatic activities. Although the development of models based on insertion, deletion, inversion, and transposition were suggested for the future, [6] was concerned most specifically with the recombination operations of double stranded DNA made possible through the application of restriction enzymes and a ligase. Although [6] also suggested the possibility of dealing with circular as well as linear DNA, this was not commented on in detail until [7], [25], [26]. By 1994 both D. Pixton [20], [21], [22] and Gh. Păun [14], [15], [16], [17], [18], [19] had begun deep investigations of splicing systems. R. Gatterdam [5] streamlined the formal definition of the splicing system concept from the original definition [6] which had kept close to the cutting patterns of restriction enzymes. Then Păun gave his more general definition of splicing [14] [15] that allows a coherent study of the action of not only finite sets of splicing rules, but also of infinite sets of splicing rules. It is Păun's definition of splicing that we adopt here in Section 2. An exposition of the splicing literature appears in [10] which includes, as an appendix, a history of the notation and the manner in which the splicing system concept was designed from the DNA biochemistry. A brief sketch of some of these topics appears in [8]. E. Laun and K. J. Reddy have recently reported the first laboratory results on *wet splicing systems* [12]. Splicing systems and their extensions have been renamed H-systems in the European literature. This change helps to prevent the confusion of the splicing operation as understood here (which has been taken from 'gene splicing') with the distinct phenomenon of RNA splicing.

Here our concern is centered on splicing systems having only *a finite number of initial strings* (molecules) and only *a finite number of rules* (enzymes). These systems generate a restricted subclass of the class of regular languages [1], [4], [20], [21], [22]. They are neither capable of generating the noncounting regular language $a^*ba^*ba^*$ nor, as pointed out by Gatterdam [4], the regular language $(aa)^*$ which is not noncounting. The problem of finding alternate characterizations of these languages has been under investigation since [6]. As yet no algorithm has been found for deciding whether a regular language is a splicing language. Here we characterize the regular languages that can be generated by splicing systems having rules that test context on only one side. An algorithm is given for deciding whether a regular language can be generated by a splicing system in which all rules are required to test context on the same side. As yet we do not have an algorithm that will decide whether a given regular language can be generated by a splicing system with each rule is required to test context on only one side, but allowing some rules to test context on the left and others to test context on the right.

Lila Kari has provided a paper [11] which is highly recommended as an introduction to current research in the use of the techniques of molecular biology in the theory, and potential practice, of computation. It is in this context that splicing systems are currently being examined by several researchers. Readers interested in the cut and paste method of generating DNA molecules will also wish to read the paper by Paul Rothemund [23].

2 Definitions

Let A be a finite set to be used as an alphabet. Let $\#$ and $\$$ be symbols not in A. A *rule set* R for an alphabet A, is a set of strings of the form $p\#q\$u\#v$ with p, q, u, and v in A^*. For ease of reading we will write such rules less formally as: $p\#q\$u\#v$. The combination of an *initial language* I contained in A^* with a rule set R yields a *splicing system* $G = (A, I, R)$. G is regarded as a formal device that generates from I a language $L(G)$ that is defined recursively. $L(G)$ is the smallest language in A^* that contains I and possesses the following closure property with respect to R: For each pair of strings $wpqx$ and $yuvz$ in $L(G)$, where w, x, y, and z are in A^* and $p\#q \$ u\#v$ is in R, $wpvz$ is also in $L(G)$. We say that $wpvz$ is obtained from $wpqx$ and $yuvz$ by *splicing* at the points of contact between p and q and between u and v. We say that a rule set R is *reflexive* if, whenever $p\#q \$ u\#v$ is in R, both $p\#q \$ p\#q$ and $u\#v \$ u\#v$ are in R. We will say that R is *symmetric* if, whenever $p\#q \$ u\#v$ is in R, $u\#v \$ p\#q$ is also in R. The rule sets that model the behavior of restriction enzymes (with an accompanying ligase) are necessarily reflexive and symmetric.

A language L with alphabet A is said to be a *splicing language* if there is a splicing system $G = (A, I, R)$ for which $L = L(G)$. A splicing language L is said to be reflexive if there is a splicing system $G = (A, I, R)$ for which $L = L(G)$ where R is reflexive. L is said to be symmetric if $L = L(G)$ for a $G = (A, I, R)$ with R symmetric.

For a detailed explanation of how these definitions have evolved from the corresponding original definitions which were derived from the actions of a set of restriction enzymes (providing the rule set R) and a ligase enzyme on an initial set of double stranded DNA molecules (providing the initial language I), see [10].

3 Splicing Systems with One Sided Context

Recall that a string x in A^* is a *prefix* of a string w in A^* if $w = xz$ for some z in A^* and that x is a *suffix* of w if $w = yx$ for some y in A^*. Recall also that x is a *factor* of w if $w = uxv$ for some strings u and v. The following concept, established by M.P. Schutzenberger [23], is very helpful: A string w is *constant* with respect to a language L if, whenever strings xwy and uwv are in L, the strings xwv and uwy are also in L. In the above definitions any of the strings is allowed to be the null string which is denoted: 1.

Let L be a regular language and let $M = (A, S, \{q_0\}, F, E)$ be the minimal finite automaton that recognizes L, where A is the alphabet of L, S is the set of states of M, q_0 is the initial state of M, F is the set of final states of M, and E is the set of directed edges of M labeled by symbols in A. Suppose that all states not accessible from q_0 have been deleted. Suppose also that all states from which F is not accessible have been deleted. After these deletion processes have been carried out we say that M has been *trimmed*. Note that each constant factor w with respect to L has the following remarkable property with respect to M: Every path in M labeled by w terminates at the same vertex. This is an immediate consequence of the definition of a constant and the fact that, in the minimal automaton of a language, states p and q must be identical if the automata $M(p) = (A, S, \{p\}, F, E)$ and $M(q) = (A, S, \{q\}, F, E)$ recognize the same language.

Note that the concept of a constant possesses a left-right symmetry: c is a constant with respect to L if and only if c^r is a constant relative to L^r, where the exponent r indicates the reversal of strings. We make fundamental use of this symmetry here. We treat L and its reversal L^r in an exactly parallel fashion in the theorem below. For this purpose we let $M' = (A, S', \{q_0'\}, F', E')$ be the trimmed minimal automaton recognizing L^r. Now we may think of M' as reading the strings of L^r from left to right – or we may think of M' as reading the strings of L from right to left. We prefer the latter interpretation as we believe it emphasizes the important left-right symmetry. Note that M and M' need not have the same number of states: For $L = \{a, ba, bb\}$ M has three states and M' has four.

Theorem 3.1 *For each regular language, L, the following two conditions are equivalent:*

(i) *L is a splicing language that is generated by a system $G = (A, I, R)$, where I is finite, where R is reflexive, and where each rule has either the form $u\#1\$v\#1$ or the form $1\#u\$1\#v$;*

(ii) *there is a finite set F of constant factors of L for which the set $X = L \backslash (\cup \{A^*cA^* \mid c \text{ in } F\})$ is finite.*

Proof. Suppose that (i) holds. Let F be the set of all those strings that appear as either the u or the v in a rule of either the form $u\#1\$v\#1$ or the form $1\#u\$1\#v$. Since there are only finitely many rules, F is finite. Since R is reflexive, each string in F is constant. Observe that when strings wux and yvz are spliced, by means of a rule $u\#1\$v\#1$ (respectively, $1\#u\$1\#v$) the resulting wuz (respectively, wvz) has the constant factor u (respectively, v). Thus any word in L that does not have a constant factor must be contained in the finite set I. We conclude that (ii) holds.

Suppose now that (ii) holds. Let M (respectively, M') be the trimmed minimal finite automaton recognizing L (respectively, L^r). Let K be the larger of the number of states in M and the number of states in M'. Let N be the length of the longest word in F[1]. Let $I = X \cup \{s \in L \mid |s| < N + 4K - 1\}$. From each of the finite number of factors of L having the form cw (respectively, wc), with c in F and $|w| = K$, we create a rule to place in R: Since M (respectively, M') has at most K states, $cw = cxyz$ (respectively, $wc = zyxc$) where $|y| \geq 1$ and the state arrived at by M on reading cxy (respectively, by M' on reading yxc) repeats the state arrived at on reading cx (respectively, xc). There may be several such factorizations of cw (respectively, wc). From among these factorizations choose one for which cxy (respectively, yxc) has minimum length. From this factorization construct the rule $cxy\#1\$cx\#1$ (respectively, $1\#yxc\$1\#xc$). Define R to be the reflexive closure of the finite set of rules obtained in this way from factors having either the form cw or the form wc, with c in F and $|w| = K$.

Let $G = (A, I, R)$ with I and R as defined in the previous paragraph. Suppose that L is not contained in $L(G)$. Let s be a shortest string in $L \backslash L(G)$. Since s is not in I, $|s| \geq N + 4K - 1$. Then s has a factorization of the form $s = pcq$ where c is in F and either $|p| \geq 2K$ or $|q| \geq 2K$. In these two cases contradictions arise in a symmetric fashion. We derive the contradiction for only one case: Suppose $|q| \geq 2K$. Since K is the number of states in the automaton M, q has a factorization $q = vwxyz$ which satisfies the following conditions: (1) $|w| \geq 1$ and cvw is the shortest prefix of cq for which M arrives at the same state on reading cvw as it does on reading cv; and (2) $|y| \geq 1$ and M arrives at the same state on reading $cvwxy$ as it does on reading $cvwx$. Then both $pcvwxz$ and $pcvxyz$ are in L and, being shorter than s, are necessarily in $L(G)$. However, condition (1) guarantees that R contains the rule $cvw\#1\$cv\#1$. But $s = pcvwxyz$ is obtained by splicing $pcvwxz$ and $pcvxyz$ by using the rule $cvw\#1\$cv\#1$. The contradiction that s is in $L(G)$ follows from the closure of $L(G)$ under

---

[1] We denote by $|x|$ the length of a string x.

splicing. Thus L is contained in $L(G)$.

Since, from the definition of I, I is contained in L, to conclude the proof we need only show that L is closed under splicing by rules in R. Each rule in R has either the form $cxy\#1\$cx\#1$ or $1\#yxc\$1\#xc$. In these two cases the closure demonstrations are symmetric. We demonstrate closure for only one of these forms. Let $pcxyq$ and $rcxs$ be strings in L for which $cxy\#1\$cx\#1$ is in R. We must show that $pcxys$ is in L. By the construction of R, M reaches the same state on reading cxy as on reading cx. Thus M arrives at the same state when reading either $pcxy$ or rcx. Then M arrives at the same state whether reading $pcxys$ or $rcxs$. Since $rcxs$ is in L, $pcxys$ is also in L as required. Thus $L(G)$ is contained in L. We conclude that $L = L(G)$ and consequently that (ii) holds. $\qquad\square$

This theorem provides for the construction of numerous simple examples of splicing languages: If L is any regular language and c is a symbol not in the alphabet A of L, then the theorem provides splicing representations of such languages as cL, Lc, LcL, LcA^*, etc. A review of the proof of the theorem confirms the following:

Corollary 3.1 *If $L = L(G)$ for a reflexive splicing system for which each rule has either the form $u\#1\$v\#1$ or the form $1\#u\$1\#v$, then $L = L(G')$ for a reflexive splicing system for which each rule has either the form $xy\#1\$x\#1$ or the form $1\#yx\$1\#x$.*

Theorem 3.2 *For a regular language L:*

1. *L is a splicing language that is generated by a system $G = (A, I, R)$, where I is finite, where R is reflexive, and where each rule has the form $u\#1 \$ v\#1$, if and only if all but finitely many prefixes of strings in L are constant;*

2. *L is a splicing language that is generated by a system $G = (A, I, R)$, where I is finite and each rule in R has the form $u\#1\$u\#1$, if and only if all but finitely many factors of strings in L are constant; and*

3. *L is a splicing language that is generated by a system $G = (A, I, R)$, where I is finite, where R is reflexive, and where each rule has the form $1\#u\$1\#v$, if and only if all but finitely many suffixes of strings in L are constant.*

Proof. (1). Let $L = L(G)$ where $G = (A, I, R)$, I is finite, and the rule set R is reflexive and has rules only of the form $u\#1\$v\#1$. Let C be the set of all those words that appear in some rule $u\#1\$v\#1$ in R as either the u or the v. Since R is reflexive, each string in C is constant. Observe that when strings wux and yvz are spliced, by means of a rule $u\#1\$v\#1$, the result wuz has a prefix, namely wu, which has a member of C, namely u, as a factor. It follows that every string in L that arises from a splicing operation has a constant prefix. Moreover, the length of the shortest constant prefix is precisely the length of the shortest constant prefix of the first member of the pair that was spliced. For each string s in I let $N(s)$ be the length of the longest prefix of s that does not contain a member of C as a factor. Let $N = \max\{N(s) \mid s \text{ in } I\}$. Then each prefix of length $N + 1$ or more of each string in L must be a constant. Thus all but finitely many prefixes are constant.

Suppose now that all but finitely many prefixes of strings in L are constant. Let N be the length of the longest non-constant prefix of a string in L. Let $F = \{w \text{ in } A^* \mid |w| = N + 1 \text{ and is a prefix of a string in } L\}$. Then $L\backslash(\cup\{cA^* \mid c \text{ in } F\})$ is finite. From this finiteness and the statement of Theorem 3.1, the proof of (1) is complete except for confirming the lack of need for rules of the form $1\#u\$1\#v$ which can done by a review of the construction of R given in Theorem 3.1.

(2) This is a consequence of the theorem appearing in [6].

(3) This result is obtained from (1) by reversal of strings in the statement and in the proof – provided each rule $xy\#1\$x\#1$ in the proof of (1) is reversed into the rule $1\#yx\$1\#x$ in the proof of (3). □

We have included (2) in the statement of Theorem 3.2 because we believe that the comparison with (1) and (3) is valuable. Result (2) is a corollary of the central theorem of [6], where the languages in (2) were called *null context languages*. It was observed in [6] that, from a theorem of A. DeLuca and A. Restivo [3], it follows that the null context languages are precisely the strictly locally testable (SLT) languages as defined originally by R. McNaughton [13]. Thus Theorem 3.2 picks up just where [6] left off. See [9] for a detailed investigation of splicing representations for SLT languages. We note that Gatterdam's example [4] $b(aa)^*$ is of type (1) but neither of type (2) nor type (3). Likewise $(aa)^*b$ is of type (3) but neither of type (1) nor type (2). A regular language that satisfies (2) must satisfy both (1) and (3), but $b(aa)^*b$ satisfies both (1) and (3) without satisfying (2). Note that there is no asymmetry in the relation of (2) to (1) and (3) since each rule of the form $u\#1\$u\#1$ can just as well be expressed in the form $1\#u \$ 1\#u$, and vice versa.

Let L be a regular language and let $M = (A, S, \{q_0\}, F, E)$ be the trimmed minimal automaton recognizing L. Then the automata Pre $= (A, S, \{q_0\}, S, E)$, Suf $= (A, S, S, F, E)$, and Fac $= (A, S, S, S, E)$ recognize the language $P(L)$ of all prefixes of L, $S(L)$ of all suffixes of L, and $F(L)$ of all factors of L, respectively. Thus $P(L)$, $S(L)$, and $F(L)$ are all constructable regular languages. The language $C(L)$ of all factors of L that are constant with respect to L is also a constructable regular language: For each p in S, let $M(p) = (A, S, S, \{p\}, E)$. For each p in S, let $C(p) = L(M(p)) \backslash (\cup \{L(M(q)) \mid q \in S \backslash \{p\}\})$. Then $C(L)$ is the regular language $\cup \{C(p) \mid p \in S\}$.

Theorem 3.3 *The following questions for a regular language L are decidable:*

1. *Does $L = L(G)$ for some reflexive splicing system $G = (A, I, R)$ where I is finite and all rules in R are of the form $u\#1\$v\#1$?*

2. *Does $L = L(G)$ for some splicing system $G = (A, I, R)$ where I is finite and all rules in R are of the form $u\#1\$u\#1$?*

3. *Does $L = L(G)$ for some reflexive splicing system $G = (A, I, R)$ where I is finite and all rules in R are of the form $1\#u\$1\#v$?*

Proof. Applying Theorem 3.2 we see that decisions (1), (2), and (3) can be made by deciding the finiteness of the intersection of the complement of $C(L)$ with $P(L)$, $F(L)$, and $S(L)$, respectively. $\qquad\square$

For the regular language $L = ba^*ba^*$, $L \backslash (\cup \{A^*wA^* \mid w \in F\})$ is empty for $F = \{ab, bb\}$ and both ab and bb are constants. For the regular language $L = a^*ba^*ba^*$ there does not exist a finite set F of constants for which $L \backslash (\cup \{A^*wA^* \mid w \in F\})$ is finite. For the language $L = (aa)^*$ the set of constants is empty. We propose the following:

Problem. Construct an algorithm for deciding whether, for an arbitrarily given regular language L, there is a finite set F of constants of L for which $L \backslash (\cup \{A^*wA^* \mid w \in F\})$ is finite.

4   Molecular Considerations

When a catalog of commercially available restriction enzymes is consulted (in 1998) one finds that approximately one third of the enzymes listed appear to act on the basis of one sided context. In fact, the sites at which these enzymes act may be listed in such a way that the context is entirely on the left. On the other hand they can equally well be listed so that the context is entirely on the right. This is a consequence of the fundamental fact [6] that (non-palindromic) double stranded DNA molecules possess two equally desirable representations as strings over the alphabet $D = \{A/T, C/G, G/C, T/A\}$ consisting of four compound symbols A/T, C/G, G/C, and T/A. Note, for example, that the following two string over this alphabet D represent the same molecule:

$$5\text{'}-\text{AAAAAAAAAA}-3\text{'} \qquad 5\text{'}-\text{TTTTTTTTTT}-3\text{'}$$
$$3\text{'}-\text{TTTTTTTTTT}-5\text{'} \qquad 3\text{'}-\text{AAAAAAAAAA}-5\text{'}$$

Of several possible enzymes for our further illustration here, we choose to focus exclusively on *Bpm*I. The sites at which this enzyme acts can equally well be expressed in either of the following two ways:

$$5\text{'}-\text{CTGGAGNNNNNNNNNNNNNNNN}-3\text{'}$$
$$3\text{'}-\text{GACCTCMMMMMMMMMMMMMMMM}-5\text{'}$$

$$5\text{'}-\text{MMMMMMMMMMMMMMMMCTCCAG}-3\text{'}$$
$$3\text{'}-\text{NNNNNNNNNNNNNNNNGAGGTC}-5\text{'}$$

Think of *each N/M* and *each M/N* as representing *any one* of the four compound alphabet symbols. In the version of the site expressed on left above, it is natural to regard the context as being entirely on the left, since the cut (in the upper strand) occurs at the extreme right end. However, in the version of this site expressed on the right above, it is natural to regard the context as being entirely on the right, since the cut (in the lower strand) occurs at the extreme left end. Moreover, since *Bpm*I produces 2-base pair overhangs, the two versions above can interact if the overhangs are compatible. Thus to represent the *full* generative capacity of *Bpm*I in the form of splicing rules we must include not only the apparent one sided rules but also the set of rules:

$$\begin{array}{ccccccc} \text{CTGGAG[14N]} & \# & \text{WX} & \$ & \text{WX} & \# & \text{[14M]CTCCAG} \\ \text{GACCTC[14M]} & & \text{YZ} & & \text{YZ} & & \text{[14N]GAGGTC} \end{array}$$

where W, X, Y, and Z satisfy the requirement that both W/Y and X/Z are in $D = \{A/T, C/G, G/C, T/A\}$. What might appear on first inspection of the usual representation of the site at which *Bpm*I operates to be 'left context only' is not even a case of 'one sided context only (with both right and left allowed)'.

Our molecular considerations above have been concerned only with the representation via splicing rules of the *full recombinant capacity* of a restriction enzyme accompanied by a ligase. When we restrict attention to the action of a set of restriction enzyme(s) and a ligase on a specific initial language I, it may be apparent that certain splicing rules in the repertoire R of the enzyme system will never have the opportunity to be exercised. An example follows that will allow us to recover contact with the message of Theorem 3.1.

We continue with the single restriction enzyme *Bpm*I and a ligase. As the initial language I we choose the single molecule m (but in literally trillions of copies, as is usual in such experiments) where the representation of m is subdivided into small groups of bases for ease in grasping the structural significance:

```
GCCTGCGCCT 2 GCGCCT 1 [8T]CTCCAG CTCCAG
CGGACGCG 2 GACGCG 1 GA[8A]GAGGTC GAGGTC
                       <---B2 <---B1

             T1---> T2--->
             CTGGAG CTGGAG[8A]GC 1 GCCTGC 2 GCCTGC
             GACCTC GACCTC[8T] 1 CGCGGA 2 CGCGGACG
```

$T1$ indicates the six base pair recognition site that allows the cuts at the positions marked with the numeral 1 at the right. $T2$ indicates the site that allows the cuts at the positions marked with the numeral 2 at the right. $B1$ shows the site that allows the cuts at the positions marked with the numeral 1 at the left. $B2$ shows the site that allows the cuts at the positions marked with the numeral 2 at the left.

We will construct as an example a splicing system $G = (D, I, R)$, where $I = \{m\}$, where m is the 70 base pair molecule listed above, and R be the reflexive symmetric closure of the set that consists of the following pair of rules which represent two of the recombinant capabilities allowed by *Bpm*I:

```
CTGGAG [8A]GC GCCTGC # 1   $   CTGGAG CTGGAG [8A]GC # 1
GACCTC [8T]CG CGGACG           GACCTC GACCTC [8T]CG
```

and

```
1 # CT[8T] CTCCAG CTCCAG   $   1 # CTGC GCCT[8T] CTCCAG
    GA[8A] GAGGTC GAGGTC           GACG CGGA[8A] GAGGTC.
```

As a *formal* splicing system, $G = (D, I, R)$ can be verified (mentally, perhaps) to generate the language $L(G)$ given by the regular expression:

```
GCCTGC (GCCTGC)* GCCT[8T] CTCCAG CTCCAG
CGGACG  CGGACG   CGGA[8A] GAGGTC GAGGTC

             CTGGAG CTGGAG [8A]GC (GCCTGC)* GCCTGC
             GACCTC GACCTC [8T]CG  CGGACG    CGGACG
```

For

```
u = GCCTGC   x = GCCT[8T] CTCCAG CTCCAG CTGGAG CTGGAG [8A]GC
    CGGACG       CGGA[8A] GAGGTC GAGGCC GACCTC GACCTC [8T]CG
```

we have $L(G) = uu^*xu^*u = u^*uxuu^*$. It is easily verified that x is a constant for $L(G)$ and that no factor of u^* is a constant. We compare $L = L(G)$ with the conditions appearing in Theorems 3.1 and 3.2: There is a finite set F of constant factors of L for which L is contained in D^*FD^*, namely the singleton set $F = \{x\}$. Infinitely many prefixes of strings in L fail to be constant as do infinitely many suffixes. Thus L cannot be generated by any splicing system for which every rule tests context on the same side, whether right or left. In particular, L is not strictly locally testable and cannot be generated by a null context splicing system as defined in [6] and [9].

If a buffer can be found that allows the parallel activity of both *BpmI* and a ligase to take place in the same test tube, is it reasonable to expect the set of well formed fully double stranded linear DNA molecules that can potentially arise from the initial set $I = \{m\}$ to be representable as $L(G)$? Although the repertoire of *BmpI* is vastly more extensive (as indicated above) than we have encoded in the small rule set R of our example G, a thoughtful examination suggests that no molecule will be generated that will allow any further rules to be applicable to molecules that will arise. Thus we predict that $L = L(G)$ will properly model the language of linear DNA molecules that would potentially arise. Three provisos follow: (1) If, after a cut of type $T2$, a cut of type $T1$ can occur in the resulting left segment then fragments of the form:

```
GCGCCT
GACGCG
```

will arise. If sufficiently many of such fragments arise then circular molecules of the form

```
^(GCGCCT)*
CGCGGA
```

may form in addition to the predicted linear molecules [7]. (2) Due to the specific manner in which *BpmI* finds and attaches to its site before cutting, it may be necessary to add additional length at the center of m, between base pairs 36 and 37, and also at each end of m. Any such extra length that might be added preserves the significance of our example, assuming

the insertions are chosen not to create new sites at which *Bpm*I can act.
(3) Since the formation of long molecules in $L(G)$ requires the formation
of many short molecules, in an actual experiment it would be well to insert
several extra copies of $[G/C][C/G][C/G][G/C][T/A][C/G]$ at each end of
m.

We invite laboratory testing of models of this type, whether based on
*Bpm*I or any of several other similar possibilities.

Acknowledgement. I wish to thank Elizabeth Laun for her careful reading of this manuscript. Partial support for this research through NSF CCR-9509831 and DARPA/NSF CCR-9725021 is gratefully acknowledged.

References

[1] K. Culik II, T. Harju, Splicing semigroups of dominoes and DNA, *Discrete Applied Mathematics*, 31 (1991), 261 – 277.

[2] K. L. Denninghoff, R. W. Gatterdam, On the undecidability of splicing systems, *Intern. J. Computer Math.*, 27 (1989), 133 – 145.

[3] A. DeLuca, A. Restivo, A characterization of strictly locally testable languages and its application to subsemigroups of a free semigroup, *Information and Control*, 44 (1980), 300 – 319.

[4] R. W. Gatterdam, Splicing systems and regularity, *Intern. J. Computer Math.*, 31 (1989), 63 – 67.

[5] R. W. Gatterdam, DNA and Twist Free Splicing Systems, in *Words, Languages and Combinatorics II* (M. Ito, H. Jurgensen, eds.), World Scientific, Singapore, 1994, 170 – 178.

[6] T. Head, Formal language theory and DNA: An analysis of the generative capacity of specific recombinant behaviors, *Bulletin of Mathematical Biology*, 49 (1987), 737 – 759.

[7] T. Head, Splicing schemes and DNA, in *Lindenmayer Systems - Impacts on Theoretical Computer Science, Computer Graphics, and Developmental Biology* (G. Rozenberg, A. Salomaa, eds.), Springer-Verlag, Berlin, 371 – 383, 1992 (also in *Nanobiology*, 1 (1992), 335 – 342).

[8] T. Head, Splicing systems and molecular processes, in *Proc. ICEC'97 Special Session on DNA Based Computation*, IN, IEEE Press, 1997, 203 – 205.

[9] T. Head, Splicing representations of strictly locally testable languages, *Discrete Applied Math.*, to appear.

[10] T. Head, Gh. Păun, D. Pixton, Language theory and molecular genetics: generative mechanisms suggested by DNA recombination, in *Handbook of Formal Languages* (G. Rozenberg, A. Salomaa, eds.), Springer-Verlag, 1997, chapter 7, vol. 2, 295 – 360.

[11] L. Kari, DNA computing: arrival of biological mathematics, *The Mathematical Intelligencer*, 19, 2 (Spring 1997), 9 – 22.

[12] E. Laun, K. J. Reddy, Wet splicing systems, *DIMACS Series in Discrete Math. and Theor. Computer Sci.*, Amer. Math. Soc., to appear.

[13] R. McNaughton, S. Papert, *Counter-Free Automata*, MIT Press, Cambridge, Massachusetts, 1971.

[14] Gh. Păun, The splicing as an operation on formal languages, *Proc. Intelligence in Neural & Biological Systems*, IEEE Press, 1995, 176 – 180.

[15] Gh. Păun, On the splicing operation, *Discrete Applied Math.*, 70 (1996), 57 – 79.

[16] Gh. Păun, Regular extended H systems are universal, *J. Automata, Languages, Combinatorics*, 1 (1996), 27 – 36.

[17] Gh. Păun, Five (plus two) universal DNA computing models based on the splicing operation, *DIMACS Series in Discrete Math. and Theor. Computer Sci.*, Amer. Math. Soc., to appear.

[18] Gh. Păun, G. Rozenberg, A. Salomaa, Computing by splicing, *Theor. Computer Sci.*, 168 (1996), 321 – 336.

[19] Gh. Păun, A. Salomaa, DNA computing based on the splicing operation, *Mathematica Japonica*, 43 (1996), 607 – 632.

[20] D. Pixton, Linear and circular splicing systems, *Proc. Intelligence in Neural & Biological Systems*, IEEE Press, 1995, 181 – 188.

[21] D. Pixton, Regularity of splicing systems, *Discrete Applied Math.*, 69 (1996), 101 – 124.

[22] D. Pixton, Splicing in abstract families of languages, *Theor. Computer Sci.*, to appear.

[23] P. W. K. Rothemund, A DNA and restriction enzyme implementation of Turing machines, *DIMACS Series in Discrete Math. and Theor. Computer Sci.*, 27, Amer. Math. Soc., 1996, 75 – 119.

[24] M. P. Schützenberger, Sur certaines operations de fermeture dans les langages rationels, *Symposium Mathematica*, 15 (1975), 245 – 253.

[25] R. Siromoney, K. G. Subramanian, V. R. Dare, Circular DNA and splicing systems, in *Parallel Image Analysis* (*Lect. Notes in Computer Sci.* 654), Springer-Verlag, Berlin, 1992, 260 – 273.

[26] T. Yokomori, S. Kobayashi, C. Ferretti, On the power of circular splicing systems and DNA computability, in *Proc. ICEC'97 Special Session on DNA Based Computation*, IN, IEEE Press, 1997, 219 – 224.

Self Cross-over Systems

Jürgen DASSOW

Faculty of Computer Science, University of Magdeburg
P.O.Box 4120, D-39016, Magdeburg, Germany
dassow@cs.uni-magdeburg.de

Victor MITRANA[1]

Faculty of Mathematics, University of Bucharest
Str. Academiei 14, 70109, Bucharest, Romania
mitrana@funinf.math.unibuc.ro

Abstract. We consider the following way of generating a language, by using the cross-over operation: start with a finite set of strings and a given set of cross-over rules which are applicable only to two identical strings. By applying a cross-over rule to two copies of an initial string we get two new strings. Iteratively, we get a language. Some properties (closure, decidability, etc.) of these languages are investigated.

1 Introduction

One of the most recent suggestions in developing new types of computers consists of considering computers based on molecular interactions, which, under some circumstances, might be an alternative to the classical Turing/von Neumann notion of a computer.

This paper deals with a very particular case of cross-over despite that this is not the only biologically significant case. One has asserted [2] that in vivo, cross-over takes place just between homologous chromosomes (chromosomes of the same type and of the same length). A first attempt to

---

[1]Research supported by Alexander von Humboldt Foundation

283

model the homologous recombination was made in [5], where cross-over between strings of equal length, which exchange each other segments of equal length, is proposed.

Roughly speaking, in the present paper, we try to model cross-over between a DNA molecule and a its replicated version. Thus, our approach appears as a model for cross-over between "sister" chromatids. In our opinion this restriction makes up a theoretical aspect of molecular biology that deserves to be investigated. What would happen if cross-over occurred only between a chromosome and its replica ?

A similar operation specific to DNA recombination in vitro is *splicing*. This operation was firstly considered in [3] and has been intensively studied in a series of paper as a model for biological computing [7], [8] and the references thereof. A survey may be found in [4].

The main idea of our approach is schematically presented in the figure below:

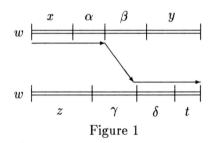

Figure 1

One gives a starting finite set of string and a finite set of cross-over rules $(\alpha, \beta, \gamma, \delta)$. One considers that every starting string is replicated so that, we have two identical copies for every initial string. The first copy is cut between the segments α and β and the other one is cut between γ and δ. Now, the last segment of the second string adheres to the first segment of the first string, and a new string is obtained. More generally, another string is also generated, by linking the first segment of the second string with the last segment of the first string. Iterating the procedure, we get a language.

We want to point out, at this moment, some connections between our approach and other large scale operations in genome and in formal language theory as well. If the situation is as in the Figure 1, then we have the deletion of a certain substring of w which may be viewed as the deletion of a segment of a chromosome.

If the situation is as in Figure 2, then we have the insertion of a substring of w in w; this appears as duplication of chromosomes in genomes.

As a matter of fact, we have to mention here that the self cross-over operation was already considered in [9], [6], as a particular case of splicing with clusters, but it was studied only as a non-iterated operation on languages.

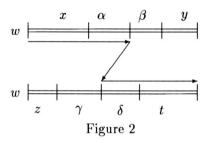

Figure 2

2 Definitions and Examples

We denote: V^* = the free monoid generated by V under catenation, λ = the empty string, $V^* = V^+ - \{\lambda\}$, $|x|$ = the length of the string x. For further details in formal language theory we refer to [11].

A self cross-over system is a triple $SCO = (V, A, R)$, where V is an alphabet, A is a finite subset of V^*, R is a finite commutative relation, $R \subset (V^* \times V^*)^2$.

With respect to a self cross-over system as above, for $x \in V^+$, we define:

$$x \bowtie y \quad \text{iff} \quad (1) x = x_1 \alpha \beta x_2 = x_3 \gamma \delta x_4$$
$$(2) \quad y = x_1 \alpha \delta x_4$$
$$(3) \quad (\alpha, \beta) R(\gamma, \delta).$$

Note that $x \bowtie x_3 \gamma \beta x_2$ follows from the definition of R. Moreover, the strings $x_1 \alpha \delta x_4$ and $x_3 \gamma \beta x_2$ are somehow "conjugated" namely, there exists $u \in V^*$ such that $x = v_1 u v_2$, for some $v_1, v_2 \in V^*$, such that $x_1 \alpha \delta x_4 = v_1 u u v_2$, $x_3 \gamma \beta x_2 = v_1 v_2$.

We denote by $\bowtie^*$ the reflexive and transitive closure of the relation $\bowtie$.

The language generated by a self cross-over system as above is

$$L(SCO) = \{x \in V^* \mid w \bowtie^* x, \ w \in A\}.$$

Example 1. Take $V = \{a, b\}$, $A = \{bab\}$, $R = \{(a, b; b, a), (b, a; a, b)\}$.

We have
$$L(SCO) = \{bb\} \cup \{ba^{2^n}b \mid n \geq 0\}.$$

Indeed, $bab \bowtie ba^2b$ and $bab \bowtie bb$. Assuming that $bab \bowtie^* ba^{2^n}b$, by applying the rule $(a, b; b, a)$ to this string, we get $ba^{2^n}b \bowtie ba^{2^{n+1}}b$. By using the other splicing rule, we get $ba^{2^n}b \bowtie bb$. □

Example 2. Consider $V = \{a_1, a_2, \ldots, a_n\}$, $A = \{a_1 a_2 \ldots a_n\}$, $R = \{(\lambda, \lambda; \lambda, \lambda)\}$.

We state that $L(SCO) = V^*$. We are going to prove our assertion by induction on n. It is obvious that the statement is true for $n = 1$.

Let $z \in V^+$,
$$z = z_1 a_n^{t_1} z_2 a_n^{t_2} \ldots z_k a_n^{t_k} z_{k+1},$$

with $t_1, t_2, \ldots, t_k \geq 1$, $z_1, z_{k+1} \in (V - \{a_n\})^*$, $z_1, z_2, \ldots, z_k \in (V - \{a_n\})^+$.

By the induction hypothesis and following the cross-over rule, we can perform the sequence of cross-over operations below:

$$w = a_1 a_2 \ldots a_n \bowtie a_1 a_2 \ldots a_n w \bowtie^* z_1 a_n w \bowtie z_1 a_n^2 w \bowtie^* z_1 a_n^{t_1} w.$$

Going on, one obtains

$$z_1 a_n^{t_1} w \bowtie z_1 a_n^{t_1} ww \bowtie^* z_1 a_n^{t_1} z_2 a_n w \bowtie^* z_1 a_n^{t_1} z_2 a_n^{t_2} w \bowtie^* \ldots$$
$$\ldots \bowtie^* z_1 a_n^{t_1} z_2 a_n^{t_2} \ldots z_k a_n^{t_k} w \bowtie^* z_1 a_n^{t_1} z_2 a_n^{t_2} \ldots z_k a_n^{t_k} z_{k+1} a_n \bowtie z.$$

Therefore, every string in V^* can be generated by the above system, which concludes the proof. □

Denote by $\mathcal{L}(SCO)$ the family of languages generated by self cross-over systems. The elements of $\mathcal{L}(SCO)$ will be referred to as self cross-over languages.

Since our operation is closely related with two operations modelling the rearrangements at the chromosome level in genomes, let us recall them following [12], [1]. Let V be an alphabet and $Dupl$ and Del be two finite subsets of V^*. For a string $w \in V^*$ we define

$$Dupl(w) = \{xyyz \mid w = xyz, \ y \in Dupl\},$$
$$Del(w) = \{xz \mid w = xyz, \ y \in Del\}.$$

They are naturally extended to languages:

$$Dupl(L) = \bigcup_{w \in L} Dupl(w), \qquad Del(L) = \bigcup_{w \in L} Del(w).$$

Let us consider now a (simplified) *evolutionary system*, in the form $ES = (V, Dupl, Del, A)$, where V is an alphabet, A is a finite subset of V^*, *Dupl* and *Del* have the above meaning; we define recursively the sets

$$L_0(ES) = A,$$
$$L_{i+1}(ES) = Dupl(L_i(ES)) \cup Del(L_i(ES)), \ i \geq 0.$$

The language produced by an evolutionary system is

$$L(ES) = \bigcup_{i \geq 0} L_i(ES).$$

The class of all languages $L(ES)$ is denoted by $\mathcal{L}(ES)$.

3 Self Cross-over Languages

The next theorem characterizes the self cross-over languages over the one letter alphabet.

Theorem 3.1 *Every self cross-over language L, over $\{a\}$, is either a finite set or a finite set $F \in V^*$ and $k > 0$ exist such that $L = F \cup \{a^n \mid n \geq k\}$.*

Proof. Let $SCO = (\{a\}, A, R)$ be a self cross-over system. Since

$$L(SCO) = \bigcup_{x \in A} L(SCO_x), \ SCO_x = (\{a\}, x, R),$$

it suffices to show that any language $L(SCO_x)$ is either a singleton or of the form $\{a^n \mid n \geq k\}$, for some $k > 0$. Clearly, every language $L(SCO_x)$ is either a singleton or an infinite language.

Let $L(SCO_x)$ be an infinite language and

$$k = min\{|\alpha\delta| \mid (\alpha, \beta)R(\gamma, \delta)\},$$
$$p = max(\{|\alpha| \mid (\alpha, \beta)R(\gamma, \delta)\} \cup \{|\beta| \mid (\alpha, \beta)R(\gamma, \delta)\}).$$

Moreover, let $(\alpha, \beta; \gamma, \delta)$ be the cross-over rule which fulfils the minimal value of k. Of course, $L(SCO_x) \subseteq \{a^n \mid n \geq k\}$. We show that any $z = a^n, n \geq k$ is a string of $L(SCO_x)$. Because $L(SCO_x)$ is an infinite set, exists $m > n + p$ such that $a^m \in L(SCO_x)$.

Then, we have the following decompositions

$$a^m = a^{n-|\alpha|-|\delta|}a^{|\alpha|}/a^{m-n+|\delta|},$$
$$a^m = a^{m-|\delta|}/a^{|\delta|},$$

where the cross-over sites are indicated by the symbol $/$, which results in generating of the string a^n.   □

Lemma 3.1 *The language* $L = a^*b^*a^*b^*$ *cannot be generated by a self cross-over system.*

Proof. Assume that L can be generated by a system $SCO = (\{a, b\}, A, R)$. Let $z = a^n b^m a^k b^p$ be a string in L such that n, m, p, k are bigger than the length of the longest string over $\{a, b\}$ which occurs in a rule of R.

Please note that, at each cross-over between two identical strings of the above form, only the number of occurrences of only one symbol, in just one of its two segments is modified. According to this remark, in order to get z, we have to produce a cross-over on a string of the form $a^n b^m a^k b^r, 0 < r < p$, at the sites indicated below by the symbol $/$:

$$a^n b^m a^k b^{r_1} / b^{r_2}, r_1 > 0,$$
$$a^n b^m a^k b^{r_3} / b^{r_4}.$$

But, we can choose also the following sites for cross-over: $a^n b^m a^k b^{r_1} / b^{r_2}$, $a^n b^{m_1} / b^{m_2} a^k b^r$, and we get the string $a^n b^m a^k b^{r_1+m_2} a^k b^r$, which leads to a contradiction.   □

The next result is a consequence of above examples and of Lemma 3.1.

Theorem 3.2 *The family* $\mathcal{L}(SCO)$ *is incomparable with the families of regular and context-free languages.*

Theorem 3.3 *The family* $\mathcal{L}(SCO)$ *is incomparable with each of the following families: D0L languages, 0L languages, E0L languages.*

Proof. The only non-trivial assertion is $\mathcal{L}(SCO) - E0L \neq \emptyset$. A language in the difference above is:

$$L = \{c(ab^m)^{2^n} c \mid m, n \geq 0\} \cup \{cab^m d^p c \mid m, p \geq 0\}.$$

A self cross-over system which generates L is

$$SCO = (\{a, b, c, d\}, \{cabdc\}, R),$$

where

$$R = R_1 \cup R_2 \cup R_3 \cup R_4 \cup R_5,$$
$$R_1 = \{(b, d; \lambda, bd), (\lambda, bd; b, d)\},$$
$$R_2 = \{(b, d; d, c), (d, c; b, d)\},$$
$$R_3 = \{(a, d; d, c), (d, c; a, d)\},$$
$$R_4 = \{(b, c; c, a), (c, a; b, c)\},$$
$$R_5 = \{(a, c; c, a), (a, c; c, a)\}.$$

It is obvious that:

$$cabdc \bowtie^*_{R_1} x, \ x \in \{cab^m dc \mid m \geq 0\},$$
$$cab^m dc \bowtie^*_{R_2} x, \ x \in \{cab^m d^p c \mid p \geq 0\}, m \geq 1,$$
$$cadc \bowtie^*_{R_3} x, \ x \in \{cad^p c \mid p \geq 0\},$$
$$cab^m c \bowtie^*_{R_4} x, \ x \in \{ca(b^m)^{2^p} c \mid p \geq 0\} \cup \{cc\}, m \geq 1,$$
$$cac \bowtie^*_{R_5} x, \ x \in \{ca^{2^p} c \mid p \geq 0\} \cup \{cc\}.$$

Consequently, $L(SCO) = L$.

On the other hand, the language

$$L' = \{(ab^m)^{2^n} \mid m, n \geq 0\} \cup \{\lambda\}$$

is not an $E0L$ language [10]. Because the family $E0L$ is closed under arbitrary morphismes, we deduce that L does not belong to $E0L$, otherwise, $h(L) = L'$ would be in $E0L$, where $h : \{a, b, c, d\}^* \longrightarrow \{a, b\}^*, h(a) = a$, $h(b) = b, h(c) = h(d) = \lambda$. □

Theorem 3.4 *The families $\mathcal{L}(ES)$ and $\mathcal{L}(SCO)$ are incomparable.*

Proof. We prove that the self cross-over language $L = \{bb\} \cup \{ba^{2^n}b \mid n \geq 1\}$ cannot be generated by any evolutionary system. Assume the contrary and let $ES = (\{a, b\}, A, R)$ be an evolutionary system generating L. Since the set $Dupl$ has to be nonempty there exists $a^k \in Dupl$. All strings $ba^{2^n}b, ba^{2^{n+1}}b, ba^{2^{n+2}}b$ are in $L(ES)$; consequently $Dupl(\{ba^{2^n}b, ba^{2^{n+1}}b, ba^{2^{n+2}}b\}) \subseteq L(ES)$. Therefore, there are integers $p < q < r$ such that $2^n + k = 2^p, 2^{n+1} + k = 2^q, 2^{n+2} + k = 2^r$, which leads to $2^{q+1} - 2^p = 2^r - 2^q$ or equivalently $2^p(2^{q+1-p} - 1) = 2^q(2^{r-q} - 1)$. It follows that $p = q$ that is contradictory.

Conversely, we observe that the language $L_1 = a^*b^*a^*b^*$ can be generated starting from $abab$ by iterating the duplication and deletion of both letters a and b, respectively. □

4   Closure Properties

In this section we shall prove that the family $\mathcal{L}(SCO)$ has very poor properties concerning the closure under usual operations in formal language theory.

Theorem 4.1 *The family $\mathcal{L}(SCO)$ is an anti-AFL and it is not closed under left/right derivatives and complement, too.*

Proof. Union: The languages $L_1 = \{bb\} \cup \{ba^{2^n}b \mid n \geq 0\}$ and $L_2 = \{baaab\}$ are self cross-over languages but not their union. We omit the simple proof of this fact.

Catenation: The language $L = \{a^n b^m \mid n, m \geq 0\}$ can be generated by the self cross-over system (a detailed proof is left to the reader):

$$SCO = (\{a, b\}, \{ab\}, \{(a, \lambda; \lambda, a), (b, \lambda; \lambda, b), (\lambda, a; a, \lambda), (\lambda, b; b, \lambda)\}).$$

From Lemma 3.1 it follows that L^2 is not a self cross-over language.

Intersection with regular sets: Consider the intersection between the self cross-over language $\{a\}^*$ and the regular language $\{a^{2n} \mid n \geq 1\}$, which is not in $\mathcal{L}(SCO)$ due to Theorem 3.1.

Morphisms: Take the morphism $h : \{a, b\}^* \to \{a\}^*$ defined by $h(a) = h(b) = a$, and the self cross-over language $L = \{bb\} \cup \{ba^{2^n}b \mid n \geq 0\}$. However, $h(L) = \{aa\} \cup \{a^{2^n+2} \mid n \geq 0\}$ is not in $\mathcal{L}(SCO)$, as a consequence of Theorem 3.1.

Inverse morphisms: Take the self cross-over language $L = \{a^n \mid n \geq 3\}$ and the morphism $k : \{a, b\}^* \to \{a\}^*$ defined by $h(a) = a$, $h(b) = \lambda$. Clearly, $h^{-1}(L) = \{x \in \{a, b\}^* \mid |x|_a \geq 3\}$. We are going to prove that $h^{-1}(L) \notin \mathcal{L}(SCO.$ Assume the contrary but notice that exists $k > 0$ such that at least a cross-over rule is applicable to the string $x = b^k ab^k ab^k ab^k$. Thus, x may be split: (1) between two a's, (2) between an a and a b, (3) between a b and an a.

Therefore, we should consider nine cases, but the reader can easily find out some appropiate sites such that each case leads to strings containing less than three a's, a contradiction.

Kleene $*$: Consider the self cross-over language $L = \{bb\} \cup \{ba^{2^n}b \mid n \geq 0\}$ and assume that L^* can be generated by a self cross-over system. Following the same idea as for proving Lemma 3.1, we want to obtain the string

$$z = ba^{2^{n_1}}bba^{2^{n_2}}\ldots bba^{2^{n_k}}b$$

with large enough $n_i, 1 \leq i \leq k$, pairwise different. For increasing the number of a's occurrences in the last segment, we must split one copy of z somewhere on its last segment of a's. But, by choosing another segment, we will obtain a string which has a substring of the form $ba^r b$, and r is not a power of 2, a contradiction.

Left derivatives: Take L the previous language. We shall prove that $\partial_b^l(L) = \{b\} \cup \{a^{2^n} b \mid n \geq 0\}$ is not in $\mathcal{L}(SCO)$.

Assume the contrary; in order to get a string $a^{2^n} b$, with large enough n, we must apply a cross-over rule to a string, say $a^{2^m} b, n \neq m$.

By crossing-over, the string $a^{2^m} b$ may give the strings $a^{2^m+k} b$ and $a^{2^m-k} b$, for some $k > 0$. For both strings must be in $\partial_b^l(L)$, it follows that exist $i \neq j$ such that $2^{m+1} = 2^i + 2^j$, a contradiction.

The case of the right derivatives is symmetric.

Complement: Take the self cross-over system

$$SCO = (\{a, b\}, \{aaabbabb\}, R),$$

where

$$R = \{(\lambda, \lambda; x, y), (x, y; \lambda, \lambda) \mid x, y \in \{aa, ab, ba, bb\}\}$$

In order to prove that $L(SCO) = \{a, b\}^* - \{\lambda, a, b\}$ we mention the following two facts: (1) All strings of length two over $\{a, b\}$ are in $L(SCO)$ (this can be easily checked; for instance, the string ba can be obtained by $aaabbabb \bowtie babb \bowtie ba$), (2) Both strings $aaabbabba, aaabbabbb$ are in $L(SCO)$ (the cross-over steps for generating these strings are $aaabbabb \bowtie aaabbabbabb \bowtie aaabbabba$ and $aaabbabb \bowtie aaabbabbbabb \bowtie aaabbabbb$).

According to the first fact, we can assume, by induction, that all strings of length $n \geq 2$ over $\{a, b\}$ are in $L(SCO)$. Let z be a string of length $n + 1$ over $\{a, b\}$ and $z = ua$. From the inductive hypothesis we have $aaabbabba \bowtie^* ua$, hence, by combining with the second fact, we get $z \in L(SCO)$. Analogously, if $z = ub$.

Therefore, the complement of $L(SCO)$ is the language $\{\lambda, a, b\}$ which, obviously, is not in $\mathcal{L}(SCO)$. $\qquad \square$

Theorem 4.2 *The family $\mathcal{L}(SCO)$ is not closed under duplications and deletions.*

Proof. For $Dupl = \{a\}$, $Dupl(\{bb\} \cup \{ba^{2^n} b \mid n \geq 0\}) = \{ba^{2^n+1} b \mid n \geq 0\}$ is not a self cross-over language. Indeed, let us assume that the string $ba^{2^n+1} b$

produces two conjugated strings, say $ba^{2^m+1}b$ and $ba^{2^k+1}b$. Observe that $2^{n+1} + 2 = 2^m + 2^k + 2$ implies $n = m = k$. Consequently, $Dupl(\{bb\} \cup \{ba^{2^n}b \mid n \geq 0\})$ is a finite set, a contradiction.

The same reasoning is valid for deletions.   □

5 Decidability

Theorem 5.1 *Let SCO be a self cross-over system over the alphabet V. If the membership problem is decidable for the system SCO, then the problem "$L(SCO) = V^*$?" is decidable as well.*

Proof. Let $SCO = (V, A, R)$ be the given self cross-over system and $k = max\{|x| \mid x \in A\}$. Assume that the problem "$w \in L(SCO)$?" is decidable for any $w \in V^*$. Consider the set

$$F = \{w \in V^* \mid |w| \leq k + 1\}.$$

Obviously, if $F \not\subseteq L(SCO)$ (which is decidable), then $L(SCO) \neq V^*$. Therefore, we check whether or not $F \not\subseteq L(SCO)$. If the answer is positive, we claim that $L(SCO) = V^*$ holds. In order to prove our claim we take a string $x = wy$, $w \in V^+$, $y \in V$, $|w| = k + 1$. There exists a string $z \in A$ such that $z \bowtie^* w$. Moreover, $|z| \leq k$. The string zy is of length at most $k + 1$, hence $zy \in L(SCO)$. By applying the same rules used in the sequence of crossing-over $z \bowtie^* w$, to the same sites, we get that $zy \bowtie^* wy$. By induction one proves that $V^* \subseteq L(SCO)$.   □

Theorem 5.2 *It is not decidable whether or not an arbitrarily given context-free language is in $\mathcal{L}(SCO)$.*

Proof. The proof is a usual reduction to the Post's Correspondence Problem. Take two arbitrary n-tuples of nonempty strings over the alphabet $\{a, b\}$, $x = (x_1, x_2, \ldots, x_n)$, $y = (y_1, y_2, \ldots, y_n)$. Then, consider the languages

$$L_z = \{ba^{t_1}ba^{t_2} \ldots ba^{t_k}cz_{t_k} \ldots z_{t_2}z_{t_1} \mid k \geq 1\}, \ z \in \{x, y\},$$
$$L_s = \{w_1cw_2cw_2^Rcw_1^R \mid w_1, w_2 \in \{a, b\}^*\},$$
$$L(x, y) = \{a, b, c\}^* - (L_x\{c\}L_y^R \cap L_s),$$

(w^R denotes the mirror image of w).

It is known that $L(x, y)$ is a context-free language. For every solution $(i_1, i_2, \ldots, i_k)$ of $PCP(x, y)$ the strings

$$ba^{i_1}ba^{i_2}\ldots ba^{i_k}cx_{i_k}\ldots x_{i_2}x_{i_1}cy_{i_1}^R y_{i_2}^R \ldots y_{i_k}^R ca^{i_k}b\ldots ba^{i_2}ba^{i_1}b$$

are not in $L(x, y)$. On the other hand, $\{a, b\}^* \subseteq L(x, y)$.

Clearly, when $L(x, y) = \{a, b, c\}^*$, then $L(x, y)$ is a self cross-over system according to Example 2.

Now, it is sufficient to prove that when $L(x, y) \neq \{a, b, c\}^*$, then $L(x, y)$ is not a self cross-over language and we will do that in the sequel.

Let us suppose that $L(x, y)$ is a self cross-over language, i.e, $L(x, y) = L(SCO)$, $SCO = (\{a, b, c\}, A, R)$. We choose a solution $(i_1, i_2, \ldots, i_k)$ such that

$$|x_{i_k}x_{i_{k-1}}\ldots x_{i_1}| > max\{|w| \mid w \in A\}.$$

For $\{a, b\}^* \subseteq L(SCO)$, there exists a word $w \in A$ such that

$$w \bowtie^* y_{i_1}^R y_{i_2}^R \ldots y_{i_k}^R \in L(SCO).$$

By the choice of the solution $(i_1, i_2, \ldots, i_k)$ the word

$$z = ba^{i_1}ba^{i_2}\ldots ba^{i_k}cx_{i_k}\ldots x_{i_2}x_{i_1}cwca^{i_k}b\ldots ba^{i_2}ba^{i_1}b$$

is in $L(EG)$. Therefore, we get

$$z \bowtie^* ba^{i_1}ba^{i_2}\ldots ba^{i_k}cx_{i_k}\ldots x_{i_2}x_{i_1}cy_{i_1}^R y_{i_2}^R \ldots y_{i_k}^R ca^{i_k}b\ldots ba^{i_2}ba^{i_1}b,$$

a contradiction, and the proof is over. □

It is left for the further study to investigate the status of other decidability questions such as the membership, the inclusion and the equivalence problems.

6 Final Remarks

In this paper, a very simple and very natural restriction on the splicing operation is investigated: the cross-over rules are applicable only to identical strings. We have tried to capture two features of the recombination of genes in a chromosome:

1. The exchange of segments occurs after the chromosomes have been replicated.

2. The recombination process means exchanging of segments between homologous chromosomes.

Even under this very simple limitation, the context-freeness barrier is overpasses by using a finite set of cross-over rules, iteratively, starting from a finite set of strings.

References

[1] J. Dassow, V. Mitrana, On some operations suggested by genome evolution, *Proc. Pacific Symposium on Biocomputing*, Hawaii, 1997.

[2] D. L. Hartl, D. Freifelder, L. A. Snyder, *Basic Genetics*, Jones and Bartlett Publ., Boston, Portola Valley, 1988.

[3] T. Head, Formal language theory and DNA: An analysis of the generative capacity of specific recombinant behaviors, *Bull. Math. Biology*, 49 (1987), 737 – 759.

[4] T. Head, Gh. Păun, D. Pixton, Language theory and molecular genetics, chapter 7 in vol. 2 of *Handbook of Formal Languages* (G. Rozenberg, A. Salomaa, eds.), Springer-Verlag, Berlin, 1997, 295 – 360.

[5] L. Ilie, V. Mitrana, Crossing-over on languages. A formal representation of the recombination of genes in a chromosome, *Proc. of German Conf. on Bioinformatics*, Leipzig, 1996, 87 – 93.

[6] L. Kari, Gh. Păun, A. Salomaa, The power of restricted splicing with rules from a regular set, *J. Universal Computer Sci.*, 2, 4 (1996), 224 – 240.

[7] Gh. Păun, On the power of the splicing operation, *Intern. J. Computer Math.*, 59 (1995), 27 – 35.

[8] Gh. Păun, Regular extended H systems are computationally universal, *J. Automata, Languages, and Combinatorics*, 1, 1 (1996), 27 – 36.

[9] Gh. Păun, G. Rozenberg, A. Salomaa, Restricted use of the splicing operation, *Intern. J. Computer Math.*, 60 (1996), 17 – 32.

[10] G. Rozenberg, A. Salomaa, *The Mathematical Theory of L Systems*, Academic Press, New York, 1980.

[11] A. Salomaa, *Formal Languages*, Academic Press, New York, 1973.

[12] D. B. Searls, The linguistics of DNA, *Scientific American*, 80 (1992), 579 – 591.

Some Results on Array Splicing

Kamala KRITHIVASAN, Shri Raghav KAUSHIK

Department of Computer Science & Engineering
I. I. T. M., Chennai - 600 036, India
kamala@iitm.ernet.in

Abstract. Array splicing systems were considered in [1]. In this paper, we consider persistent and permanent array splicing systems and the concept of constants for such systems. We prove some results about these systems and some reduction theorems. In addition it is shown that there are strictly locally testable array languages that are not splicing languages at all.

1 Introduction

In the last few years, there has been a lot of interest in formal language theory applied to DNA computing. A specific model of DNA recombination is the splicing operation which consists of cutting DNA sequences and then pasting the fragments again, under the influence of restriction enzymes and ligases. In [2], Tom Head defined splicing systems motivated by this behavior of DNA sequences. Languages generated by splicing systems and their sub-classes were studied. Păun, in [4], extended the definition to Extended H-systems (EH systems) and proved that they are computationally universal.

In [1], array splicing systems were defined. The motivation was more from formal language theory, rather than from biology, though such systems may find applications for describing splicing of several DNA sequences. Some subclasses were also defined in [1]. Shearing imposes some restrictions in the case of arrays. Because of this, some results which hold for string splicing systems do not hold for array splicing systems.

In this paper, we define persistent and permanent splicing systems. We find that there are strictly locally testable array languages that cannot be generated by any array splicing system. Also, persistent systems need not have the property that all images of some given size are constants, and persistent languages are strictly higher than null context languages.

These results are different from the results for string splicing systems, but we find that some of the reduction results of Gatterdam [3] can be extended to array splicing languages. However, here too, there is one result that does not extend from strings.

In the next section, we give the basic definitions. In Section 3, we give an example of a strictly locally testable language that is not a splicing language. In Section 4, we give the definition of persistent and permanent array splicing systems and prove some hierarchy results. In Section 5, we define constants for array languages and show that there is a persistent language which has arbitrarily large non-constants. Section 6 explores the concept of reduction and some results about reduction are shown there.

2   Basic Definitions

2.1   Array Splicing Systems

Definition 1. Let $S = \langle x_{1,1}, x_{1,2}, \ldots, x_{1,n}, x_{2,1}, x_{2,2}, \ldots, x_{2,n}, \ldots, x_{m,1}, x_{m,2}, \ldots, x_{m,n} \rangle$ be a sequence of mn images defined over the alphabet Σ. Let,

$$
\begin{aligned}
I_1 \; = \; & (x_{1,1}\Phi x_{1,2}\Phi x_{1,3}\Phi \ldots \Phi x_{1,n})\Theta \\
& (x_{2,1}\Phi x_{2,2}\Phi x_{2,3}\Phi \ldots \Phi x_{2,n})\Theta \\
& \dotfill \\
& (x_{m,1}\Phi x_{m,2}\Phi x_{m,3}\Phi \ldots \Phi x_{m,n}),
\end{aligned}
$$

$$
\begin{aligned}
I_2 \; = \; & (x_{1,1}\Theta x_{2,1}\Theta x_{3,1}\Theta \ldots \Theta x_{m,1})\Phi \\
& (x_{1,2}\Theta x_{2,2}\Theta x_{3,2}\Theta \ldots \Theta x_{m,2})\Phi \\
& \dotfill \\
& (x_{1,m}\Theta x_{2,m}\Theta x_{3,m}\Theta \ldots \Theta x_{m,n}),
\end{aligned}
$$

where Φ and Θ are the column and row concatenations respectively.

If I_1 and I_2 are legal and $I_1 = I_2$ then the sequence S is said to be a proper sequence of cardinality $\langle m, n \rangle$.

Definition 2. Let $S = \langle x_{1,1}, x_{1,2}, \ldots, x_{1,n}, x_{2,1}, x_{2,2}, \ldots, x_{2,n}, \ldots, x_{m,1}, x_{m,2}, \ldots, x_{m,n} \rangle$ be a proper sequence of cardinality $\langle m, n \rangle$. Let,

$$I = (x_{1,1}\Phi x_{1,2}\Phi x_{1,3}\Phi \ldots \Phi x_{1,n})\Theta$$
$$(x_{2,1}\Phi x_{2,2}\Phi x_{2,3}\Phi \ldots \Phi x_{2,n})\Theta$$
$$\ldots\ldots\ldots\ldots\ldots\ldots\ldots\ldots\ldots$$
$$(x_{m,1}\Phi x_{m,2}\Phi x_{m,3}\Phi \ldots \Phi x_{m,n}).$$

Then, I is called the matrix-image of S and is represented by $MI(S, \langle m, n \rangle)$.

Definition 3. A sequence S of cardinality $\langle m, n \rangle$ is said to be a matrix split of an image I, if and only if S is a proper sequence and I is the matrix-image of S.

Note that for an image more than one matrix-split can exist.

Definition 4. Let I be an image and $M = \langle x_{1,1}, x_{1,2}, x_{1,5}, x_{2,1}, x_{2,2}, x_{2,5}, x_{5,1}, x_{5,2}, x_{5,5} \rangle$ be a matrix split of I of cardinality $\langle 5, 5 \rangle$. We define several prefixes and suffixes of I with respect to M, grouped under four types as follows.

TYPE 1 :

    Prefix: $MI((x_{1,1}, x_{1,2}, x_{1,3}, x_{2,1}, x_{2,2}, x_{2,3}, x_{3,1}, x_{3,2}, x_{3,3}), \langle 3, 3 \rangle)$.

    LRSuffix: $MI((x_{1,4}, x_{1,5}, x_{2,4}, x_{2,5}, x_{3,4}, x_{3,5}), \langle 3, 2 \rangle)$.

    BUSuffix: $MI((x_{4,1}, x_{4,2}, x_{4,3}, x_{5,1}, x_{5,2}, x_{5,3}), \langle 2, 3 \rangle)$.

    CSuffix: $MI((x_{4,4}, x_{4,5}, x_{5,4}, x_{5,5}), \langle 2, 2 \rangle)$.

TYPE 2 :

    Prefix: $MI((x_{1,3}, x_{1,4}, x_{1,5}, x_{2,3}, x_{2,4}, x_{2,5}, x_{3,3}, x_{3,4}, x_{3,5}), \langle 3, 3 \rangle)$.

    LRSuffix: $MI((x_{1,1}, x_{1,2}, x_{2,1}, x_{2,2}, x_{3,1}, x_{3,2}), \langle 3, 2 \rangle)$.

    BUSuffix: $MI((x_{4,3}, x_{4,4}, x_{4,5}, x_{5,3}, x_{5,4}, x_{5,5}), \langle 2, 3 \rangle)$.

    CSuffix: $MI((x_{4,1}, x_{4,2}, x_{5,1}, x_{5,2}), \langle 2, 2 \rangle)$.

TYPE 3 :

    Prefix: $MI((x_{3,1}, x_{3,2}, x_{3,3}, x_{4,1}, x_{4,2}, x_{4,3}, x_{5,1}, x_{5,2}, x_{5,3}), \langle 3, 3 \rangle)$.

    LRSuffix: $MI((x_{3,4}, x_{3,5}, x_{4,4}, x_{4,5}, x_{5,4}, x_{5,5}), \langle 3, 2 \rangle)$.

    BUSuffix: $MI((x_{1,1}, x_{1,2}, x_{1,3}, x_{2,1}, x_{2,2}, x_{2,3}), \langle 2, 3 \rangle)$.

    CSuffix: $MI((x_{1,4}, x_{1,5}, x_{2,4}, x_{2,5}), \langle 2, 2 \rangle)$.

TYPE 4 :

    Prefix: $MI((x_{3,3}, x_{3,4}, x_{3,5}, x_{4,3}, x_{4,4}, x_{4,5}, x_{5,3}, x_{5,4}, x_{5,5}), \langle 3, 3 \rangle)$.

    LRSuffix: $MI((x_{3,1}, x_{3,2}, x_{4,1}, x_{4,2}, x_{5,1}, x_{5,2}), \langle 3, 2 \rangle)$.

    BUSuffix: $MI((x_{1,3}, x_{1,4}, x_{1,5}, x_{2,3}, x_{2,4}, x_{2,5}), \langle 2, 3 \rangle)$.

    CSuffix: $MI((x_{1,1}, x_{1,2}, x_{2,1}, x_{2,2}), \langle 2, 2 \rangle)$.

In the above definition LR stands for left-or-right, BU stands for bottom-or-up, C stands for corner.

It can be easily seen that the image I can be expressed in terms of these prefixes and suffixes. If P, L, B, and C represent Prefix, LRsuffix, BUsuffix, and CSuffix, respectively, then I is

$$\text{TYPE 1: } I = (P\Phi L)\Theta(B\Phi C) = (P\Theta B)\Phi(L\Theta C),$$
$$\text{TYPE 2: } I = (L\Phi P)\Theta(C\Phi B) = (L\Theta C)\Phi(P\Theta B),$$
$$\text{TYPE 3: } I = (B\Phi C)\Theta(P\Phi L) = (B\Theta P)\Phi(C\Theta L),$$
$$\text{TYPE 4: } I = (C\Phi B)\Theta(L\Phi P) = (C\Theta L)\Phi(B\Theta P).$$

These are represented by prefixing with their type. For example, 2-BUSuffix represents a Type-2 BUSuffix.

Definition 5. A 2D splicing system S is a 4 tuple $S = \langle \Sigma, I, B, f \rangle$. Σ is the set of symbols used by S, $\Sigma = A \cup A', A \cap A' = \emptyset$ (A is the alphabet of the language generated by the splicing system, $L(S)$; A' is called the set of special symbols), f is a mapping $f : A' \longrightarrow A$, I is the set of initial images, B is a 4 tuple, $B = \langle B_1, B_2, B_3, B_4 \rangle$, where B_i is the set of Type-i patterns. A pattern p is a 9 tuple $\langle x_1, x_2, x_3, x_4, x_5, x_6, x_7, x_8, x_9 \rangle$, $x_1, x_2, x_3, x_4, x_6, x_7, x_8, x_9 \in \Sigma^{**}$, $x_5 \in \Sigma^{++}$, subject to the condition that p is a proper sequence of cardinality $\langle 3, 3 \rangle$. The middle term x_5 is called the crossing of p. The matrix image of p is called the site of the pattern p.

Splicing products. Four types of splicing operations are defined for images and a splicing operation between two images is uniquely specified by giving the two images, the type of splicing (1, 2, 3, or 4) and two matrix splits of cardinality $\langle 5, 5 \rangle$, one for each of the two images. The result of the splicing operation consists of two resultants or splicing products. For the splicing to take place, certain conditions have to be satisfied. Let X and Y be the two images. Let

$$MX = \langle x_{1,1}, x_{1,2}, \ldots, x_{1,5}, x_{2,1}, x_{2,2}, \ldots, x_{2,5}, \ldots, x_{5,1}, x_{5,2}, \ldots, x_{5,5} \rangle,$$
$$MY = \langle y_{1,1}, y_{1,2}, \ldots, y_{1,5}, y_{2,1}, y_{2,2}, \ldots, y_{2,5}, \ldots, y_{5,1}, y_{5,2} \ldots, y_{5,5} \rangle$$

be the two matrix splits of cardinality $\langle 5, 5 \rangle$ of X and Y, respectively. Type-i splicing between X and Y with respect to these matrix splits MX and MY can take place if and only if the following conditions hold:

1. $\langle x_{2,2}, x_{2,3}, x_{2,4}, x_{3,2}, x_{3,3}, x_{3,4}, x_{4,2}, x_{4,3}, x_{4,4} \rangle$ and
   $\langle y_{2,2}, y_{2,3}, y_{2,4}, y_{3,2}, y_{3,3}, y_{3,4}, y_{4,2}, y_{4,3}, y_{4,4} \rangle$ are Type-i patterns.

2. $x_{3,3} = y_{3,3}$. That is, the crossings are the same.

3. For the four types of splicing operations, define R_1 and R_2 as follows (XP and YP represent the Type-i prefixes of X and Y with respect to MX and MY, respectively). Similarly for suffixes.

TYPE 1: $R_1 = (XP\Phi YL)\Theta(YB\Phi YC) = (XP\Theta YB)\Phi(YL\Theta YC)$,

$\qquad R_2 = (YP\Phi XL)\Theta(XB\Phi XC) = (YP\Theta XB)\Phi(XL\Theta XC)$,

TYPE 2: $R_1 = (YL\Phi XP)\Theta(YC\Phi YB) = (YL\Theta YC)\Phi(XP\Theta YB)$,

$\qquad R_2 = (XL\Phi YP)\Theta(XC\Phi XB) = (XL\Theta XC)\Phi(YP\Theta XB)$,

TYPE 3: $R_1 = (YB\Phi YC)\Theta(XP\Phi YL) = (YB\Theta XP)\Phi(YC\Theta YL)$,

$\qquad R_2 = (XB\Phi XC)\Theta(YP\Phi XL) = (XB\Theta YP)\Phi(XC\Theta XL)$,

TYPE 4: $R_1 = (YC\Phi YB)\Theta(YL\Phi XP) = (YC\Theta YL)\Phi(YB\Theta XP)$,

$\qquad R_2 = (XC\Phi XB)\Theta(XL\Phi YP) = (XC\Theta XL)\Phi(XB\Theta YP)$.

Type-i splicing between X and Y with respect to these matrix splits is said to be defined/legal if the corresponding R_1 and R_2 are legal, i.e., the concatenation operations performed in the above expressions are legal. The two images X and Y are i-compatible with respect to these matrix splits if and only if the Type-i splicing is defined as per the above rules. We can see that, if none of the prefixes and suffixes are Λ, then, the images will be i-compatible if and only if the i-prefixes of the images are of same size.

Auxiliary language L'. The language of the splicing system $S, L(S)$, is obtained as follows. An auxiliary language L' is defined inductively as follows.

1. $L' \leftarrow I$.

2. Add images in I to L'.

3. Select any two images I_1 and I_2 from L'.
   Splice them in all the possible four types, using all matrix splits.
   If the resultants are rectangular and are not in L', add them to L'.

4. If no new images are added in step 3, then exit, else goto step 3.

Language $L(S)$. For an image X, define $F(X)$, for $X \in L'$, to be an image obtained by replacing every special symbol s in X by $f(s)$. Thus, $F : \Sigma^{++} \rightarrow A^{++}$ and $L(S) = \{F(X) \mid X \in L'\}$.

2.2 Classes of 2D Splicing Systems

Let $S = \langle \Sigma, I, B, f \rangle$ be a splicing system with $\Sigma = A \cup A'$. S is said to be a *simple* splicing system if $A' = \emptyset$. That is there are no special symbols used. Consequently, f is not defined for a simple splicing system. Hence, a simple splicing is represented as $S = \langle A, I, B \rangle$ and A is the alphabet of $L(S)$.

A splicing system $S = \langle \Sigma, I, \langle B_1, B_2, B_3, B_4 \rangle, f \rangle$, is said to be a *finite* splicing system if I, B_1, B_2, B_3 and B_4 are all finite.

A splicing system S is said to be *null-context* if all the patterns of S are of the form, $\langle \Lambda, \Lambda, \Lambda, \Lambda, c, \Lambda, \Lambda, \Lambda, \Lambda \rangle$, and $c \in \Sigma^{++}$. Note that in such a system, the crossing itself is a site. The patterns of this form are called null-context patterns. In a null-context pattern of the 9 images, only the crossing c is a nonempty image. The null-context patterns are represented by $\langle c \rangle$, where, c is the crossing.

A null-context splicing system S is said to an *equal* splicing system, if the four sets of patterns B_1, B_2, B_3, and B_4 are equal to each other, i.e., $B_1 = B_2 = B_3 = B_4$. Such a splicing system can be expressed by a 4 tuple $\langle \Sigma, I, B_1, f \rangle$. B_1 is the set of patterns, and the patterns can be used for any type of splicing.

A null-context splicing system S is said to be *uniform*, if the set of patterns are such that, $B_1 = B_2 = B_3 = B_4 = \{ \langle a \rangle \mid a \in \Sigma^{m,n}$, for some $m, n \geq 0 \}$. That is, every image in $\Sigma^{m,n}$ is a site of all the four types and these are the only sites. An uniform splicing system is represented as $S = \langle A, I, m, n \rangle$. Note that every uniform splicing system is also null-context.

An array language L is said to be a finite splicing language, if it can be generated by a finite splicing system. L is said to be a simple splicing language if it can be generated by a simple splicing system. Similarly, null-context, equal and uniform splicing languages are defined.

A language L defined over an alphabet A is said to be p,q-*strictly locally testable* (SLT), if the sets $U, V, Y, Z, W \subseteq A^{p,q}$ can be constructed such that,

$$L' = (L_U \cap L_V \cap L_Y \cap L_Z) - L_W,$$

where

$$L' = L \cap ((A^{p,q} \Phi A^{**}) \Theta A^{**}),$$
$$L_U = ((U \Phi A^{**}) \Theta A^{**}),$$
$$L_V = ((A^{**} \Phi V) \Theta A^{**}),$$
$$L_Y = (A^{**} \Theta (Y \Phi A^{**})),$$
$$L_Z = (A^{**} \Theta (A^{**} \Phi Z)),$$
$$L_W = (A^{**} \Phi (A^{**} \Theta W \Theta A^{**}) \Phi A^{**}).$$

It can be easily seen that L' is the set of images in L with size greater than or equal to $\langle p, q \rangle$, L_U is the set of images over A, with an image in U

as a top-left sub-image; L_V is the set of images over A, with an image in V as a top-right sub-image; L_Y is the set of images over A, with an image in Y as bottom-left sub-image; L_Z is the set of images over A, with an image in Z as a bottom-right sub-image. L_W is the set of images with an image in W as their sub-image.

Note. Unless otherwise specified, in this paper, the terms splicing system and array splicing system mean finite-simple array splicing systems. Here, we extend the hierarchy of array splicing systems and prove the hierarchy shown in Figure 1, where U, E, NC, P, A, SLT are the families of uniform splicing languages, equal splicing languages, null context splicing languages, persistent splicing languages, all splicing languages, and strictly locally testable languages, respectively.

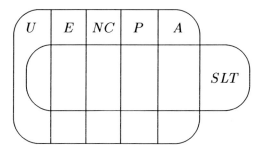

Figure 1: Hierarchy of array splicing languages

3 A Strictly Locally Testable Language that is not a Splicing Language

Here we further contribute to the hierarchy in [1]. We show that there are Strictly Locally Testable languages that are not splicing languages at all.

The proof follows by considering the following language which is given in [1].

Let L be a language over the alphabet $A = \{1, 0\}$ defined as follows. L consists of all images of 1's and 0's which do not have $0^{3,3}$ as a sub-image. That is, the images do not contain the sub-image $\begin{pmatrix} 0 & 0 & 0 \\ 0 & 0 & 0 \\ 0 & 0 & 0 \end{pmatrix}$. This language is obviously 3,3-strictly locally testable with $W = \{0^{3,3}\}$ and $U = V = Y = Z = A^{3,3} - W$.

Now let us show that this language L cannot be produced by any FS-array splicing system. On the contrary, let there be an FS-array splicing system $S = \langle A, I, B_1, B_2, B_3, B_4 \rangle$ with $L(S) = L$. As I is finite, the language L has to be produced by splicing only. So there should be at least one pattern in S. Let it be a Type-1 pattern $X = \langle Y_{11}, Y_{12}, Y_{13}, Y_{21}, Y, Y_{23}, Y_{31}, Y_{32}, Y_{33} \rangle$. Note that X cannot contain $0^{3,3}$ as its subimage. Let $Size(X) = \langle m, n \rangle$. Let Y be the crossing. Consider two images I_1 and I_2 as in Figure 2.

I_1 and I_2 do not contain $0^{3,3}$ as a subimage. Hence, both belong to L. A splicing as shown in Figure 2 results in a resultant image with $0^{3,3}$ as its subimage. This image by the rule of splicing should belong to $L(S)$ but does not belong to L. Thus, $L(S) \neq L$. Notice that we have assumed in the figure that m is sufficiently large. But it is clear that the proof for small m is similar.

$$
\begin{array}{ccc|cc}
1\ 1 & \cdots & 1\ 1\ 1 & 0\ 1 \cdots 1 \\
1\ 1 & \cdots & 1\ 1\ 1 & 0\ 1 \cdots 1 \\
1\ 1 & \cdots & 1\ 1\ 1 & 0\ 1 \cdots 1 \\
1\ 1 & \cdots & 1\ 1\ 1 & 1\ 1 \cdots 1 \\
1 & Y_{11} & Y_{12} & Y_{13} & 1 \\
\cdots & Y_{21} & Y & Y_{33} & \cdots \\
1 & Y_{31} & Y_{32} & Y_{33} & 1
\end{array}
\qquad
\begin{array}{ccc|cc}
1\ 1 \cdots 1\ 0\ 0 & 1\ 1 \cdots 1 \\
1\ 1 \cdots 1\ 0\ 0 & 1\ 1 \cdots 1 \\
1\ 1 \cdots 1\ 0\ 0 & 1\ 1 \cdots 1 \\
1\ 1 \cdots 1\ 1\ 1 & 1\ 1 \cdots 1 \\
1 & Y_{11} & Y_{12} & Y_{13} & 1 \\
\cdots & Y_{21} & Y & Y_{33} & \cdots \\
1 & Y_{31} & Y_{32} & Y_{33} & 1
\end{array}
\qquad
\begin{array}{ccc|cc}
1\ 1 \cdots 1\ 0\ 0 & 0\ 1 \cdots 1 \\
1\ 1 \cdots 1\ 0\ 0 & 0\ 1 \cdots 1 \\
1\ 1 \cdots 1\ 0\ 0 & 0\ 1 \cdots 1 \\
1\ 1 \cdots 1\ 1\ 1 & 1\ 1 \cdots 1 \\
1 & Y_{11} & Y_{13} & & 1 \\
\cdots & Y_{21} & Y & Y_{33} & \cdots \\
1 & Y_{31} & Y_{32} & Y_{33} & 1
\end{array}
$$

$$I_1 \qquad\qquad\qquad I_2 \qquad\qquad\qquad\qquad I$$

Figure 2: Example of a SLT language that is not a splicing language

A similar proof can be given for all the other 3 types of splicing.

Thus we have shown the existence of languages that are SLT but not splicing languages at all. This contrasts with string splicing systems, as defined by Tom Head in [2] where the family of SLT languages is equal to the family of persistent splicing languages, which is a subset of the family of all splicing languages.

4   Persistence and Permanence

4.1   Definition of Persistence

An FS array splicing system $S = \langle A, I, B_1, B_2, B_3, B_4 \rangle$ is said to be type-1 persistent if for any two images in A^{**}, $I_1 =$

$$\begin{pmatrix} X_{1,1} & X_{1,2} & X_{1,3} & X_{1,4} & X_{1,5} \\ X_{2,1} & X_{2,2} & X_{2,3} & X_{2,4} & X_{2,5} \\ X_{3,1} & X_{3,2} & X_{3,3} & X_{3,4} & X_{3,5} \\ X_{4,1} & X_{4,2} & X_{4,3} & X_{4,4} & X_{4,5} \\ X_{5,1} & X_{5,2} & X_{5,3} & X_{5,4} & X_{5,5} \end{pmatrix}, \text{ where } \begin{pmatrix} X_{2,2} & X_{2,3} & X_{2,4} \\ X_{3,2} & X_{3,3} & X_{3,4} \\ X_{4,2} & X_{4,3} & X_{4,4} \end{pmatrix} \in B_1$$

and $X_{3,3}$ is the crossing of this site, and $I_2 =$

$$\begin{pmatrix} X'_{1,1} & X'_{1,2} & X'_{1,3} & X'_{1,4} & X'_{1,5} \\ X'_{2,1} & X'_{2,2} & X'_{2,3} & X'_{2,4} & X'_{2,5} \\ X'_{3,1} & X'_{3,2} & X_{3,3} & X'_{3,4} & X'_{3,5} \\ X'_{4,1} & X'_{4,2} & X'_{4,3} & X'_{4,4} & X'_{4,5} \\ X'_{5,1} & X'_{5,2} & X'_{5,3} & X'_{5,4} & X'_{5,5} \end{pmatrix}, \text{ where } \begin{pmatrix} X'_{2,2} & X'_{2,3} & X'_{2,4} \\ X'_{3,2} & X_{3,3} & X'_{3,4} \\ X'_{4,2} & X'_{4,3} & X'_{4,4} \end{pmatrix} \in B_1$$

and $X_{3,3}$ is the crossing of this site, if I_1 and I_2 are Type-1 splicing compatible with respect to the (above occurrences of) the above patterns and if Y

is a sub-image of $\begin{pmatrix} X_{1,1} & X_{1,2} & X_{1,3} \\ X_{2,1} & X_{2,2} & X_{2,3} \\ X_{3,1} & X_{3,2} & X_{3,3} \end{pmatrix}$ which is the crossing of a site in I_1,

then the same occurrence of Y in $I_3 = \begin{pmatrix} X_{1,1} & X_{1,2} & X_{1,3} & X'_{1,4} & X'_{1,5} \\ X_{2,1} & X_{2,2} & X_{2,3} & X'_{2,4} & X'_{2,5} \\ X_{3,1} & X_{3,2} & X_{3,3} & X'_{3,4} & X'_{3,5} \\ X'_{4,1} & X'_{4,2} & X'_{4,3} & X'_{4,4} & X'_{4,5} \\ X'_{5,1} & X'_{5,2} & X'_{5,3} & X'_{5,4} & X'_{5,5} \end{pmatrix}$

which is obtained by splicing I_1 and I_2 by type-1 splicing contains an occurrence of the crossing of some site in I_3, i.e., has a subimage which is a crossing of some site in I_3.

The definitions for Type-2, Type-3 and Type-4 persistence are similar.

An FS-array splicing system is *persistent* if it is Type-i persistent for all $i = 1, 2, 3, 4$.

The difference between this definition of persistence and the definition for string splicing systems is clear. In the definition of persistence for string splicing systems, y can be a substring of either uax or xbv (where $uaxbv \in A^*$ corresponds to I_1 above and y corresponds to Y above). This is not extended to array splicing systems. Here, Y can be a sub-image of only the prefix of I_1.

4.2 Definition of Permanence

The definition of permanence is obtained by replacing the word "contains" in the definition of persistence with "is", as in string splicing systems. The formal definition is given below.

An FS array splicing system $S = \langle A, I, B_1, B_2, B_3, B_4 \rangle$ is said to be Type-1 permanent if for any two images in A^{**}, $I_1 =$

$$
\begin{pmatrix}
X_{1,1} & X_{1,2} & X_{1,3} & X_{1,4} & X_{1,5} \\
X_{2,1} & X_{2,2} & X_{2,3} & X_{2,4} & X_{2,5} \\
X_{3,1} & X_{3,2} & X_{3,3} & X_{3,4} & X_{3,5} \\
X_{4,1} & X_{4,2} & X_{4,3} & X_{4,4} & X_{4,5} \\
X_{5,1} & X_{5,2} & X_{5,3} & X_{5,4} & X_{5,5}
\end{pmatrix}
, \text{ where }
\begin{pmatrix}
X_{2,2} & X_{2,3} & X_{2,4} \\
X_{3,2} & X_{3,3} & X_{3,4} \\
X_{4,2} & X_{4,3} & X_{4,4}
\end{pmatrix}
\in B_1
$$

and $X_{3,3}$ is the crossing of this site, and $I_2 =$

$$
\begin{pmatrix}
X'_{1,1} & X'_{1,2} & X'_{1,3} & X'_{1,4} & X'_{1,5} \\
X'_{2,1} & X'_{2,2} & X'_{2,3} & X'_{2,4} & X'_{2,5} \\
X'_{3,1} & X'_{3,2} & X_{3,3} & X'_{3,4} & X'_{3,5} \\
X'_{4,1} & X'_{4,2} & X'_{4,3} & X'_{4,4} & X'_{4,5} \\
X'_{5,1} & X'_{5,2} & X'_{5,3} & X'_{5,4} & X'_{5,5}
\end{pmatrix}
, \text{ where }
\begin{pmatrix}
X'_{2,2} & X'_{2,3} & X'_{2,4} \\
X'_{3,2} & X_{3,3} & X'_{3,4} \\
X'_{4,2} & X'_{4,3} & X'_{4,4}
\end{pmatrix}
\in B_1
$$

and $X_{3,3}$ is the crossing of this site, if I_1 and I_2 are type-1 splicing compatible with respect to the same occurrences of the given patterns as given above and if Y is a sub-image of $\begin{pmatrix} X_{1,1} & X_{1,2} & X_{1,3} \\ X_{2,1} & X_{2,2} & X_{2,3} \\ X_{3,1} & X_{3,2} & X_{3,3} \end{pmatrix}$ which is the crossing of a site in I_1, then the same occurrence of Y in $I_3 =$

$$
\begin{pmatrix}
X_{1,1} & X_{1,2} & X_{1,3} & X'_{1,4} & X'_{1,5} \\
X_{2,1} & X_{2,2} & X_{2,3} & X'_{2,4} & X'_{2,5} \\
X_{3,1} & X_{3,2} & X_{3,3} & X'_{3,4} & X'_{3,5} \\
X'_{4,1} & X'_{4,2} & X'_{4,3} & X'_{4,4} & X'_{4,5} \\
X'_{5,1} & X'_{5,2} & X'_{5,3} & X'_{5,4} & X'_{5,5}
\end{pmatrix}
\text{ which is obtained by splicing } I_1 \text{ and }
$$

I_2 by type-1 splicing is the crossing of some site in I_3.

The definitions for Type-2, Type-3 and Type-4 permanence are similar.

An FS-array splicing system is *permanent* if it is Type-i permanent for all $i = 1, 2, 3, 4$.

Note. It can be seen that there is a variation from the definition for strings; while Gatterdam [3] defined permanence only for crossing disjoint systems, nothing like that is done here.

Thus, from the definitions it is clear that a permanent system is always persistent. The converse is not true.

Theorem 4.1 *A persistent array splicing system needs not be permanent.*

Proof. Consider the FS array splicing system $S = \langle A, I, B_1, B_2, B_3, B_4 \rangle$ where $A = \{a, b, c\}$, $I = \{cabc\}$, $B_2 = B_3 = B_4 = \emptyset$, $B_1 = \{\langle \lambda, \lambda, \lambda, c, ab,$

$c, \lambda, \lambda, \lambda\rangle, \langle b\rangle\}$. This system is persistent because any crossing must contain b, which by itself is a null-site. However this system is not permanent. Let

$$I_1 = \begin{pmatrix} c & a & b & c \\ a & a & b & c \end{pmatrix},$$

$$I_2 = \begin{pmatrix} a & a & a & a \\ a & a & b & c \end{pmatrix}.$$

I_1 and I_2 splice by Type-1 splicing at null site $\langle b\rangle$ to give

$$I = \begin{pmatrix} c & a & b & a \\ a & a & b & c \end{pmatrix}$$

$I_1, I_2 \in A^{**}$. In I_1, ab is a sub-image of the prefix corresponding to site b. But when I_1 and I_2 are spliced at b, the resulting image I_3 does not have the same ab as the crossing of a site. □

4.3 The Place of Persistent and Permanent Systems in the Hierarchy

From the definition of null-context array splicing systems, it is clear that the family of null-context array languages is contained in the family of permanent (and so, persistent) array languages.

However, unlike with string splicing systems, the converse is not true.

Theorem 4.2 *There are persistent splicing languages that are not null-context and are not strictly locally testable either.*

Proof. The following is an example of a persistent array language that cannot be generated by a null-context array splicing system and is not SLT.

Define an array splicing system $S = \langle A, I, B_1, B_2, B_3, B_4\rangle$, where

$$A = \{0, b, a, c, e, f\}$$

$$I = \{ \begin{pmatrix} 0 & b & 0 \\ a & 0 & c \\ a & 0 & c \end{pmatrix}, \begin{pmatrix} 0 & b & b & 0 \\ a & 0 & 0 & c \\ a & 0 & 0 & c \end{pmatrix}, \begin{pmatrix} 0 & b & 0 \\ e & 0 & f \\ e & 0 & f \end{pmatrix},$$

$$\begin{pmatrix} 0 & b & b & 0 \\ e & 0 & 0 & f \\ e & 0 & 0 & f \end{pmatrix}, \begin{pmatrix} 0 & b & b & b & 0 \\ e & 0 & 0 & 0 & f \\ e & 0 & 0 & 0 & f \end{pmatrix}\},$$

$$B_1 = \emptyset,$$
$$B_2 = \{\langle b \rangle, \langle bb \rangle\},$$
$$B_3 = \{\langle \lambda, \lambda, \lambda, \lambda, b, 0, \lambda, 0, c \rangle, \langle \lambda, \lambda, \lambda, 0, b, \lambda, a, 0, \lambda \rangle,$$
$$\langle \lambda, \lambda, \lambda, \lambda, bb, 0, \lambda, 00, f \rangle, \langle \lambda, \lambda, \lambda, 0, bb, \lambda, e, 00, \lambda \rangle\},$$
$$B_4 = \emptyset.$$

It is clear that the language generated by S is given by

$$L = \begin{pmatrix} 0 \\ a \\ a \end{pmatrix} \left(\begin{pmatrix} b \\ 0 \\ 0 \end{pmatrix} \right)^+ \begin{pmatrix} 0 \\ c \\ c \end{pmatrix} \bigcup \begin{pmatrix} 0 \\ e \\ e \end{pmatrix} \left(\begin{pmatrix} b \\ 0 \\ 0 \end{pmatrix} \right)^+ \begin{pmatrix} 0 \\ f \\ f \end{pmatrix}.$$

S is persistent, in fact permanent, because the only possible crossings are b and bb which are also the crossings of a null-context site, albeit in B_2. Hence, no crossing can be destroyed.

Claim 1: L is not null-context.

Proof. On the contrary, suppose a null-context array splicing system N generates L. First of all, N must have some sites for array splicing, because the language is infinite and the initial images set must be finite.

Now, there is a splicing that increases the length of a string. Let the site (and hence crossing) be X. Suppose X has a or c or e or f. Then it clearly cannot be a length increasing site. Hence, X contains only 0 and b. But in this case, it is obvious that images in the set $\begin{pmatrix} 0 \\ a \\ a \end{pmatrix} \left(\begin{pmatrix} b \\ 0 \\ 0 \end{pmatrix} \right)^+ \begin{pmatrix} 0 \\ f \\ f \end{pmatrix}$ can be obtained by splicing, which is a contradiction.

Claim 2: L is not SLT.

Proof. It is obvious that if L is SLT, then the images in the set $\begin{pmatrix} 0 \\ a \\ a \end{pmatrix} \left(\begin{pmatrix} b \\ 0 \\ 0 \end{pmatrix} \right)^+ \begin{pmatrix} 0 \\ f \\ f \end{pmatrix}$ lie in L, which is a contradiction.   □

Thus we have the hierarchy of array splicing languages as shown in the first section.

5 Constants

Definition 6. Let L be a language over alphabet A. An image $C \in A^{**}$ is defined to be a type-1 constant if whenever $I_1 = \begin{pmatrix} X_{1,1} & X_{1,2} & X_{1,3} \\ X_{2,1} & C & X_{2,3} \\ X_{3,1} & X_{3,2} & X_{3,3} \end{pmatrix}$

$$\text{and } I_2 = \begin{pmatrix} X'_{1,1} & X'_{1,2} & X'_{1,3} \\ X'_{2,1} & C & X'_{2,3} \\ X'_{3,1} & X'_{3,2} & X'_{3,3} \end{pmatrix} \text{ are in } L \text{ and } I_{12} = \begin{pmatrix} X_{1,1} & X_{1,2} & X'_{1,3} \\ X_{2,1} & C & X'_{2,3} \\ X'_{3,1} & X'_{3,2} & X'_{3,3} \end{pmatrix}$$

is a proper image, $I_{12} \in L$.

Definitions for type-i constant for $i = 2, 3, 4$ are similar.

An image is said to be a constant with respect to a language if it is a type-i constant for some $i \in \{1, 2, 3, 4\}$.

5.1 Constants in Persistent Array Languages

In string splicing systems, we have the result that in a persistent splicing system there is a constant such that all strings in the alphabet of this length are constants with respect to the language.

The equivalent result does not hold in array splicing systems.

Theorem 5.1 *There are null context (and hence persistent) splicing languages such that there are no constants m, n such that all images of size $\langle m, n \rangle$ are constants for the language.*

Proof. The following null-context (and hence persistent) array splicing system S is such that there are no m, n such that all $I \in A^{m,n}$, where A is the alphabet I, are constant, i.e., we can have arbitrarily large non-constant images.

$$S = \langle A, I, B_1, B_2, B_3, B_4 \rangle,$$
$$A = \{0, a, b, c, d\},$$
$$I = \{I_1 = \begin{pmatrix} 0 & 0 & c \\ 0 & 0 & c \\ b & b & 0 \end{pmatrix}, I_2 = \begin{pmatrix} 0 & d & d \\ a & 0 & 0 \\ a & 0 & 0 \end{pmatrix} \},$$
$$B_1 = \{\langle b \rangle\}, \ B_2 = \{\langle a \rangle\}, \ B_3 = \{\langle c \rangle\}, \ B_4 = \{\langle d \rangle\}.$$

$L(S) = L_1 \cup L_2$ where L_1 is generated from I_1, B_1 and B_3 while L_2 is generated from I_2, B_2 and B_4. It can be easily seen that

$$L_1 = \{ \begin{pmatrix} 0 & \cdots & 0 & c \\ \vdots & & & \vdots \\ \vdots & & & \vdots \\ 0 & \cdots & 0 & c \\ b & \cdots & b & 0 \end{pmatrix} \text{ with } r \geq 2 \text{ rows and } c \geq 2 \text{ columns}\},$$

$$L_2 = \left\{ \begin{pmatrix} 0 & d & . & . & . & d \\ a & 0 & . & . & . & 0 \\ . & & & & & . \\ . & & & & . & . \\ . & & & & & . \\ a & 0 & . & . & . & 0 \end{pmatrix} \text{ with } r \geq 2 \text{ rows and } c \geq 2 \text{ columns} \right\}.$$

Now assume that every $I \in A^{m,n}$ for some m, n is a constant for some type-i, $i = 1, 2, 3, 4$.

Thus $0^{m,n}$ is also a constant for some type i. Let us assume that it is a type-1 constant. The proof for other types is similar.

Consider the two images I and I' in $L(S)$ shown in Figure 3. As I and I' are compatible and as $0^{m,n}$ is a type-1 constant, we get that the image I'' shown in the figure also lies in $L(S)$. This contradiction completes the proof for type-1 constants.

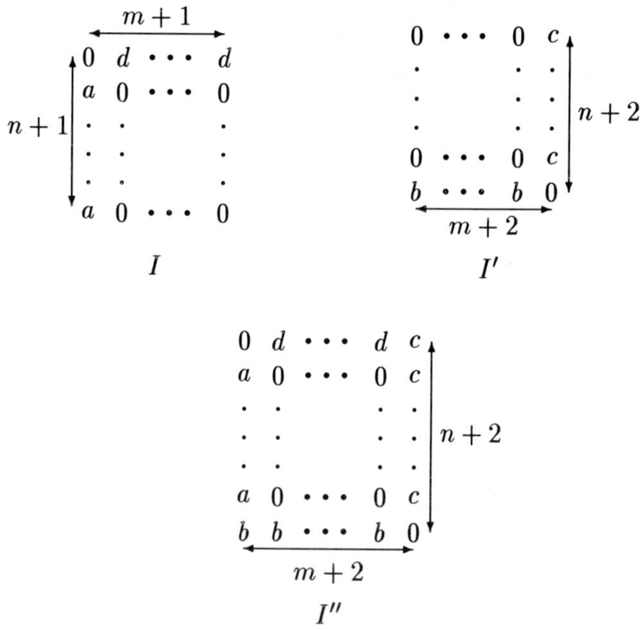

Figure 3: Arbitrarily large constants in a persistent language

Hence, it is possible to construct arbitrarily large non-constants for the above language.   □

In string splicing systems, the proof of the result that there is a (numerical) constant such that any string of this length is a constant, for persistent languages, is based on the result that in a persistent string splicing system S, any substring of a string in $L(S)$ of length $M(S)$ has the crossing of a site. Here $M(S)$ is one plus the length of the maximum substring of any string in I, the initial set of images of S, which has no crossing of any site.

This does not hold in array splicing systems. Intuitively, this can be seen because there are two degrees of freedom in arrays.

The above array splicing system S is an example for this. Any $0^{m,n}$ is the sub-image of some image in $L(S)$ but $0^{m,n}$ does not contain any crossing. And m, n can each be arbitrarily large.

6 Reduction in Array Splicing Systems

In [3], Gatterdam has defined the concept of reduction in string splicing systems. Here we extend the concept to array splicing systems. However, unlike in [3], we do not assume that the array splicing systems are crossing disjoint (i.e., if X is a type-1 crossing, then X cannot be a type-i crossing for $i = 2, 3$, or 4).

6.1 Type 1 Reduction

If there are two patterns of the same hand of the form $X = \langle x_{22}, x_{23}, x_{24},$ $x_{32}, x_{33}, x_{34}, x_{42}, x_{43}, x_{44} \rangle$ and $Y = \langle y_{22}, y_{23}, y_{24}, y_{32}, x_{33}, y_{34}, y_{42}, y_{43}, y_{44} \rangle$ such that X is a sub-image of Y, *with the x_{33} in both matching each other*, then Y is redundant from the set of patterns, i.e., the language generated by the system will be the same if pattern Y is removed.

6.2 Type 2 Reduction

If there are two images x_{33} and y_{33} such that y_{33} is a proper *right bottom* sub-image of x_{33} and for each pattern of type-1 $X = \langle x_{22}, x_{23}, x_{24},$ $x_{32}, x_{33}, x_{34}, x_{42}, x_{43}, x_{44} \rangle$, there is a type-1 pattern $Y = \langle y_{22}, y_{23}, y_{24}, y_{32},$ $y_{33}, y_{34}, y_{42}, y_{43}, y_{44} \rangle$ such that Y is a sub-image of X, *with the y_{33} matching the corresponding y_{33} which is a part of x_{33}*, then all type-1 patterns containing x_{33} as a crossing are redundant.

The same rule applies to type-2 patterns except that y_{33} must be a bottom left subimage of x_{33}, for type-3 patterns, *top right* subimage and for type-4, top left.

An array splicing system S is reduced if all possible type-1 and type-2 reductions have been performed. Let S_0 be the reduced splicing system.

Theorem 6.1 $L(S) = L(S_0)$.

Proof. Clearly $L(S_0) \subseteq L(S)$.

Conversely, $L(S) \subseteq L(S_0)$: We show that any splicing operation that can be done in S can also be done equivalently in S_0.

The proof is by induction. Let L be a set of images. Let us define $\sigma(L) = \{$image $X \mid X$ can be obtained by splicing any two images in L according to the rules of $S\} \cup L$. Now, define $\sigma^0(L) = L$ and $\sigma^{i+1}(L) = \sigma(\sigma^i(L))$ for $i \geq 0$.

Clearly $L(S) = \bigcup_{i=0}^{\infty} \sigma^i(I)$ where I is the initial set of images of S. We show by induction on i that $\sigma^i(I)$ is contained in $L(S_0)$.

Basis: This is trivial as the initial set of images for both S and S_0 are the same.

Induction: Suppose that the hypothesis holds for $0, 1, 2, \ldots, i$. Take any image R in $\sigma^{i+1}(I)$. It should have been derived from two images X and X' in $\sigma^i(I)$. We assume that this splicing is a type-1 splicing. The proof is similar for other types. Let

$$X = \begin{pmatrix} X_{1,1} & X_{1,2} & X_{1,3} & X_{1,4} & X_{1,5} \\ X_{2,1} & X_{2,2} & X_{2,3} & X_{2,4} & X_{2,5} \\ X_{3,1} & X_{3,2} & X_{3,3} & X_{3,4} & X_{3,5} \\ X_{4,1} & X_{4,2} & X_{4,3} & X_{4,4} & X_{4,5} \\ X_{5,1} & X_{5,2} & X_{5,3} & X_{5,4} & X_{5,5} \end{pmatrix},$$

$$X' = \begin{pmatrix} X'_{1,1} & X'_{1,2} & X'_{1,3} & X'_{1,4} & X'_{1,5} \\ X'_{2,1} & X'_{2,2} & X'_{2,3} & X'_{2,4} & X'_{2,5} \\ X'_{3,1} & X'_{3,2} & X'_{3,3} & X'_{3,4} & X'_{3,5} \\ X'_{4,1} & X'_{4,2} & X'_{4,3} & X'_{4,4} & X'_{4,5} \\ X'_{5,1} & X'_{5,2} & X'_{5,3} & X'_{5,4} & X'_{5,5} \end{pmatrix}$$ splice at the crossing X_{33} to give

$$R = \begin{pmatrix} X_{1,1} & X_{1,2} & X_{1,3} & X'_{1,4} & X'_{1,5} \\ X_{2,1} & X_{2,2} & X_{2,3} & X'_{2,4} & X'_{2,5} \\ X_{3,1} & X_{3,2} & X_{3,3} & X'_{3,4} & X'_{3,5} \\ X'_{4,1} & X'_{4,2} & X'_{4,3} & X'_{4,4} & X'_{4,5} \\ X'_{5,1} & X'_{5,2} & X'_{5,3} & X'_{5,4} & X'_{5,5} \end{pmatrix}$$ in S.

If both patterns $P_1 = \langle X_{22}, X_{23}, X_{24}, X_{32}, X_{33}, X_{34}, X_{42}, X_{43}, X_{44} \rangle$ and $P_2 = \langle X'_{22}, X'_{23}, X'_{24}, X'_{32}, X_{33}, X'_{34}, X'_{42}, X'_{43}, X'_{44} \rangle$ are in S_0, then the proof is trivial. If one of the two is not there, without loss of generality, say P_1, then the other must have been removed by a type-1 reduction. Thus, there is a pattern P_3 with crossing X_{33} in S_0 which is a sub-image of P_1. Clearly we can splice X and X' with patterns P_3 and P_2 to get R. This completes this case. If both P_1 and P_2 have been removed by type-1 reductions and there has been no type-2 reduction on patterns of crossing X_{33}, then there are patterns P_3 and P_4 each with crossing X_{33} in S_0 such that P_3 is a sub-image of P_1 and P_4 is a sub-image of P_2. Thus, we can use P_3 and P_4 to splice X and X' and get R.

If patterns of crossing X_{33} have been removed by type-2 reductions, then there are two patterns P_3 and P_4, each with crossing some Y_{33} such that Y_{33} is a sub-image of X_{33} and P_3, P_4 are sub-images of P_1 and P_2 respectively. Clearly, P_3 and P_4 can be used to splice X and X' to get R.

In all cases, we can see that R can be derived in S_0 also. Thus, $R \in L(S_0)$.

Thus we can conclude that $L(S) = L(S_0)$. □

Theorem 6.2 *The set S_0 is independent of the order of removal of patterns by either type of reduction.*

Proof. We first make the observation that both types of reduction satisfy the transitivity property, i.e., in a type-1 reduction if pattern P_1 causes P_2 to be removed and P_2 causes P_3 to be removed then P_1 can cause P_3 to be removed, and in a type-2 reduction, if type-i patterns of crossing z cause patterns of crossing y to be removed and type-i patterns of crossing y cause patterns of crossing x to be removed then the patterns of crossing z can cause the patterns of crossing x to be removed. Now assuming the contrary, suppose two different orders of removal yield the two minimal sets of patterns S_1 and S_2. Without loss of generality, assume that $P \in S_1 - S_2$.
Case 1: P got removed in S_2 by a type-1 reduction.

In this case, considering the transitivity of type-1 reductions and the fact that no pattern with the same crossing as P got removed in S_1 due to a type-2 reduction, we arrive at the contradiction that $P \notin S_1$.
Case 2: P got removed in S_2 by a type-2 reduction.

In this case we make the observation that if a set of patterns A is used to remove a set of patterns B and if a few patterns in A get removed by a type-1 reduction then the remaining set, say A', can also be used to remove B by a type-2 reduction. This observation coupled with the transitivity of type-2 reductions completes this case.

The two cases complete the proof. □

Definition 7. A set of type-1 patterns is *full context* if whenever $\langle x_{22}, x_{23}, x_{24}, x_{32}, x_{33}, x_{34}, x_{42}, x_{43}, x_{44} \rangle$ and $\langle x'_{22}, x'_{23}, x'_{24}, x'_{32}, x_{33}, x'_{34}, x'_{42}, x'_{43}, x'_{44} \rangle$ are two patterns of type-1, then if $\langle x_{22}, x_{23}, x'_{24}, x_{32}, x_{33}, x'_{34}, x'_{42}, x'_{43}, x'_{44} \rangle$ is a proper sequence then it is a type-1 pattern.

A set of patterns of type-i, for $i = 2, 3, 4$, is full-context if the condition satisfied is similar.

An array splicing system is full-context if the set of patterns of each type is full context.

Theorem 6.3 *There are array splicing systems that are permanent and reduced, but not full context, unlike in string splicing systems.*

Proof. Define the array splicing system S with alphabet $\{0, 1, 2\}$ and with the following patterns:

Type-1 :

$$P_1 = \langle 11, 1, 1, 11, 0, 1, 11, 1, 1 \rangle,$$
$$P_2 = \langle 22, 2, 2, 22, 0, 2, 22, 2, 2 \rangle;$$

Type-2 $= \{\langle 0 \rangle\}$, Type-3 $=$ Type-4 $= \emptyset$.

Now, S is permanent because the only possible crossing is 0 which is a null site of type 2, and so cannot be destroyed. It is also reduced: Since there is only one crossing in the patterns of S, there can be no type-2 reduction. And on inspection, we see that no pattern is a sub-image of another.

Moreover, S is not full context: We can see that while $\langle 22, 2, 2, 22, 0, 2, 22, 2, 2 \rangle$ and $\langle 11, 1, 1, 11, 0, 1, 11, 1, 1 \rangle$ are compatible patterns, $\langle 22, 2, 1, 22, 0, 1, 11, 1, 1 \rangle$ is not a pattern, which contradicts the definition of full contextuality.

Thus S is an example of an array splicing system that is permanent and reduced but not full context.   □

7   Conclusions

In this paper, we have proved some hierarchy results over array splicing systems defined in [1]. We have also considered some reduction results. We find that the hierarchy for arrays is different from the hierarchy for string splicing systems, whereas some reduction results are similar for both.

It would be interesting to find out whether the family of persistent languages forms a subclass of the family of all array splicing languages. It is possible to give a different definition of array splicing systems and consider the hierarchy there. Another possible topic which could be pursued is distributed array splicing systems. Yet another direction of research could be to extend the definition of splicing to graphs. This might have more biological significance. We are working on these lines.

References

[1] K. Krithivasan, V. T. Chakaravarthy, R. Rama, Array splicing systems, in *New Trends in Formal Languages. Control, Cooperation, and Combinatorics* (Gh. Păun, A. Salomaa, eds.), *Lecture Notes in Computer Science* 1218, Springer-Verlag, Berlin, (1997), 346 – 365.

[2] T. Head, Formal language theory and DNA; An analysis of the generative capacity of specific recombinant behaviours, *Bulletin of Mathematical Biology*, 49 (1987), 737 – 759.

[3] R. W. Gatterdam, Algorithms for splicing systems, *SIAM Journal of Computing*, 21 (1992), 507 – 520.

[4] Gh. Păun, Regular extended H-systems are computationally universal, *Journal of Automata, Languages and Combinatorics*, 1 (1996), 27 – 36.

Splicing on Routes and Recognizability[1]

Alexandru MATEESCU

Faculty of Mathematics, University of Bucharest
Str. Academiei 14, 70109, Bucharest, Romania, and
Turku Centre for Computer Science,
Lemminkäisenkatu 14 A, 20520 Turku, Finland
mateescu@utu.fi

Abstract. Assume that L_1 and L_2 are (regular) languages and that M_1, M_2 are (finite) monoids, such that M_i recognizes L_i, $i = 1, 2$. Moreover, assume that $\oplus$ is an operation with languages, such that $L_1 \oplus L_2$ is a language. The following problem is of a special interest: find a function Ψ such that the language $L_1 \oplus L_2$ is recognized by the monoid $\Psi(M_1, M_2)$. In this paper we investigate this problem for the case when the operation $\oplus$ is the operation of splicing on routes.

1   Introduction

The operation of splicing on routes was introduced in [9]. Splicing on routes defines new formal ways to specify restrictions on the crossover operation used in the theory of DNA computation. The crossover operation models the recombination of DNA sequences. The recombination produces a new sequence starting from two parent sequences. The resulting sequence is formed by starting at the left end of one parent sequence, copying a substring, crossing over to some site in the other parent sequence, copying a substring, crossing back to some site in the first parent sequence and so on.

The constraints that occur in the operation of splicing on routes involve the general strategy to switch from one word to another word. Once such a strategy is defined, the structure of the words that are spliced does not play any role.

---

[1]This work has been supported by the Project 137358 of the Academy of Finland.

It was shown in [9] that, if two regular languages are spliced on a set of regular routes, then the resulting language is also regular.

The following problem is central in the algebraic theory of automata: let L_1 and L_2 be two (regular) languages and assume that M_1, M_2 are (finite) monoids, such that M_i recognizes L_i, $i = 1, 2$. Moreover, assume that $\oplus$ is an operation with languages, such that $L_1 \oplus L_2$ is a language. The following problem is of a special interest: find a function Ψ such that the language $L_1 \oplus L_2$ is recognized by the monoid $\Psi(M_1, M_2)$. In this paper we investigate this problem for the case when the operation $\oplus$ is the operation of splicing on routes.

Note that for several, well-known operations with languages, such as: union, intersection, complementation, catenation, Kleene star, shuffle, the above problem was widely investigated, [21], [2], [17], [18].

The method of finding such a function Ψ, that we describe in this paper is very general and covers most of the well-known operations with languages. Additionaly, this method can be applied to some other operations with languages, like: crossover, simple splicing, various shuffle-like operations, etc. This method can be used also for a large number of unary operations.

2 Routes and Splicing on Routes

In this section we recall the notions of *route* and of *splicing on routes*. For other facts and results concerning these notions, see [9].

The reader is referred to [20], [12] for general results in formal languages and to [6], [5], [1], [15], [16] for results on splicing and on DNA computation.

Consider the alphabet $V = \{r, \bar{r}, u, \bar{u}\}$. Elements in V are referred to as *versors*. Denote $V_1 = \{r, \bar{r}\}$ and $V_2 = \{u, \bar{u}\}$.

Definition 2.1 *A route is an element $t \in V^*$. A set of routes is a subset of V^*.*

Let Σ be an alphabet and let α, β be words over Σ. Assume that $d \in V$ and $t \in V^*$.

Definition 2.2 *The splicing of α with β on the route dt, denoted $\alpha \bowtie_{dt} \beta$, is defined as follows:*

if $\alpha = au$ and $\beta = bv$, where $a, b \in \Sigma$ and $u, v \in \Sigma^*$, then:

$$au \bowtie_{dt} bv = \begin{cases} a(u \bowtie_t bv), & \text{if } d = r, \\ b(au \bowtie_t v), & \text{if } d = u, \\ (u \bowtie_t bv), & \text{if } d = \bar{r}, \\ (au \bowtie_t v), & \text{if } d = \bar{u} . \end{cases}$$

If $\alpha = au$ and $\beta = \lambda$, $a \in \Sigma, u \in \Sigma^*$, then

$$au \bowtie_{dt} \lambda = \begin{cases} a(u \bowtie_t \lambda), & \text{if } d = r, \\ (u \bowtie_t \lambda), & \text{if } d = \bar{r}, \\ \emptyset, & \text{otherwise.} \end{cases}$$

If $\alpha = \lambda$ and $\beta = bv$, $b \in \Sigma, v \in \Sigma^*$, then

$$\lambda \bowtie_{dt} bv = \begin{cases} b(\lambda \bowtie_t v), & \text{if } d = u, \\ (\lambda \bowtie_t v), & \text{if } d = \bar{u}, \\ \emptyset, & \text{otherwise.} \end{cases}$$

Finally,

$$\lambda \bowtie_t \lambda = \begin{cases} \lambda, & \text{if } t = \lambda , \\ \emptyset, & \text{otherwise.} \end{cases}$$

Remark 2.1 *One can easily notice that, if $|\alpha| \neq |t|_{V_1}$ or $|\beta| \neq |t|_{V_2}$, then $\alpha \bowtie_t \beta = \emptyset$.*

The operation of splicing on a route is extended in a natural way to the operation of splicing on a set of routes as well as to an operation between languages.

If T is a set of routes, the *splicing of α with β on the set T of routes*, denoted $\alpha \bowtie_T \beta$, is:

$$\alpha \bowtie_T \beta = \bigcup_{t \in T} \alpha \bowtie_t \beta.$$

If $L_1, L_2 \subseteq \Sigma^*$, then:

$$L_1 \bowtie_T L_2 = \bigcup_{\alpha \in L_1, \beta \in L_2} \alpha \bowtie_T \beta.$$

Example 2.1 *Let α and β be the words $\alpha = a_1 a_2 a_3 a_4 a_5 a_6 a_7 a_8$, $\beta = b_1 b_2 b_3 b_4 b_5$ and assume that $t = ur\bar{r}\bar{r}ru\bar{u}uru\bar{r}r$. The splicing of α with β on the route t is:*

$$\alpha \bowtie_t \beta = \{b_1 a_1 a_4 a_5 b_2 b_4 a_6 b_5 a_8\}.$$

The result has the following geometrical interpretation (see Figure 1): the route t defines a line starting in the origin and continuing one unit to the right or up, depending of the definition of t. If the current symbol in t is r, then the line is continuing one unit to the right by a continuous line, if the current symbol in t is $\bar{r}$, then the line is continuing one unit also to the right, but by a dot line. Similarly, if the current symbol in t is u, then the line is continuing one unit up by a continuous line, if the current symbol in t is $\bar{u}$, then the line is continuing one unit also up, but by a dot line.

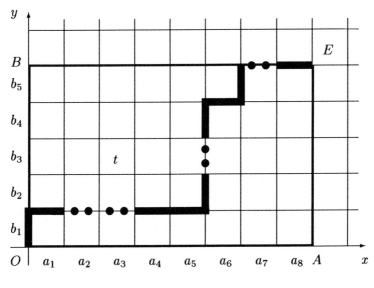

Figure 1

In our case, first there is a unit up, then one unit right, then two units right depicted by a dot line, etc. Assign α on the Ox axis and β on the Oy axis of the plane. Observe that the route ends in the point with coordinates $(8, 5)$ (denoted by E in Figure 1) that is exactly the upper right corner of the rectangle defined by α and β, i.e., the rectangle $OAEB$ in Figure 1. Hence, the result of the splicing of α with β on the route t is nonempty. The result can be read following the line defined by the route t: that is, when being in a lattice point of the route, with the route going to the right by a continuous line, one should pick up the corresponding letter from α, otherwise, if the route is going up by a continuous line, then one should add to the result the corresponding letter from β. The dot lines indicate that the corresponding symbols should be omitted.

Hence, the route t defines a line in the rectangle $OAEB$, on which one has "to walk" starting from the corner O, the origin, and ending in the corner E, the exit point. In each lattice point one has to follow one of the versors from V, according to the definition of t.

Assume now that t' is another route, say:

$$t' = rru\bar{u}\bar{u}rru\bar{r}\bar{r}rur.$$

Observe that:

$$\alpha \bowtie_{t'} \beta = \{a_1a_2b_1a_3a_4b_4a_7b_5a_8\}.$$

Consider the set of routes, $T = \{t, t'\}$. The splicing of α with β on the set T of routes is:

$$\alpha \bowtie_T \beta = \{b_1a_1a_4a_5b_2b_4a_6b_5a_8, \, a_1a_2b_1a_3a_4b_4a_7b_5a_8\}.$$

In the sequel we consider some particular cases of the operation of splicing on routes. This will prove that splicing on routes is a very general operation with a great power of expressibility.

• Examples of binary operations that are instances of splicing on routes:

1. Let T be the set $T = \{r, \bar{r}, u, \bar{u}\}^*$. Observe that $\bowtie_T$ is the crossover, [8], (pure recombination, [13], or iterated splicing, [15]).

2. Assume that $T = r^*\bar{r}^*u^*\bar{u}^*$. Notice that in this case $\bowtie_T$ is the simple splicing operation, [10].

3. Let T be the set $T = \{r^n\bar{u}^nu^m\bar{r}^m \mid n, m \geq 0\}^*$. Observe that $\bowtie_T$ is the equal-length crossover, [8], [13], [14].

4. Assume that $T = \{r, u\}^*$. Note that in this case $\bowtie_T$ is the shuffle operation.

5. Consider that $T = (ru)^*(r^* \cup u^*)$. Note that in this case $\bowtie_T$ is the literal shuffle.

6. Assume that $T \subseteq \{r, u\}^*$ and observe that $\bowtie_T$, is the shuffle on the set T of trajectories, [11].

7. Consider that $T = r^*u^*$. It follows that $\bowtie_T = \cdot$, the catenation operation.

8. Let T be the following set of routes: $T = r^*u^* \cup u^*r^*$. In this case $\bowtie_T = \bullet$, the bi-catenation operation.

9. Assume that $T = u^*r^*$ and observe that $\bowtie_T$ is the anti-catenation operation.

10. Define $T = r^*u^*r^*$ and note that $\bowtie_T = \longleftarrow$, the insertion operation, [7].

• **Examples of unary operations that are instances of splicing on routes:**

Let Σ be an alphabet. Assume that T is a set of routes such that $T \subseteq \{r, \bar{r}, \bar{u}\}^*$. Note that for all languages $L, L_1, L_2 \subseteq \Sigma^*$, with L_1, L_2 nonempty, it follows that:

$$L \bowtie_T L_1 = L \bowtie_T L_2.$$

Therefore, in this case, the operation $\bowtie_T$ does not depend of the second argument. Similarly, if $T \subseteq \{\bar{r}, u, \bar{u}\}^*$, the operation $\bowtie_T$ does not depend of the first argument, assuming that the first argument is nonempty.

Consequently, several well-know unary operations of words and languages are particular cases of the operation of splicing on routes. In the sequel we denote by ∇_T the unary operation defined by $\bowtie_T$, in the case $T \subseteq \{r, \bar{r}, \bar{u}\}^*$.

1. Let T be the set $T = r^*\bar{r}^*\bar{u}^*$. Note that $\nabla_T(L) = Pref(L)$, that is, ∇_T is the unary operation $Pref$.

2. Assume that $T = \bar{r}^*r^*\bar{u}^*$. Note that $\nabla_T(L) = Suf(L)$, i.e., ∇_T is the unary operation Suf.

3. Let $T = \bar{r}^*r^*\bar{r}^*\bar{u}^*$. It is easy to see that $\nabla_T(L) = Sub(L)$, that is, ∇_T is the unary operation Sub.

4. Assume that $T = \{r, \bar{r}\}^*\bar{u}^*$. Note that $\nabla_T(L) = Scatt(L)$, that is, ∇_T is the unary operation $Scatt$.

5. Consider that $T = \{r^k\bar{r}^k | k \geq 0\}\bar{u}^*$. Note that $\nabla_T(L) = \frac{1}{2}(L)$, that is, ∇_T is the unary operation $\frac{1}{2}$, [4].

6. Assume that $T = r^*\bar{r}^*r^*\bar{u}^*$. Note that $\nabla_T(L) = L \longrightarrow \Sigma^*$, that is, ∇_T is the unary operation of sequential deletion of L with Σ^*, [7].

Consider the following set P of rewriting rules:

$$\bar{r}\bar{u} \longrightarrow \bar{u}\bar{r}, \; \bar{u}\bar{r} \longrightarrow \bar{r}\bar{u}, \; r\bar{u} \longrightarrow \bar{u}r,$$

$$u\bar{r} \longrightarrow \bar{r}u, \; \bar{r}u \longrightarrow u\bar{r}, \; \bar{u}r \longrightarrow r\bar{u}.$$

Denote by π the transitive and reflexive closure of the following relation θ:

$$\alpha u \beta \; \theta \; \alpha v \beta \text{ iff } (u \longrightarrow v) \in P.$$

where $\alpha, \beta, u, v \in V^*$.

Note that π is an equivalence relation on V^*. Let T be a set of routes. Denote by $\pi(T)$ the set:

$$\pi(T) = \{t' \mid \text{ there is } t \in T \text{ such that } t\pi t'\}.$$

For the following result see [9].

Theorem 2.1 *Let t, t' be in V^* and let T be a set of routes.*
(i) If $t\pi t'$, then $\bowtie_t = \bowtie_{t'}$.
(ii) If T' is a set of routes such that $T \subseteq T' \subseteq \pi(T)$, then $\bowtie_T = \bowtie_{T'}$.

3   The Problem of Recognizability

We start this section by recalling some general facts about languages recognized by monoids. Of a special interest is the case of languages of the form $L_1 \bowtie_T L_2$, where T is a set of routes.

A monoid M_1 is *embedded* in a monoid M_2 iff there exists an injective morphism from M_1 to M_2. A monoid M_1 *divides* a monoid M_2, denoted $M_1 < M_2$, iff M_1 is isomorphic with a quotient of a submonoid of M_2. Clearly, if M_1 is embedded in M_2, then M_1 divides M_2. The division relation is transitive.

The unit element of a monoid is denoted by 1. If M is a monoid, then the set $\mathcal{P}(M)$ is a monoid with the multiplication defined by $AB = \{xy \mid x \in A, y \in B\}$, where $A, B \subseteq M$.

Let M be a monoid and assume that C is a binary relation over M, i.e., $C \subseteq M \times M$. The relation C is a *congruence* iff C is an equivalence relation and, moreover, whenever $(u, v) \in C$ and $(u', v') \in C$ then also $(uv, u'v') \in C$.

It is well-known that, if C is a congruence, then the set of all equivalences classes with respect to C is a monoid, called the factor monoid and denoted by M/C.

The equivalence class (with respect to the congruence C) of an element u, $u \in M$, is denoted by $\langle u \rangle_C$ or simply by $\langle u \rangle$, if C is understood from context. The fact that u, v are in the relation C, i.e., $(u, v) \in C$, will be also denoted as uCv, where $u, v \in M$.

If C, C' are binary relation over M, then $C^{-1} = \{(v, u) \mid (u, v) \in C\}$,

$$C \circ C' = \{(u, w) \mid (u, v) \in C \text{ and } (v, w) \in C', \text{ for some } v \in M\},$$

and

$$CC' = \{(uu', vv') \mid (u, v) \in C , (u', v') \in C'\}.$$

If C is a binary relation over M, then $\bar{C}$ denotes the conguence generated by C, i.e., the smallest congruence on M that contains C. Note that $\bar{C}$ is the intersection of all those congruences over M that contain C.

Remark 3.1 *Let C be a relation over M. Consider the following sequence of relations:*

$$C_0 = C \cup \{(u, u) \mid u \in M\},$$
$$C_{k+1} = C_k \cup C_k^{-1} \cup C_k \circ C_k \cup C_k C_k, \text{ where } k \geq 0.$$

One can easily see that:

$$\bar{C} = \bigcup_{k \geq 0} C_k.$$

This construction of the congruence generated by a relation will be used in the sequel.

Definition 3.1 *Let L be a language, $L \subseteq \Sigma^*$. A monoid M recognizes L iff there exists a morphism $\varphi : \Sigma^* \longrightarrow M$, a subset F of M, $F \subseteq M$, such that $L = \varphi^{-1}(F)$.*

For each language L, $L \subseteq \Sigma^*$, there exists a monoid M that recognizes L. An example of such a monoid is the syntactic monoid of L. The *syntactic congruence* defined by L is the congruence $\approx_L$ on Σ^* defined as: $x \approx_L y$ iff for all $\alpha, \beta \in \Sigma^*$

$$\alpha x \beta \in L \quad \text{iff} \quad \alpha y \beta \in L,$$

where $x, y \in \Sigma^*$. The *syntactic monoid* of L, denoted by M_L, is the quotient monoid $\Sigma^* / \approx_L$. One can easily verify that M_L recognizes L. A monoid M recognizes L iff M_L divides M. If a monoid M_1 recognizes L and if M_1 divides M_2, then M_2 recognizes L, too.

The following theorem is a fundamental result that goes back to Kleene:

Theorem 3.1 *A language L is regular iff L is recognized by some finite monoid.*

Let L_1, L_2 be languages $L_1, L_2 \subseteq \Sigma^*$. Let M_1, M_2 be monoids, such that M_i recognizes L_i, $i = 1, 2$. Assume that $\bowtie$ is an operation with languages such that $L_1 \bowtie L_2 \subseteq \Sigma^*$.

The following problem has been widely investigated: find a function $\Psi_{\bowtie}$ such that the language $L_1 \bowtie L_2$ is recognized by $\Psi_{\bowtie}(M_1, M_2)$.

For more details on this problem, as well as for a large bibliography, the reader is referred to [2], [17], or more recently, [18].

In the sequel we solve this problem for the operation $\bowtie_T$, where T is an arbitrary set of routes. The solution offers a uniform method to find monoids that recognize a large number of operations with languages.

Let L_1, L_2 be languages over an alphabet Σ. Assume that M_i is a monoid that recognize L_i, $i = 1, 2$. That means that there exists a morphism $\varphi_i : \Sigma^* \longrightarrow M_i$ and $F_i \subseteq M_i$, such that $L_i = \varphi_i^{-1}(F_i)$, $i = 1, 2$.

Let T be a set of routes, $T \subseteq V^*$ and assume that T is recognized by the monoid M_T, i.e., there exists a morphism $\varphi_T : V^* \longrightarrow M_T$, and there exists $F_T \subseteq M_T$ such that $T = \varphi_T^{-1}(F_T)$.

Let M be the following monoid: $M = M_1 \times M_2 \times M_T$. Elements in M will be denoted by $[x_1, x_2, x_3]$, where $x_1 \in M_1$, $x_2 \in M_2$ and $x_3 \in M_T$.

Now consider the following relation:

$$\rho = \{([\varphi_1(a), 1, \varphi_T(\bar{r})], [1, 1, 1]) \mid a \in \Sigma\}$$
$$\cup \ \{([1, \varphi_2(a), \varphi_T(\bar{u})], [1, 1, 1]) \mid a \in \Sigma\}.$$

Let $\bar{\rho}$ be the congruence generated by the relation ρ on the monoid M.

Let M' be the factor monoid $M/\bar{\rho}$, i.e., $M' = M/\bar{\rho}$ and denote by $\mathcal{M}$ the monoid of all subsets of M', i.e., $\mathcal{M} = \mathcal{P}(M')$.

Remark 3.2 *Without loss of generality we may assume that all three morphisms φ_1, φ_2 and φ_T are surjective morphisms. (If one of these morphisms, say φ_1, is not surjective, then the corresponding monoid M_1 can be replaced with the image monoid $\varphi(M_1)$, etc.).*

We are now in the position to state the main result of this paper:

Theorem 3.2 *Using the above notations and definitions, it follows that the monoid $\mathcal{M}$ recognizes the language $L_1 \bowtie_T L_2$.*

Proof. Consider the morphism ψ, $\psi : \Sigma^* \longrightarrow \mathcal{M}$, defined as follows:

$$\psi(a) = \{\langle \varphi_1(a), 1, \varphi_T(r)\rangle, \langle 1, \varphi_2(a), \varphi_T(u)\rangle \mid a \in \Sigma\}.$$

Also, define the subset F of $\mathcal{M}$ as being:

$$F = \{K \subseteq \mathcal{M} \mid \text{ there exist } f_1 \in F_1, f_2 \in F_2, f_T \in F_T, \text{ such that} \\ \langle f_1, f_2, f_T \rangle \in K\}.$$

Now we prove the following:

Claim A. For all $w \in \Sigma^*$, the morphism ψ satisfies the equality:

$$\psi(w) = \{\langle \varphi_1(\alpha), \varphi_2(\beta), \varphi_T(t)\rangle \mid w \in \alpha \bowtie_t \beta\}.$$

Proof of Claim A. The proof is by induction on the length of w. Assume that $|w| = 0$, i.e., $w = \lambda$. Since ψ is a morphism, it follows that $\psi(\lambda) = \langle 1, 1, 1 \rangle$.

Now consider that $\langle \varphi_1(w_1), \varphi_2(w_2, \varphi_T(t)\rangle = \langle 1, 1, 1\rangle$. From the definition of the congruence $\bar{\rho}$, we conclude that $t \in \{\bar{r}, \bar{u}\}$, and, moreover, $|w_1| = |t|_{\bar{r}}$, $|w_2| = |t|_{\bar{u}}$. Therefore, $w_1 \bowtie_t w_2 = \lambda$ and hence, $\lambda = w \in \alpha \bowtie_t \beta$. Now assume that Claim A is true for all words x with $|x| \leq n$ ant let w be a word of length $n + 1$. Consider that $w = w'a$, where $a \in \Sigma$ and $|w'| = n$. Using the inductive assumption and the definition of the congruence $\bar{\rho}$ we obtain that:

$$\psi(w) = \psi(w'a)$$
$$= \{\langle \varphi_1(\alpha'), \varphi_2(\beta'), \varphi_T(t')\rangle \mid w' \in \alpha' \bowtie_{t'} \beta'\}$$
$$\{\langle \varphi_1(a), 1, \varphi_T(r)\langle, \rangle 1, \varphi_2(a), \varphi_T(u)\rangle\}.$$

Now, there are two cases:

Case 1.

$$\psi(w) = \psi(w'a)$$
$$= \{\langle \varphi_1(\alpha'), \varphi_2(\beta'), \varphi_T(t')\rangle \mid w' \in \alpha' \bowtie_{t'} \beta'\}\langle \varphi_1(a), 1, \varphi_T(r)\rangle.$$

Therefore, $\psi(w) = \{\langle \varphi_1(\alpha'), \varphi_2(\beta'), \varphi_T(t')\rangle\langle \varphi_1(a), 1, \varphi_T(r)\rangle \mid w' \in \alpha' \bowtie_{t'} \beta'\}$. Thus, $\psi(w) = \{\langle \varphi_1(\alpha'a), \varphi_2(\beta'), \varphi_T(t'r)\rangle \mid w' \in \alpha' \bowtie_{t'} \beta'\}$. Notice that $w = w'a \in \alpha'a \bowtie_{tr} \beta'$, and hence Claim A holds in this case.

Case 2.

$$\psi(w) = \psi(w'a)$$
$$= \{\langle \varphi_1(\alpha'), \varphi_2(\beta'), \varphi_T(t')\rangle \mid w' \in \alpha' \bowtie_{t'} \beta'\}\langle 1, \varphi_2(a), \varphi_T(u)\rangle.$$

The proof is similar with the proof of Case 1.

This ends of proof of Claim A.

We end the proof of this theorem showing that

$$\psi^{-1}(F) = L_1 \bowtie_T L_2.$$

Assume first that $w \in \psi^{-1}(F)$. Hence $\psi(w) \in F$. Now, from Claim A, it follows that

$$\psi(w) = \{\langle \varphi_1(\alpha), \varphi_2(\beta), \varphi_T(t)\rangle \mid w \in \alpha \bowtie_t \beta\}.$$

Using the definition of the set F we conclude that there are $f_1 \in F_1$, $f_2 \in F_2$ and $f_T \in F_T$ such that for some class from $\psi(w)$, $\langle f_1, f_2, f_T\rangle = \langle \varphi_1(\alpha), \varphi_2(\beta), \varphi_T(t)\rangle$, and, moreover, $w \in \alpha \bowtie_t \beta$. One can easily show that in this case, $\alpha \in L_1$, $\beta \in L_2$, and $t \in T$. Therefore, $w \in L_1 \bowtie_T L_2$ and thus

$$\psi^{-1}(F) \subseteq L_1 \bowtie_T L_2.$$

For the converse inclusion, assume that $w \in L_1 \bowtie_T L_2$. Therefore, there are $\alpha \in L_1$, $\beta \in L_2$ and $t \in T$, such that $w \in \alpha \bowtie_t \beta$. Notice that $\varphi_1(\alpha) = f_1 \in F_1$, $\varphi_2(\beta) = f_2 \in F_2$ and, moreover, $\varphi_T(t) = f_T \in F_T$. Hence, using Claim A, i.e.,

$$\psi(w) = \{\langle \varphi_1(\alpha), \varphi_2(\beta), \varphi_T(t)\rangle \mid w \in \alpha \bowtie_t \beta\},$$

we conclude that $\langle f_1, f_2, f_T\rangle \in \psi(w)$. Hence, from the definition of F, we deduce that $\psi(w) \in F$ and thus, $w \in \psi^{-1}(F)$. Therefore, $\psi^{-1}(F) = L_1 \bowtie_T L_2$., i.e., the monoid $\mathcal{M}$ recognizes the language $L_1 \bowtie_T L_2$. $\square$

The above theorem has the following direct consequence:

Theorem 3.3 *If L_1, L_2 are regular languages, and if T is a regular set of routes, then $L_1 \bowtie_T L_2$ is a regular language.*

Proof. Note that by Theorem 3.1, we can consider M_1, M_2 and M_T are finite monoids. One can easily see that in this case also the monoid $\mathcal{M}$ is finite. Therefore, again using Theorem 3.1 we conclude that $L_1 \bowtie_T L_2$ is a regular language. $\square$

For a different proof of the above theorem, as well as for other results of this type, the reader is referred to [9].

From Theorem 3.3 we conclude several closure properties (well-known or new) of the class of regular languages:

Corollary 3.1 *The class of regular languages is closed under the following operations: crossover, simple splicing, shuffle, literal shuffle, shuffle on a regular set of trajectories, catenation, bi-catenation, anti-catenation, insertion, taking prefixes, taking suffixes, taking subwords, taking scattered subwords.*

Proof. Note that all these operations are particular instances of the operation of splicing on a regular set of routes, as shown in Section 2. □

Remark 3.3 *Note that if the set T of routes does not contain occurrences of $\bar{r}$ or of $\bar{u}$, i.e., $T \subseteq \{r, u\}^*$, then the congruence $\bar{\rho}$ is the identity relation. Therefore, the construction of the monoid $\mathcal{M}$ from Theorem 3.2 is much simpler, i.e., $\mathcal{M} = \mathcal{P}(M_1 \times M_2 \times M_T)$.*

4 Conclusion

Splicing on routes is a very general operation with words and languages. In view of Theorem 3.2 it will be interesting to find the function Ψ for all the operations with languages listed in Section 2. This will provide a comprehensive way to compute monoids that recognize a large variety of operations with languages. Also, Theorem 3.2 opens new possibilities to investigate many new problems related to the Theory of Algebraic Varieties of Formal Languages, see [2], [17] and [18]. Finally, we like to mention that DNA computation offers a new area of research that leads also to new theoretical problems. Splicing on routes is such an example.

References

[1] K. Culik II, T. Harju, Splicing semigroups of dominoes and DNA, *Discrete Appl. Math.*, 31 (1991), 261 – 277.

[2] S. Eilenberg, *Automata, Languages and Machines*, Academic Press, New York, vol. A, 1974, vol. B, 1976.

[3] T. Harju, A. Mateescu, A. Salomaa, *Shuffle on trajectories: a simplified approach to the Schützenberger product and related operations families of languages*, TUCS Technical Report 163, 1998.

[4] J. Hartmanis, R. E. Stearns, Regularity-preserving modifications of regular expressions, *Information and Control*, 6 (1963), 55 – 69.

[5] T. Head, Splicing schemes and DNA, in *Lindenmayer Systems: Impacts on Theoretical Computer Science and Developmental Biology* (G. Rozenberg, A. Salomaa, eds.), Springer, Heidelberg, 1992, 371 – 383.

[6] T. Head, Gh. Păun, D. Pixton, Language Theory and Molecular Genetics, Chapter 7, Vol. 2, in *Handbook of Formal Languages* (G. Rozenberg, A. Salomaa, eds.), Springer, Heidelberg, 1997, 295 – 360.

[7] L. Kari, *On Insertion and Deletion in Formal Languages*, PhD Thesis, University of Turku, Turku, Finland, 1991.

[8] J. Kececioglu, D. Gusfield, Reconstructing a history of recombinations from a set of sequences, *Proc. of the 5-th ACM-SIAM Symposium on Discrete Algorithms*, 1994, 471 – 480.

[9] A. Mateescu, Splicing on Routes: A Framework of DNA Computation, in *Unconventional Models of Computation* (C. S. Calude, J. Casti, M. J. Dinneen, eds.), Springer, Singapore, 1998, 273 – 285.

[10] A. Mateescu, Gh. Păun, G. Rozenberg, A. Salomaa Simple splicing systems, to appear in *Discrete Applied Mathematics*.

[11] A. Mateescu, G. Rozenberg, A. Salomaa, Shuffle on Trajectories: Syntactic Constraints, *Theoretical Computer Science*, 197, 1-2 (1998), 1 – 56.

[12] A. Mateescu, A. Salomaa, Formal Languages: An Introduction and a Synopsis, Chapter 1, Vol. 1, in *Handbook of Formal Languages* (G. Rozenberg, A. Salomaa, eds.), Springer, Heidelberg, 1997, 1 – 40.

[13] V. Mitrana, On the interdependence between shuffle and crossing-over operations, *Acta Informatica*, 34 (1997), 257 – 266.

[14] Gh. Păun, G. Rozenberg, A. Salomaa, Restricted use of the splicing operation, *Intern. J. Computer Math.*, 60 (1996), 17 – 32.

[15] Gh. Păun, G. Rozenberg, A. Salomaa, Computing by Splicing, *Theoretical Computer Science* 168, 2 (1996), 321 – 336.

[16] Gh. Păun, A. Salomaa, DNA computing based on the splicing operation, *Mathematica Japonica*, 43, 3 (1996), 607 – 632.

[17] J. E. Pin, *Varieties of Formal Languages*, North Oxford Academic, 1986.

[18] J.E. Pin, Syntactic Semigroups, Chapter 10, Vol. 1, in *Handbook of Formal Languages* (G. Rozenberg, A. Salomaa, eds.), Springer, Heidelberg, 1997, 679 – 746.

[19] D. Pixton, Regularity of splicing languages, *Discrete Applied Math.*, 69 (1996), 101 – 124.

[20] A. Salomaa, *Formal Languages*, Academic Press, New York, 1973.

[21] M. P. Schützenberger, On finite monoids having only trivial subgroups, *Information and Control*, 8 (1965), 190 – 194.

Algebraic Properties of DNA Operations

Zhuo LI

Department of Computer Science
University of Western Ontario
London, Ontario, N6A 5B7 Canada
zli@csd.uwo.ca

Abstract. Any DNA strand can be identified with a word in the language X^* where $X = \{A, C, G, T\}$. By encoding A as 000, C as 010, G as 101, and T as 111, we treat the DNA operations concatenation, union, reverse, complement, annealing and melting, from the algebraic point of view. The concatenation and union play the roles of multiplication and addition over some algebraic structures, respectively. Then the rest of the operations turn out to be the homomorphisms or anti-homomorphisms of these algebraic structures. Using this technique, we find the relationship among these DNA operations.

1 Introduction

This paper is a first attempt to treat the DNA operations from an algebraic point of view. It offers a new formalization of DNA operations, based on the notations defined in [2].

In 1994, L. Aldeman [1] successfully solved an instance of the Directed Hamiltonian Path Problem solely by manipulating DNA strands. Following [1], DNA algorithms have been found to solve other problems: expansion of symbolic determinants [7], matrix multiplication [8], addition [4], exascale computer algebra [9], and so on. The basic idea behind the DNA computing is to use DNA strands to encode information and employ enzymes to simulate simple computations. This shows that the tools of biology could be widely used in solving mathematical problems. Conversely, it is natural and necessary to study the DNA strands and the operations of the DNA strands in a mathematically precise way. This is the motivation of this paper.

Let us consider the DNA strings which are the DNA strands with polarity ignored. Then it is natural to relate these strings with formal language and power series. More precisely, let $X = \{A, C, G, T\}$, where A, C, G, T represent the building blocks of a DNA strand, called nucleotides: *adenine, guanine, cytosine*, and *thymine*. The single nucleotides are linked together end-to-end to form DNA strands. The DNA sequence has a *polarity*: a sequence of DNA is distinct from its reverse. The two ends are known under the name of the $5'$ end and $3'$ end, respectively. Taken as pairs, the nucleotides C and G and the nucleotides A and T are said to be *complementary*. Two complementary single strands with opposite orientation will join together to form a *double helix* in a process called *annealing*. The reverse process is called *melting* (refer to [5] for more details).

We encode A by 000, C by 010, G by 101, and T by 111. The reason for the choice of our encoding is the following. It would seem natural to encode nucleotides A, C, G, and T as 00, 01, 10, and 11, respectively. However, this encoding that satisfies the condition that $\hat{A} = \hat{}(00) = 11 = T$ and $\hat{C} = \hat{}(01) = 10 = G$, does not satisfy the natural requirement that the reverse of a nucleotide is the nucleotide itself. Indeed, $C^R = (01)^R = 10 = G$, which is not what we expect. In contrast, our encoding satisfies both natural constraints mentioned above.

Let $\Sigma = \{0, 1\}$. Let S be any commutative *semiring* (Definition 2.2). Let $S\langle\langle\Sigma^*\rangle\rangle$ and $S\langle\langle X^*\rangle\rangle$ be the corresponding sets of *power series* (Definition 2.3) over Σ^* and X^*, respectively. Then $S\langle\langle\Sigma^*\rangle\rangle$ and $S\langle\langle X^*\rangle\rangle$ are semirings (Example 2.2 in Section 2), where the multiplication on both semirings is the generalization of concatenation of two words over Σ^* and X^*, respectively, and the addition on both semirings is the generalization of the union of two words over Σ^* and X^*, respectively. Then under our encoding scheme. $S\langle\langle X^*\rangle\rangle$ is subsemiring of $S\langle\langle\Sigma^*\rangle\rangle$. Next, we generalize the operations of DNA strings, *reverse*, denoted by R, *complement*, denoted by $\hat{}$ to *anti-automorphism* and *automorphism* (Definition 2.6) of the semiring $S\langle\langle\Sigma^*\rangle\rangle$. The subsemiring $S\langle\langle X^*\rangle\rangle$ of $S\langle\langle\Sigma^*\rangle\rangle$ is invariant under these two morphisms. That is, the automorphism $\hat{}$ is still an automorphism of $S\langle\langle X^*\rangle\rangle$ and the anti-automorphism is still an anti-automorphism of $S\langle\langle X^*\rangle\rangle$.

To consider the polarity of the DNA strands, we introduce *semimodules* (Definition 2.4) and *direct sum* of two semimodules (Definition 2.5). To get a rough idea of what a semimodule and a direct sum are, the reader could think that the semimodule is a vector space and the direct sum is a Cartesian product of two vector spaces. Each element of the direct sum has two components. In our interpretation, the first component could be thought as the DNA strand which starts with 5′ and ends with 3′, while the second component could be thought as the DNA strand which starts with 3′ and end with 5′. This provides a mathematical model to simulate the double DNA strands. The operations of DNA strands denoted by notations ↑, ↓, and ↕ in [2] can then be generalized to homomorphisms from the semimodule $S\langle\langle X^*\rangle\rangle$ to the direct sum of $S\langle\langle X^*\rangle\rangle \oplus S\langle\langle X^*\rangle\rangle$. This brings about the relationship among these operators.

2 Background of Linear Algebra

We introduce some background in mathematics. The interested readers should refer to [6] for more detail.

Definition 2.1 *A monoid* $(M, \cdot, 1)$ *consists of a set* M, *an associative operation on* M *and of an identity* 1 *such that* $1 \cdot a = a \cdot 1 = a$. *A monoid is commutative if and only if* $a \cdot b = b \cdot a$ *for every* a *and* b *in* M.

Definition 2.2 *A semiring is a set* S *together with two binary operations* $+$ *and* $\cdot$ *and two elements* 0 *and* 1 *such that:*

1. $(S, +, 0)$ *is a commutative monoid,*

2. $(S, \cdot, 1)$ *is a monoid,*

3. *the distribution laws* $a \cdot (b + c) = a \cdot b + a \cdot c$ *and* $(b + c) \cdot a = b \cdot a + c \cdot a$ *hold for every* a, b, *and* c *in* S,

4. $0 \cdot a = a \cdot 0 = 0$ *for every* a.

If $a \cdot b = b \cdot a$ *for any* $a, b \in S$, *then* S *is called a commutative semiring.*

Example 2.1 (i) $(\mathbf{B}, +, \cdot)$ *is a semiring, where* " $+$ " *is* "*OR*", "$\cdot$" *is* "*AND*", *and* $B = \{0, 1\}$.
(ii) $(X^*, \cdot)$ *is the free monoid generated by a nonempty countable set* X. *It has all the finite strings, also referred to as words,* $x_1 x_2 \cdots x_n$, $x_i \in X$, *as its elements and the product* $w_1 \cdot w_2$ *is formed by writing the string* w_2 *immediately after the string* w_1. *The identity element is the empty word, denoted by* ε.

Definition 2.3 *Let S be a semiring. Let X be an alphabet, i.e., a finite non-empty set. A map $r : X^* \to S$ is called a formal power series; r itself is written as a formal sum*

$$r = \sum_{w \in X^*} r(w)w.$$

The values $r(w)$ are referred to as the coefficients of the series. The collection of all power series over X^* on S is denoted by $S\langle\langle X^* \rangle\rangle$. Given $r \in S\langle\langle X^* \rangle\rangle$ the subset of X^* defined by

$$\{w \mid r(w) \neq 0\}$$

is termed the support of r and denoted by supp(r). The subset of $S\langle\langle X^* \rangle\rangle$ consisting of all series with finite support is denoted by $S\langle X^* \rangle$. Series of $S\langle X^* \rangle$ are referred to as polynomials.

Example 2.2 *For $r_1, r_2 \in S\langle\langle X^* \rangle\rangle$, define $r_1 + r_2$ by $(r_1 + r_2)(w) = r_1(w) + r_2(w)$ and $r_1 r_2$ by $(r_1 r_2)(w) = \sum_{w_1 w_2 = w} r_1(w_1) r_2(w_2)$ for all $w \in X^*$. Then $S\langle\langle X^* \rangle\rangle$ is a semiring whose zero element is the same as the zero element of S and whose identity element is the empty word ε.*

Definition 2.4 *A left S-semimodule M is a commutative monoid $(M, +, 0)$ together with a left scalar multiplication satisfying for $a, b, 1 \in S$ and $x, y \in M$:*

(i) $a(x + y) = ax + ay$,

(ii) $(a + b)x = ax + bx$,

(iii) $(ab)x = a(bx)$,

(iv) $1x = x$, $0x = 0$, *and*

(v) $a0 = 0$,

One can define the right S-semimodule in a similar way.

Example 2.3 *For $a \in S$, $r \in S\langle\langle X^* \rangle\rangle$, if we define the left scalar product ar by $ar = \sum_{w \in X^*} ar(w)w$ then $S\langle\langle X^* \rangle\rangle$ is a left S-semimodule. For $a \in S$, $r \in S\langle\langle X^* \rangle\rangle$, if we define the right scalar product ra by $ra = \sum_{w \in X^*} r(w)aw$ then $S\langle\langle X^* \rangle\rangle$ is a right S-semimodule.*

If S is commutative then $S\langle\langle X^*\rangle\rangle$ is both left S-module and right S-module. From now on, we assume that the semiring S is commutative.

Definition 2.5 *Let M, N be two left S-semimodules. Then the direct sum of M and N, denoted by $M \oplus N$, is a Cartesian product set $M \times N$ of tuples (m, n) where $m \in M$ and $n \in N$, together with an addition, a 0 element, and a left scalar multiplication satisfying the following for $a \in R$, $m_i \in M$, and $n_i \in N$ for $i = 1, 2$:*

(i) $(m_1, n_1) + (m_2, n_2) = (m_1 + m_2, n_1 + n_2)$,

(ii) $0 = (0, 0)$, *and*

(iii) $a(m_1, n_1) = (am_1, an_1)$.

Any element in the direct sum $M \oplus N$ is denoted by (m, n) for $m \in M$ and $n \in N$.

Example 2.4 *The direct sum $S\langle\langle X^*\rangle\rangle \oplus S\langle\langle X^*\rangle\rangle$ is a left S- semimodule. If we define the multiplication on the direct sum by*

$$(r_1, r_2)(r_3, r_4) = (r_1 r_3, r_2 r_4)$$

and addition on the direct sum by

$$(r_1, r_2) + (r_3, r_4) = ((r_1 + r_3), (r_2 + r_4)),$$

then $S\langle\langle X^*\rangle\rangle \oplus S\langle\langle X^*\rangle\rangle$ is a semiring whose 0 element is $(0, 0)$ and whose identity element $(\varepsilon, \varepsilon)$, where ε is the empty word of X^*. Similarly, since S itself is a left S-semimodule $S \oplus S$ is also a semiring whose 0 element is $(0, 0)$ and whose identity element is $(1, 1)$. Note that we can think of elements of $S\langle\langle X^*\rangle\rangle$ as single DNA strands, and of elements of $S\langle\langle X^*\rangle\rangle \oplus S\langle\langle X^*\rangle\rangle$ as double DNA strands. Then, the multiplication on $S\langle\langle X^*\rangle\rangle \oplus S\langle\langle X^*\rangle\rangle$ is the ligation of two double DNA strands, while the addition is just union of two double DNA strands.

A map $r : X^* \times X^* \to S \oplus S$ is a power series; r can be written as a formal sum

$$\sum_{w_1 \in X^*, w_2 \in X^*} r(w_1, w_2)(w_1, w_2).$$

The collection of all power series over $X^* \times X^*$ on the semiring $S \oplus S$ is denoted by $(S \oplus S)\langle\langle X^* \times X^*\rangle\rangle$. One can define the support of the power series by

$$supp(r) = \{(w_1, w_2) \mid r(w_1, w_2) \neq (0, 0)\} \subseteq (X^* \times X^*).$$

Definition 2.6 *A map f from a semiring R to a semiring S is a homomorphism if it satisfies for any $a, b \in R$,*

(i) $f(a + b) = f(a) + f(b)$,

(ii) $f(ab) = f(a) f(b)$.

If a homomorphism is both injective and surjective, then it is called an isomorphism. A homomorphism from S to itself is called an endomorphism. An isomorphism from S to itself is called an automorphism.

A map f from a semiring R to a semiring S is an anti-homomorphism if it satisfies

(i) $f(a + b) = f(a) + f(b)$,

(ii') $f(ab) = f(b) f(a)$.

Similarly, we can define anti-isomorphism, and anti-automorphism.

Proposition 2.1 $S\langle\langle X^* \rangle\rangle \oplus S\langle\langle X^* \rangle\rangle$ *is a subsemiring of* $(S \oplus S)\langle\langle X^* \times X^* \rangle\rangle$ *up to an isomorphism.*

Proof. Given two maps (or power series) $r_1, r_2 : X^* \to S$, define a map f from $S\langle\langle X^* \rangle\rangle \oplus S\langle\langle X^* \rangle\rangle$ to $(S + S)\langle\langle X^* \times X^* \rangle\rangle$ by

$$f((r_1, r_2)) = r, \text{ where } r(w_1, w_2) = (r_1(w_1), r_2(w_2)).$$

It is easy to check that the map f is an injection. We now need to prove that f is an homomorphism. In fact, for any $(r_i, s_i) \in S\langle\langle X^* \rangle\rangle \oplus S\langle\langle X^* \rangle\rangle$ for $i = 1, 2$,

$$
\begin{aligned}
f((r_1, s_1)(r_2, s_2)) &= f(r_1 r_2, s_1 s_2) \\
&= \sum_{w_1, w_2 \in X^*} ((r_1 r_2)(w_1), (s_1 s_2)(w_2))(w_1, w_2), \\
f((r_1, s_1)) f((r_2, s_2)) &= \Big(\sum_{w_{11}, w_{12} \in X^*} (r_1(w_{11}), s_1(w_{12}))(w_{11}, w_{12}) \Big) \cdot \\
&\qquad \Big(\sum_{w_{21}, w_{22} \in X^*} (r_2(w_{21}), s_2(w_{22}))(w_{21}, w_{22}) \Big) \\
&= \sum_{\substack{w_{11} w_{21} = w_1 \\ w_{12} w_{22} = w_2}} (r_1(w_{11}) r_2(w_{21}), s_1(w_{12}) s_2(w_{22}))(w_1, w_2) \\
&= \sum_{w_1, w_2 \in X^*} ((r_1 r_2)(w_1), (s_1 s_2)(w_2))(w_1, w_2).
\end{aligned}
$$

Therefore, $f(r_1r_2, s_1s_2) = f(r_1, s_1)f(r_2, s_2)$. It is easy to show that

$$f(r_1 + r_2, s_1 + s_2) = f(r_1, s_1) + f(r_2, s_2).$$

Hence, f is a homomorphism. □

By Proposition 2.1, we can define the support of a power series $r \in S\langle\langle X^* \rangle\rangle \oplus S\langle\langle X^* \rangle\rangle$. Let $r = (r_1, r_2)$, where $r_1, r_2 \in S\langle\langle X^* \rangle\rangle$. Then

$$\text{supp}(r) = \{(w_1, w_2) \in X^* \times X^* \mid r(w_1, w_2) = (r_1(w_1), r_2(w_2)) \neq (0, 0)\}.$$

3   DNA Strings and Regular Language

DNA strings are words over $\{A, C, G, T\}^*$, the free monoid generated by the alphabet $\{A, C, G, T\}$ under the concatenation operation. (e.g., $x = ACCTGAC$). The *DNA strands* are DNA strings with a polarity (e.g. $3' - ACCTGAC - 5'$). Now, let us ignore the polarity for the moment, and consider all the DNA strings. We encode A by 000, C by 010, G by 101 and T by 111. Then the following proposition is easy to show.

Proposition 3.1 *Let x, y be two DNA string encodings. Then x and y are complementary if and only if $|x| = |y|$ and x XOR $y = 1 \cdots 1$. Here, the XOR represent the bitwise exclusive OR.*

Let S be an arbitrary semiring and let $\Sigma = \{0, 1\}$. We can build a regular grammar $G = (N, \Sigma, R, Q)$ such that the set of all the DNA strings is the language of $L(G)$: we take $N = \{Q\}$ and the rewriting rules of R are:

$$Q \to 000|010|101|111|000Q|010Q|101Q|111Q|.$$

Given the $S\langle\langle \Sigma^* \rangle\rangle$-left linear system

$$Q = 000 + 010 + 101 + 111 + (000 + 010 + 101 + 111)Q \qquad (1)$$

corresponding to the above regular grammar, there exists a unique *quasiregular* solution (Theorem 14.11 in [6]), where a *solution* of equation 1 is a power series $r \in S\langle\langle \Sigma^* \rangle\rangle$ such that $r = 000 + 010 + 101 + 111 + (000 + 010 + 101 + 111)r$ and is quasiregular if $r(\varepsilon) = 0$.

Theorem 3.1 *Let $r = \sum_{w \in L(G)} w$. Then the power series r is the only quasiregular solution.*

Proof. Replacing Q in (1) by r, we can easily show that the equation (1) holds. □

4   DNA Operations

The notations ˆ, R, ↓, ↑, and ↕ for biological operations have been defined in [2]. Let x be any DNA string. Then:

1. $\hat{x}$ is the string that results when each character of x has been replaced by its complement. By Proposition 3.1, $\hat{x}$ is obtained by flipping the bits of x if the string x is thought as a string in Σ^* under our encoding scheme.

2. x^R is the reverse of a string x.

3. ↑ x denotes the DNA strand whose corresponding DNA string is x and whose polarity is $5' - 3'$.

4. ↓ x denotes the $3' - 5'$ DNA strand complementary to ↑ x.

5. ↕ x denotes the double strand that results when ↑ x and ↓ x anneal in solution.

Example 4.1 *The table below summarizes these operators on a given DNA string:*

$$x = ACCTGAC$$
$$\hat{x} = TGGACTG,$$
$$x^R = CAGTCCA,$$
$$x^{\hat{R}} = GTCAGGT \quad (= \hat{x}^R = x^{\hat{R}}),$$
$$\uparrow x = 5' - ACCTGAC - 3',$$
$$\downarrow x = 3' - TGGACTG - 5',$$
$$\updownarrow x = \begin{array}{l} 5' - ACCTGAC - 3' \\ 3' - TGGACTG - 5'. \end{array}$$

We generalize these operators to homomorphisms of semirings. For every $r \in S\langle\langle\Sigma^*\rangle\rangle$, we define

$$\hat{r} = \sum_{w \in \Sigma^*} r(w)\hat{w},$$
$$r^R = \sum_{w \in \Sigma^*} r(w)w^R,$$
$$r^{\hat{R}} = \sum_{w \in \Sigma^*} r(w)w^{\hat{R}}.$$

Theorem 4.1 *The operators R and $\hat{R}$ are anti-automorphisms of $S\langle\langle\Sigma^*\rangle\rangle$. The operator $\hat{}$ is an automorphism of $S\langle\langle\Sigma^*\rangle\rangle$.*

Proof. For any two power series $r_1, r_2 \in S\langle\langle\Sigma^*\rangle\rangle$,

$$\hat{}(r_1 + r_2) = \sum_{w\in\Sigma^*} (r_1(w) + r_2(w))\hat{w}$$

$$= \sum_{w\in\Sigma^*} r_1(w)\hat{w} + \sum_{w\in\Sigma^*} r_2(w)\hat{w} = \hat{r_1} + \hat{r_2}, \text{ and}$$

$$\hat{}(r_1 r_2) = \sum_{w\in\Sigma^*, w_1 w_2 = w} (r_1(w_1)r_2(w_2))\hat{w}$$

$$= \sum_{w\in\Sigma^*} r_1(w)\hat{w} \sum_{w\in\Sigma^*} r_2(w)\hat{w} = \hat{r_1}\hat{r_2}.$$

Since $\hat{}$ is equivalent to the bitwise exclusive OR, it is one to one. So $\hat{}$ is an automorphism of $S\langle\langle\Sigma^*\rangle\rangle$.

Similarly, we can show that $(r_1 + r_2)^R = r_1^R + r_2^R$, and, since S is commutative,

$$(r_1 r_2)^R = \sum_{w\in\Sigma^*, w_1 w_2 = w} (r_1(w_1)r_2(w_2))w^R$$

$$= \sum_{w\in\Sigma^*, w_1 w_2 = w} (r_1(w_1)r_2(w_2))w_2^R w_1^R$$

$$= \sum_{w\in\Sigma^*} r_2(w)w^R \sum_{w\in\Sigma^*} r_1(w)w^R = r_2^R r_1^R.$$

So R is an anti-automorphism of $S\langle\langle\Sigma^*\rangle\rangle$. Since $\hat{R} = R\hat{} = \hat{}R$, $\hat{R}$ is an anti-automorphism of $S\langle\langle\Sigma^*\rangle\rangle$. $\square$

Let $X = \{A, C, G, T\}$. Under our encoding scheme, X^* is a subset of Σ^*. More precisely, X^* is a submonoid of Σ^*, that is, X^* is closed under the multiplication of Σ^*. From now on, we consider X^* as a submonoid of Σ^*.

Theorem 4.2 *The semiring $S\langle\langle X^*\rangle\rangle$ is a subsemiring of $S\langle\langle\Sigma^*\rangle\rangle$. It is invariant under the automorphism $\hat{}$ and anti-automorphism R of $S\langle\langle\Sigma^*\rangle\rangle$, that is, $\hat{}(S\langle\langle X^*\rangle\rangle) = S\langle\langle X^*\rangle\rangle$ and $R(S\langle\langle X^*\rangle\rangle) = S\langle\langle X^*\rangle\rangle$.*

Proof. First, we show that $S\langle\langle X^*\rangle\rangle$ is a subsemiring of $S\langle\langle\Sigma^*\rangle\rangle$. It suffices to prove that for any two power series $r_1, r_2 \in S\langle\langle X^*\rangle\rangle$, $r_1 + r_2, r_1 r_2 \in$

$S\langle\langle X^*\rangle\rangle$. The fact $r_1 + r_2 \in S\langle\langle X^*\rangle\rangle$ is clearly true. Assume that $r_i = \sum_{w \in X^*} r_i(w)w$ for $i = 1, 2$. Then

$$r_1 r_2 = \sum_{w, w_1, w_2 \in X^*, w_1 w_2 = w} (r_1(w_1)r_2(w_2))w.$$

Since w_1, w_2 are in X^* and X^* is a monoid, $w_1 w_2$ belongs to X^*. Therefore, $r_1 r_2$ is contained in $S\langle\langle X^*\rangle\rangle$

Secondly, we show that $S\langle\langle X^*\rangle\rangle$ is invariant under the morphisms $\hat{\ }$ and R. That is, for any power series $r \in S\langle\langle X^*\rangle\rangle$, r^R and $\hat{r}$ are contained in $S\langle\langle X^*\rangle\rangle$. Since the morphisms have nothing to do with the coefficients of the power series it suffices to prove that w^R and $\hat{w}$ are contained in X^* for any $w \in X^*$. By our encoding scheme, $(000)^R = 000$, $(010)^R = 010$, $(101)^R = 101$, and $(111)^R = 111$. And $\hat{A} = \hat{\ }(000) = 111 = T$, $\hat{C} = \hat{\ }(010) = 101 = G$, $\hat{G} = \hat{\ }(101) = 010 = C$, and $\hat{T} = \hat{\ }(111) = 000 = A$. Therefore, w^R, $\hat{w}$ are in X^* for any $w \in X^*$ since w is a word formed by $A(= 000)$, $C(= 010)$, $G(= 101)$ and $T(= 111)$. □

Corollary 4.1 *The compositions, $R\hat{\ }$ and $\hat{\ }R$ are anti-automorphism of $S\langle\langle X^*\rangle\rangle$. Moreover, $R\hat{\ } = \hat{\ }R$.*

5   Double Strands and Annealing

By Theorem 4.2, the semiring $S\langle\langle X^*\rangle\rangle$ is a subsemiring of $S\langle\langle X^*\rangle\rangle$. Hence it is both left and right S-semimodule (recall that we assume that S is commutative). Therefore, we can define $S\langle\langle X^*\rangle\rangle \oplus S\langle\langle X^*\rangle\rangle$ which is also a subsemiring of $S\langle\langle \Sigma^*\rangle\rangle \oplus S\langle\langle \Sigma^*\rangle\rangle$.

Now we consider the polarity of the DNA strings. Suppose that r is a power series in $S\langle\langle X^*\rangle\rangle \oplus S\langle\langle X^*\rangle\rangle$. An element $w \in \text{supp}(r)$ is of the form (w_1, w_2). We make the convention that $w_1 = \uparrow w_1 = 5' - w_1 - 3'$, and $w_2 = 3' - w_2 - 5'$. For example, assume that $(w_1, w_2) = (ACTG, AG) \in \text{supp}(r)$ then what we really mean is that $w_1 = 5' - ACTG - 3'$ and $w_2 = 3' - AG - 5'$.

Now we define $i_j : S\langle\langle X^*\rangle\rangle \to S\langle\langle X^*\rangle\rangle \oplus S\langle\langle X^*\rangle\rangle$, for $j = 1, 2$ by $i_1(r) = r \oplus 0$ and $i_2(r) = 0 \oplus r$. Note that the supports of the images of i_1 and i_2 are empty. Both maps i_1 and i_2 are homomorphisms of semirings from $S\langle\langle X^*\rangle\rangle$ to $S\langle\langle X^*\rangle\rangle \oplus S\langle\langle X^*\rangle\rangle$. Conversely, we define $p_j : S\langle\langle X^*\rangle\rangle \oplus S\langle\langle X^*\rangle\rangle \to S\langle\langle X^*\rangle\rangle$ for $j = 1, 2$ by

$$p_1(r_1 \oplus r_2) = r_1, \text{ and } p_2(r_1 \oplus r_2) = r_2.$$

Then p_1 and p_2 are homomorphisms of semirings from $S\langle\langle X^*\rangle\rangle \oplus S\langle\langle X^*\rangle\rangle$ to $S\langle\langle X^*\rangle\rangle$.

We identify the operators $\uparrow$, $\downarrow$, and $\updownarrow$ with homomorphisms from $S\langle\langle X^*\rangle\rangle$ to $S\langle\langle X^*\rangle\rangle \oplus S\langle\langle X^*\rangle\rangle$ as follows.

$$\uparrow = i_1,$$
$$\downarrow = i_{\hat{2}},$$
$$\updownarrow (r) = (r \oplus \hat{r}).$$

Theorem 5.1 *The operators* $\uparrow$, $\downarrow$, *and* $\updownarrow$ *are homomorphisms from* $S\langle\langle X^*\rangle\rangle$ *to* $S\langle\langle X^*\rangle\rangle \oplus S\langle\langle X^*\rangle\rangle$.

Proof. The proof is straightforward and similar to the proof of Theorem 4.1. □

We interpret p_1 and p_2 as melting and $\updownarrow$ as the process of annealing. Note that $\updownarrow$ describes the process of a single DNA strand looking for its matching single DNA strand to form a double DNA strand.

Theorem 5.2 *The map* $\updownarrow \hat{\ } p_2$ *is an endomorphism of* $S\langle\langle X^*\rangle\rangle \oplus S\langle\langle X^*\rangle\rangle$. *The invariant of the endomorphism is the set*

$$\updownarrow (S\langle\langle X^*\rangle\rangle) = \{\updownarrow (r) \mid \text{ for all } r \in S\langle\langle X^*\rangle\rangle\}$$

which is a subsemiring of $S\langle\langle X^*\rangle\rangle \oplus S\langle\langle X^*\rangle\rangle$.

Proof. Let us consider the following diagram:

$$
\begin{array}{ccc}
 & p_2 & \\
S\langle\langle X^*\rangle\rangle \oplus S\langle\langle X^*\rangle\rangle & \longrightarrow & S\langle\langle X^*\rangle\rangle \\
\updownarrow \hat{\ } p_2 \downarrow & & \downarrow \hat{\ } \\
S\langle\langle X^*\rangle\rangle \oplus S\langle\langle X^*\rangle\rangle & \overset{\updownarrow}{\longleftarrow} & S\langle\langle X^*\rangle\rangle
\end{array}
$$

Since p_2, $\hat{\ }$, and $\updownarrow$ are homomorphisms, the map $\updownarrow \hat{\ } p_2$ is an endomorphism of the semiring $S\langle\langle X^*\rangle\rangle \oplus S\langle\langle X^*\rangle\rangle$.

Now we should prove that the invariant of the endomorphism $\updownarrow \hat{\ } p_2$ is not empty.

For any $r \in S\langle\langle X^* \rangle\rangle$, $\updownarrow (r) = (r \oplus \hat{r})$ is in the invariant since

$$\updownarrow \hat{\ } p_2(\updownarrow (r)) = \updownarrow \hat{\ } p_2(r \oplus \hat{r}) = \updownarrow \hat{\ }(\hat{r}) = \updownarrow \hat{r} = \updownarrow (r).$$

This also proves that the invariant of the endomorphism contains the set $\{\updownarrow (r) \mid \text{ for all } r \in S\langle\langle X^* \rangle\rangle\}$.

Conversely, assume that $r_1 \oplus r_2 \in S\langle\langle X^* \rangle\rangle \oplus S\langle\langle X^* \rangle\rangle$ is an element in the invariant. Then

$$\updownarrow \hat{\ } p_2(r_1 \oplus r_2) = \updownarrow \hat{\ }(r_2) = \updownarrow (\hat{r}_2) = \hat{r}_2 \oplus \hat{\hat{r}}_2 = \hat{r}_2 \oplus r_2 = r_1 \oplus r_2.$$

So $r_1 = \hat{r}_2$. Thus $r_1 \oplus r_2 = \updownarrow (\hat{r}_2)$. This shows that the invariant is contained in the set $\{\updownarrow (r) \mid \text{ for all } r \in S\langle\langle X^* \rangle\rangle\}$.

Hence the invariant of the endomorphism is exactly the same as the set $\{\updownarrow (r) | for \ all \ r \in S\langle\langle X^* \rangle\rangle\}$. It is a subsemiring of $S\langle\langle X^* \rangle\rangle \oplus S\langle\langle X^* \rangle\rangle$ since it is the image of $S\langle\langle X^* \rangle\rangle$ under the homomorphism $\updownarrow$. $\square$

The following theorem summarizes the relationships among the operators.

Theorem 5.3   (i) $\hat{\ } R = R \hat{\ } = Rp_2 \downarrow = p_1 \uparrow \hat{R} = \hat{R}$,

(ii) $\hat{\ } p_1 \updownarrow = p_2 \updownarrow = \hat{\ }$;

(iii) $p_1 \updownarrow$ is the identity map,

(iv) $\uparrow x \oplus \downarrow x = \updownarrow x$ for $x \in S\langle\langle X^* \rangle\rangle$

Proof. For relation (i), it suffices to prove that $Rp_2 \downarrow = p_1 \uparrow \hat{R} = \hat{R}$. In fact, for any $r \in S\langle\langle X^* \rangle\rangle$,

$$Rp_2 \downarrow (r) = Rp_2(0 \oplus \hat{r}) = R\hat{r} = r^{\hat{R}}$$
$$p_1 \uparrow \hat{R}(r) = p_1(r^{\hat{R}} \oplus 0) = r^{\hat{R}}.$$

The relation (ii) is true since, for any $r \in S\langle\langle X^* \rangle\rangle$,

$$\hat{\ } p_1 \updownarrow (r) = \hat{\ } p_1(r \oplus \hat{r}) = \hat{r}$$
$$p_2 \updownarrow (r) = p_2(r \oplus \hat{r}) = \hat{r}.$$

The relation (iii) is true since, for any $r \in S\langle\langle X^* \rangle\rangle$,

$$p_1 \updownarrow (r) = p_1(r \oplus \hat{r}) = r.$$

The relation (iv) is true by the definitions of $\downarrow$, $\uparrow$ and $\updownarrow$. $\qquad \square$

Acknowledgment. The author wants to thank professor Lila Kari for her suggestion and comments.

References

[1] L. Aldeman, Molecular computation of solutions to combinatorial problems, *Science*, 266 (1994), 1021 – 1024.

[2] D. Boneh, C. Dunworth, R. J. Lipton, A notation for DNA operations, *to appear.*

[3] D. Boneh, R. J. Lipton, C. Dunworth, Breaking DES using molecular computer, *http://www.cs.princeton.edu/ dabo.*

[4] F. Guarmieri, C. Bancroft, Use of horizontal chain reaction for DNA-based addition, in *Proceedings of the 2nd DIMACS Workshop on DNA-Based Computers*, 1996, 249 – 259.

[5] L. Kari, DNA computing: Arrival of biological mathematics, *The Mathematical Intelligencer,* 2 (1997), 9 – 22.

[6] W. Kuich, A. Salomaa, *Semirings, Automata, Languages*, Springer-Verlag, Berlin, 1986.

[7] T. Leete, M. Schwartz, R. Williams, D. Wood, J. Salem, H. Rubin, Massively parallel DNA computation: Expansion of symbolic determinants, in *Proceedings of the 2nd DIMACS Workshop on DNA-Based Computers*, 1996, 49 – 66.

[8] J. Oliver, Computation with DNA: Matrix multiplication, in *Proceedings of the 2nd DIMACS Workshop on DNA-Based Computers*, 1996, 236 – 248.

[9] R. Williams, D. Wood, Exascale computer algebra problems interconnect with molecular reactions and complexity theory, in *Proceedings of the 2nd DIMACS Workshop on DNA-Based Computers*, 1996, 260 – 268.

Modeling DNA Recombinant Behavior with Fixed-Point Equations

Rodica CETERCHI

Bucharest University, Faculty of Mathematics
Academiei 14, 70109 Bucureşti, Romania
rc@funinf.math.unibuc.ro

Abstract. We introduce the concept of selective CP-function (cut-and-paste function) in order to provide a unified framework for the formal treatment of a great variety of behaviors involving cut and paste operations on strings. We show that the splicing operation can be modeled with fixed-point equations, thus adding a new and powerful tool, of a purely algebraic nature, to the existing generative devices.

1   Introduction

In [7] (and [8]), T. Head introduced a new operation on strings called *splicing*, motivated by the behavior of double-stranded DNA in the process of its replication. Since then, the field of DNA computing has rapidly grown, see the monograph [14], and [3].

The notion of *H system* was introduced in formal language theory, [13], as a generative model based on the splicing operation, and the theoretical generative and computational power of such systems was investigated thoroughly. The splicing operation on strings involves essentially cutting and pasting operations, performed in a "selective" manner: we can cut only in certain places, we can paste only certain strings.

Besides H systems, a variety of other formal generative devices were considered in order to model a great variety of behaviors in which cutting and pasting play a central role.

In [6] CR (cutting/recombination) systems, and mCR systems (CR systems working in a multiset style) were introduced. In [15] CP (cutting and

340

pasting) systems, and also CR (cutting and recombination) systems are considered.

The aim of the present paper is to show that fixed-point equations, attached to functions of a certain type, can be used as an alternative generative mechanism, to model recombinant behavior involving cut and paste operations on strings.

The concept of cut-and-paste language (CP-language) was introduced in [4]. It was shown that important languages, among which Dyck's language, are CP-languages, and also that the class of first degree 2CP-languages is incomparable with REG, LIN and CF.

One distinctive feature of CP-languages is that they use fixed-point equations as their generative device, attached to w-continuous functions P on $\mathcal{P}(V^*)$. The other distinctive feature is the way in which P acts on words: it "cuts" them in pieces, glues some "coefficients" to the parts, and finaly "pastes" the resulting pieces together. The pasting and cutting are done using nothing more spectacular than the catenation operation and its inverse. An initial approach to this kind of generative device was made in [11].

In [5] we have begun investigating the problem of generalizing the initial concept of CP-language, in order to be able to cover cut and paste operations ruled by selection mechanisms. We have introduced there the concept of ϕ-selective CP-language, based on a *selection function* $\phi : (V^*)^n \longrightarrow \mathcal{P}(C)$ which is commonly used by the contextual grammars with choice.

But ϕ-selection is not powerfull enough to model splicing, so we have been led to consider a much more general type of function.

The central concept of this paper, that of *selective CP-function*, is introduced in Section 3. Fixed-point equations attached to such functions generate the selective CP-languages.

We begin to investigate the power of our formal tool with the insertion operation in Section 4. We show that a selective CP-function can be attached to every insertion rule, and we obtain insertion languages as minimal solutions of fixed-point equations.

Our main result is in Section 5, where we show that the splicing operation can also be modeled using selective CP-functions. As a consequence, languages generated by iterated splicing are obtained as solutions of fixed-point equations. To our knowledge, this is the first attempt to model splicing with fixed-point equations.

Fixed-point techniques are powerful tools in various areas of Computer

Science. The most commonly used are Banach's contraction principle and Kleene's fixed-point theorem. According to [1], [2], the first one turned out to be a particular case of the latter. We hope that this classical algebraic tool, Kleene's fixed-point theorem, which has proved its usefulness in a great variety of domains, will be useful in the field of DNA computing as well.

2   Preliminaries

We present here the notations which will be used in the rest of the paper, together with some essential definitions and results. For all unexplained notions of formal language theory the reader is referred to [16].

Let V be an alphabet, V^* the free monoid generated by V, and let $1 \in V^*$ denote the empty word.

For a natural number $n \geq 2$, consider the product monoid $(V^*)^n$, with catenation defined componentwise and $(1, 1, \ldots, 1)$ the neutral element.

Let $c_n : (V^*)^n \to V^*$ be the n-catenation function, defined by

$$c_n(x_1, x_2, \ldots, x_n) = x_1 x_2 \ldots x_n \in V^*,$$

and consider its extension to the image function $c_n : \mathcal{P}((V^*)^n) \to \mathcal{P}(V^*)$.

Let $c_n^{-1} : \mathcal{P}(V^*) \to \mathcal{P}((V^*)^n)$ be the pre-image function, i.e., for $w \in V^*$

$$c_n^{-1}(w) = \{(x_1, x_2, \ldots, x_n) \in (V^*)^n \mid x_1 x_2 \ldots x_n = w\},$$

and for an $L \subseteq V^*$, let $c_n^{-1}(L) = \bigcup_{w \in L} c^{-1}(w)$.

$(\mathcal{P}((V^*)^n), \bigcup, \cdot, \emptyset, \{1\})$ is a semiring, where $\cdot$ denotes the usual product of sets and will usually be omitted.

For fixed $(a_1, \ldots, a_n)$, $(b_1, \ldots, b_n)$ in $(V^*)^n$, identified with their corresponding singletons, we have the product

$$(a_1, a_2, \ldots, a_n) c_n^{-1}(w)(b_1, b_2, \ldots, b_n) =$$
$$\{(a_1 x_1 b_1, a_2 x_2 b_2, \ldots, a_n x_n b_n) \mid x_1 x_2 \ldots x_n = w\}.$$

We give below the definitions of CP-functions introduced in [4].

Definition 2.1 *We call* first degree n-cut-and-paste monomial (nCP-monomial *for short*), *an expression of the form*

$$M(X) = c_n((a_1, a_2, \ldots, a_n) c_n^{-1}(X)(b_1, b_2, \ldots, b_n)).$$

We will call first degree nCP-polynomial *an expression*

$$P(X) = \bigcup_{i=1}^{k} M_i(X) \bigcup L,$$

where $M_i(X)$, $i = \overline{1,k}$, *are first-degree* nCP-monomials, and $L \subset V^*$ is *finite and is called the* free term *of P.*

The *monomial function* $M : \mathcal{P}(V^*) \to \mathcal{P}(V^*)$, is defined by the composition of functions $c_n \circ (a_1, a_2, \ldots, a_n)(\cdot) \circ (\cdot)(b_1, b_2, \ldots, b_n) \circ c_n^{-1}$ and all these functions commute with arbitrary unions. It follows that any monomial function and also the polynomial functions commute with arbitrary unions. In particular, they will be ω-continuous functions from $\mathcal{P}(V^*)$ to itself.

Applying Kleene's fix-point theorem we obtain the following result:

Theorem 2.1 *Let P be a first-degree nCP-polynomial function. Then, the fix-point equation $P(X) = X$ has a least solution, L, obtained as the limit of the Kleene sequence:*

$$L = \bigcup_{m \geq 0} P^m(\emptyset).$$

L is also the smallest $L' \in \mathcal{P}(V^*)$ such that $P(L') \subseteq L'$.

Definition 2.2 *A language $L \subseteq V^*$ which is the solution of a fix-point equation $P(X) = X$, with P a first-degree nCP-polynomial, will be called an nCP-language.*

To see the justification for the terminology, and to get a feeling of the way in which CP-functions work, let us apply a 2-CP-monomial function M to a word $w \in V^*$, identified with its corresponding singleton $\{w\} \in \mathcal{P}(V^*)$:

$$M(w) = c_2((a_1, a_2)c_2^{-1}(w)(b_1, b_2)) = \{c_2(a_1 w_1 b_1, a_2 w_2 b_2) \mid w_1 w_2 = w\}$$
$$= \{a_1 w_1 b_1 a_2 w_2 b_2 \mid w_1 w_2 = w\}.$$

We first apply c_2^{-1} to w, which means we cut it in two in all possible manners. We "multiply" every such cutting (w_1, w_2) with the left and right coefficients, obtaining $(a_1 w_1 b_1, a_2 w_2 b_2)$. Finally, we paste the two resulting strings together using catenation, and we get all possible words of the form $a_1 w_1 b_1 a_2 w_2 b_2$.

Note that cutting and pasting work "blindly": we cut anywhere, without restrictions, we paste the resulting strings without restrictions.

To illustrate the functioning of a CP-polynomial as a generative device, we have to look at its Kleene sequence:

$$\emptyset \subset P(\emptyset) \subset \cdots \subset P^n(\emptyset) \subset P^{n+1}(\emptyset) \subset \cdots$$

The first application of P produces $P(\emptyset) = L$, the free term of P. The next application produces $P(L)$, which contains L, and is enriched with all new words obtained from it by cutting and pasting done by the monomials. The union of this ascending chain of sets of words is the CP-language attached to P, the smallest language which contains L, and is closed to the operations performed by P.

3 Selective nCP-Functions and Languages

In order to introduce the main definition of the paper, the concept of *selective n*CP-polynomial and the corresponding languages, consider the following construction, which will be used as a recipe to construct particular selective CP-functions.

A partial function $f : A \longrightarrow B$ of sets is extended to its image function, denoted still by f, $f : \mathcal{P}(A) \longrightarrow \mathcal{P}(B)$ by defining it on singletons as

$$f(\{x\}) = \begin{cases} \{f(x)\}, & \text{if } x \in Dom(f), \\ \emptyset, & \text{otherwise.} \end{cases}$$

and extending it naturally to a monotonous function on the whole $\mathcal{P}(A)$ by defining $f(X) = \bigcup_{x \in X} f(\{x\})$. It is obvious that f thus defined will commute with arbitrary unions.

Recall also that the composition of partial functions leads to a partial function. For $f : A \longrightarrow B$ and $g : B \longrightarrow C$ partial functions, their composition $g \circ f : A \longrightarrow C$ has domain $Dom(g \circ f) = Dom(f) \cap \{x \mid f(x) \in Dom(g)\}$.

Let $\bar{c}_n : (V^*)^n \to V^*$ denote a partial n-catenation function, i.e., a partial function defined on its domain as the usual catenation, and denote still by $\bar{c}_n$ its corresponding image function.

Intuitively, a partial catenation is nothing more that "selective paste": certain strings get pasted, those in its domain, and nothing happens to the rest.

Let $\bar{c}_n^{-1} : V^* \to \mathcal{P}((V^*)^n)$ denote a partial function, which works on its domain as the usual catenation preimage, and denote still by $\bar{c}_n^{-1}$ its extension to $\mathcal{P}(V^*)$.

A partial catenation pre-image is nothing more that "selective cut": some strings get cut, in certain places, some do not. Specifying the cutting place is the most common type of selection mechanism. Anticipating a little on the theme of the next section, suppose that we are allowed to cut only between two given strings, u and v. If a word x contains uv, i.e. is of the form $x = x_1 uv x_2$, then it belongs to $Dom(\overline{c}_2^{-1})$, and it can be cut, producing $\overline{c}_2^{-1}(x) = (x_1 u, v x_2)$. If a word x does not contain uv, it cannot be cut, and consequently $\overline{c}_2^{-1}(x) = \emptyset$.

Let f be a partial function from $\mathcal{P}((V^*)^n)$ to itself. Partial versions of the product of sets, $\cdot$, are examples of such functions. A partial product of strings means that certain string can be catenated, others not. Other operations on strings, which can be performed only on a certain set of strings, and not on all, can be examples of such strictly partial functions.

We are interested in functions obtained by compositions of functions of type $\overline{c}_n \circ f \circ \overline{c}_n^{-1}$ because they model processes which start by cutting one (or several) strings, do some operations with them, and end by pasting some of the substrings involved. If a sort of selection mechanism is involved in the process, then at least one of the functions involved will be strictly partial, that is it cannot act on the whole set of strings. This motivates our next definition.

Definition 3.1 *We call* first degree selective nCP-monomial, *a function* $M : \mathcal{P}(V^*) \to \mathcal{P}(V^*)$, *defined by a composition of type* $\overline{c}_n \circ f \circ \overline{c}_n^{-1}$ *where at least one of the functions involved,* $\overline{c}_n^{-1}$, $\overline{c}_n$, *or* f, *is a strict partial function, and the result of the composition,* M, *is an* ω-*continuous function.*

We will call first degree selective nCP-polynomial *an expression*

$$P(X) = \bigcup_{i=1}^{k} M_i(X) \bigcup L,$$

where $M_i(X)$, $i = \overline{1, k}$, *are first-degree selective* nCP-monomials, *and* $L \subset V^*$ *is finite and is called the* free term *of* P.

Since by definition selective n-CP monomials are ω-continuous functions, so will be the respective polynomials, and applying Kleene's fixed-point theorem we obtain the following result:

Theorem 3.1 *Let* P *be a first-degree selective* nCP-polynomial function. *Then, the fix-point equation* $P(X) = X$ *has a least solution,* L, *obtained as the limit of the Kleene sequence:*

$$L = \bigcup_{m \geq 0} P^m(\emptyset).$$

L is also the smallest $L' \in \mathcal{P}(V^*)$ such that $P(L') \subseteq L'$.

Definition 3.2 *A language $L \subseteq V^*$ which is the solution of a fix-point equation $P(X) = X$, with P a selective first-degree nCP-polynomial, will be called a* selective nCP-language.

The generative mechanism of a selective CP-polynomial is identical to that of a (non-selective) CP-polynomial, as discussed at the end of Section 2. New words are added to the language at each application of P. The only difference (an important one!) is that some selection processes guide the cutting and/or pasting. Moreover, in between the initial cutting and the final pasting, a variety of other processes between strings can occur, all of them represented formally by the partial function f. We illustrate in the next sections how we model such mechanisms with partial functions which lead to selective CP-monomials and polynomials.

4   The Case of Insertion

We will now show that insertion operations can be modeled using selective nCP-functions.

Following [12], [10], we define an *insertion grammar* as a construct $G = (V, A, R)$, where V in an alphabet, A is a finite set of words over V called the *axioms*, and R is a finite set $R \subseteq (V^*)^3$ called the set of *insertion rules* of G.

For $x, y \in V^*$ the one-step derivation is defined by:

$$x \underset{G}{\Longrightarrow} y \quad \text{iff} \quad x = x_1 u v x_2, \; y = x_1 u z v x_2,$$

$$\text{for } x_1, x_2 \in V^* \text{ and } (u, z, v) \in R.$$

Denoting by $\underset{G}{\overset{*}{\Longrightarrow}}$ the reflexive and tranzitive closure of $\underset{G}{\Longrightarrow}$, the language generated by G is

$$L(G) = \{ y \in V^* \mid x \underset{G}{\overset{*}{\Longrightarrow}} y, \; x \in A \}.$$

and is called an *insertion language*.

For a fixed $(u, z, v) \in R$ we will construct a selective 2CP-monomial. Consider the partial function $\overline{c}_2^{-1} : V^* \to \mathcal{P}(V^*u \times vV^*)$ obtained by restricting the codomain of the inverse of two-catenation. This means precisely that we can cut only between u and v. Denote still by $\overline{c}_2^{-1}$ its extension to $\mathcal{P}(V^*)$.

Let $M_{(u,z,v)} : \mathcal{P}(V^*) \to \mathcal{P}(V^*)$ denote the composition $c_2 \circ (1,z)(\cdot) \circ \overline{c}_2^{-1}$. Applied to a word x we get

$$M_{(u,z,v)}(x) = c_2 \circ (1,z)(\cdot) \circ \overline{c}_2^{-1}(x) = c_2((1,z)(x_1 u, vx_2))$$
$$= c_2((x_1 u, zvx_2)) = x_1 uzvx_2.$$

$M_{(u,z,v)}$ is an ω-continuous function, and thus a selective 2CP-monomial. We have the following result:

Theorem 4.1 *Let $G = (V, A, R)$ be an insertion grammar. Then $L(G)$ is a selective 2CP-language. More precisely, it is the least solution of the fix-point equation attached to the selective 2CP-polynomial*

$$P(X) = \bigcup_{(u,z,v) \in R} M_{(u,z,v)}(X) \bigcup A.$$

Proof. Let us denote

$$L_n(G) = \{y \in V^* \mid x \xRightarrow[G]{n} y, x \in A\}.$$

the set of words obtained from axioms by applying at most n times all rules in R. One application of all rules to a word means selecting the rules which can act on it, in all possible ways, and applying them. It is clear that

$$L(G) = \bigcup_{n \geq 0} L_n(G).$$

On the other hand, construct the Kleene sequence of P, denote it by $\{X_n\}_{n \geq -1}$. Start the sequence with $X_{-1} = \emptyset$. Applying P once we get $X_0 = L$, and so $X_0 = L_0(G)$. We continue now with the classical induction argument. Suppose $X_n = L_n(G)$ for an $n \geq 1$. We show that $X_{n+1} = L_{n+1}(G)$. But for this, it is sufficient to prove that one application of a derivation is equivalent to one application of P. But this is obvious from the way P was constructed: a derivation using rule $(u, z, v) \in R$ applied to an x produces an element in $M_{(u,z,v)}(x)$; conversely, for any element in $M_{(u,z,v)}(x)$ we can show that it was derived from x by applying rule (u, z, v). $\square$

5   The Case of Splicing

In this last paragraph we show that the language generated by an H scheme (the iterated case) can be obtained using fixed-point equations and is therefore a selective 2CP-language. We confine ourselves to the case when an infinite number of copies of each string are available.

We give the definitions below following [9]. For a given alphabet V, and two special symbols, $\#, \$$, not in V, a *splicing rule* (over V) is a string of the form $r = u_1 \# u_2 \$ u_3 \# u_4$ where $u_i \in V^*, 1 \le i \le 4$.

For $x, y, z \in V^*$ we say that z *is obtained by splicing* x *and* y *according to the rule* r, and we write

$$(x, y) \vdash_r z \quad \text{iff} \quad x = x_1 u_1 u_2 x_2, \ y = y_1 u_3 u_4 y_2,$$
$$z = x_1 u_1 u_4 y_2, \text{ for some } x_1, x_2, y_1, y_2 \in V^*.$$

An H *scheme* is a pair $\sigma = (V, R)$, where V is an alphabet and $R \subseteq V^* \# V^* \$ V^* \# V^*$ is a set of splicing rules.

For a given H scheme $\sigma = (V, R)$ and a language $L \subseteq V^*$, we define

$$\sigma(L) = \{z \in V^* \mid (x, y) \vdash_r z \text{ for some } x, y \in L, r \in R\}.$$

One can also consider the case, when σ is applied iteratively to a language $L \subseteq V^*$. We define:

$$\sigma^0(L) = L,$$
$$\sigma^{i+1}(L) = \sigma^i(L) \cup \sigma(\sigma^i(L)), \ i \ge 0,$$
$$\sigma^*(L) = \bigcup_{i \ge 0} \sigma^i(L).$$

In other words, $\sigma^i(L)$ consists of the words $z \in V^*$ obtained from words in L by splicing at most i times, and $\sigma^*(L)$ is the smallest language L' which contains L, and is closed under splicing with respect to σ, $\sigma(L') \subseteq L'$.

In [9] a thorough investigation is made of the families of languages generated by applying an H scheme to a language L, taking into consideration all possible choices of types for L and R in the hierarchy $\{FIN, REG, LIN, CF, CS, RE\}$. Of particular interest is the case when both L and R are in FIN.

We intend to show in the following that, with L and R finite, $\sigma^*(L)$ is a selective 2CP-language. We will attach to σ and L a selective 2CP-polynomial, with free term L, and with selective monomials attached to every splicing rule $r \in R$.

Consider a rule $r \in R$, $r = u_1 \# u_2 \$ u_3 \# u_4$. Let c and c^{-1} denote 2-catenation and its pre-image, where we have dropped the index 2 in order to simplify the notation. We will use c^{-1} to cut strings in the cutting sites $u_1 \# u_2$ and $u_3 \# u_4$, using the same mechanism we described for insertion. For $u_1 \# u_2$ consider the partial function $\overline{c}_{1,2}^{-1} : V^* \longrightarrow \mathcal{P}(V^* u_1 \times u_2 V^*)$ obtained by restricting the codomain of c^{-1}. For an x in its domain,

$$\overline{c}_{1,2}^{-1}(x) = \{(x_1 u_1, u_2 x_2) \mid x_1 u_1 u_2 x_2 = x\}.$$

Denote by

$$C_r^{1,2} : \mathcal{P}(V^*) \longrightarrow \mathcal{P}(V^* u_1 \times u_2 V^*)$$

the total function obtained from it by defining if appropriately on singletons and making it commute with unions.

For the cutting site $u_3 \# u_4$, we construct analogously

$$C_r^{3,4} : \mathcal{P}(V^*) \longrightarrow \mathcal{P}(V^* u_3 \times u_4 V^*).$$

For $i = 1, 2$, let $\pi_i : V^* \times V^* \longrightarrow V^*$ denote the first and second projection, and then consider their corresponding image functions

$$\pi_1, \pi_2 : \mathcal{P}(V^* \times V^*) \longrightarrow \mathcal{P}(V^*).$$

Consider the following compositions of functions:

$$\mathcal{P}(V^*) \xrightarrow{C_r^{1,2}} \mathcal{P}(V^* u_1 \times u_2 V^*) \xrightarrow{\pi_1} \mathcal{P}(V^* u_1),$$
$$\mathcal{P}(V^*) \xrightarrow{C_r^{3,4}} \mathcal{P}(V^* u_3 \times u_4 V^*) \xrightarrow{\pi_2} \mathcal{P}(u_4 V^*),$$

where π_1 and π_2 work on the appropriately restricted domains.

Passing to products we can construct now the function

$$\Phi_r^{1,4} : \mathcal{P}(V^*) \times \mathcal{P}(V^*) \longrightarrow \mathcal{P}(V^* u_1) \times \mathcal{P}(u_4 V^*)$$

defined by

$$\Phi_r^{1,4}(A, B) = (\pi_1(C_r^{1,2}(A)), \pi_2(C_r^{3,4}(B))),$$

and we can further compose it with

$$P_r^{1,4} : \mathcal{P}(V^* u_1) \times \mathcal{P}(u_4 V^*) \longrightarrow \mathcal{P}(V^*).$$

which is a restriction of the usual product of sets. Note that on singletons this is precisely the 2-catenation, with restricted domain.

Now let

$$M_r = P_r^{1,4} \circ \Phi_r^{1,4} : \mathcal{P}(V^*) \times \mathcal{P}(V^*) \longrightarrow \mathcal{P}(V^*)$$

be the result of this final composition.

By carefully analizing all the functions involved and their compositions we can prove the following:

Lemma 5.1 *For a splicing rule $r \in R$, the function $M_r : \mathcal{P}(V^*) \times \mathcal{P}(V^*) \longrightarrow \mathcal{P}(V^*)$ constructed above is an w-continuous function, i.e., it is w-continuous in every argument.*

Consider now the following polynomial function attached to σ and L

$$P_{\sigma,L}(X) = \bigcup_{r \in R} M_r(L, X) \bigcup_{r \in R} M_r(X, L) \bigcup L$$

As a consequence of the lemma above, $P_{\sigma,L}$ is a selective 2CP-polynomial function.

Let us first treat the uniterated case. We have the following result.

Lemma 5.2 $\sigma(L) = P_{\sigma,L}(L)$.

Proof. For every $z \in \sigma(L)$ there exists a rule $r \in R$, and words $x, y \in L$ such that $(x, y) \vdash_r z$. Taking into account the definition of M_r let us calculate how it works on the singletons $\{x\}$ and $\{y\}$, where x and y are in its domain, i.e., $x = x_1 u_1 u_2 x_2$, $y = y_1 u_3 u_4 y_2$.

$$
\begin{aligned}
M_r(\{x\}, \{y\}) &= P_r^{1,4}(\pi_1(C_r^{1,2}(\{x\})), \pi_2(C_r^{3,4}(\{y\}))) \\
&- P_r^{1,4}(\pi_1(x_1 u_1, u_2 x_2), \pi_2(y_1 u_3, u_4 y_2)) \\
&= P_r^{1,4}(x_1 u_1, u_4 y_2) = x_1 u_1 u_4 y_2.
\end{aligned}
$$

But this means precisely that

$$(x, y) \vdash_r z \text{ iff } z \in M_r(\{x\}, \{y\}),$$

and thus $\sigma(L) = P_{\sigma,L}(L)$.   □

We can now state the main result:

Theorem 5.1 *Let* $L \subseteq V^*$ *be a finite language and* $\sigma = (V, R)$ *an H scheme. Then* $\sigma^*(L)$ *is the least solution of the fixed-point equation attached to the selective 2CP-polynomial* $P_{\sigma,L}$ *constructed above.*

Proof. Let $\{X_n\}_{n \geq -1}$, denote the Kleene sequence of $P_{\sigma,L}$. Starting with $X_{-1} = \emptyset$, we get $X_0 = L$, and thus $X_0 = \sigma^0(L)$. The proof of the next step, $X_1 = \sigma(L)$ is precisely that of the previous lemma. We continue now with the usual induction argument: suppose $X_k = \sigma^k(L)$, which means that words obtained by applying splicing at most k times are precisely those obtained by k applications of the polynomial $P_{\sigma,L}$. But, according to the lemma, one application of σ is equivalent to one application of $P_{\sigma,L}$.   □

6   Some Concluding Remarks

The selective CP-functions introduced in this paper can be used to model the insertion and the splicing operation.

There were two limitations in our treatment of splicing: the presence of an infinite number of copies at each step of the process, and the fact that we chose σ and L in LIN. We consider that they can both be overcome by slightly extending the present formalism.

We have reason to think that other operations in molecular computing and DNA computing, involving cut and paste operations on strings, can also be modeled with the aid of this concept, and we intend to continue this line of research.

We have also introduced fixed-point equations an alternative generative mechanism, of a purely algebraic nature, for languages generated by splicing. We hope that it will prove to be a useful tool in DNA computing, both for obtaining new results, and for providing alternative proofs to the ones already existing.

References

[1] A. Baranga, The contraction principle as a particular case of Kleene's fixed-point theorem, *Discrete Mathematics*, 98 (1991), 75 – 79.

[2] A. Baranga, C. Popescu, Ordering, compactness, continuity, *Bulletin Mathematique de la Soc. Sci. Math. de la R.S. de Roumanie*, 31(79), 2 (1987), 99 – 106.

[3] C. S. Calude, J. Casti, M. J. Dinneen, eds., *Unconventional Models of Computation*, Springer-Verlag, Singapore, 1998.

[4] R. Ceterchi, Cut-and-paste languages, submitted, 1998.

[5] R. Ceterchi, Marcus contextual languages and their cut-and-paste properties, *Rough Sets and Current Trends in Computing* (L. Polkowski, A. Skowron, eds.), RSCTC'98 Proceedings, Springer-Verlag, Berlin, 1998.

[6] R. Freund, F. Wachtler, Universal systems with operations related to splicing, *Computers and AI*, 15, 4 (1996), 273 – 294.

[7] T. Head, Formal language theory and DNA: An analysis of the generative capacity of specific recombinant behaviors, *Bull. Math. Biology*, 49 (1987), 737 – 759.

[8] T. Head, Splicing systems and DNA, in *Lindenmeyer Systems: Impacts on Theoretical Computer Science, Computer Graphics and Developmental Biology* (G. Rozenberg, A. Salomaa, eds.), Springer-Verlag, Berlin, 1992, 371 – 383.

[9] T. Head, Gh. Păun, D. Pixton, Language theory and molecular genetics: generative mechanisms suggested by DNA recombination, chapter 7 in vol. 2 of [16], 295 – 360.

[10] L. Kari, Gh. Păun, G. Thierrin, S. Yu, Characterizing recursively enumerable languages using insertion-deletion systems, *Proc. of the Third DIMACS Workshop on DNA Based Computers*, June 23 – 25, 1997, Philadelphia, 318 – 333.

[11] A. Mateescu, Rational sets over a product of monoids, in *Proc. 2nd Nat. Colloq. Languages, Logic, Mathematical Linguistics*, Braşov, 1988, 135 – 141.

[12] Gh. Păun, *Marcus Contextual Grammars*, Kluwer Academic Publishers, Dordrecht, Boston, London, 1997.

[13] Gh. Păun, G. Rozenberg, A. Salomaa, Computing by splicing, *Theoretical Computer Science*, 168 (1996), 321 – 336.

[14] Gh. Păun, G. Rozenberg, A. Salomaa, *DNA Computing. New Computing Paradigms*, Springer-Verlag, Berlin, Heidelberg, 1998.

[15] D. Pixton, Splicing in abstract families of languages, manuscript, May 1997.

[16] G. Rozenberg, A. Salomaa, eds., *The Handbook of Formal Languages*, 3 volumes, Springer-Verlag, Berlin, Heidelberg, 1997.

Springer Series in
Discrete Mathematics and Theoretical Computer Science